D0821613
3 3604 00021 7687

Numerical Weather Prediction

Numerical Weather Prediction

GEORGE J. HALTINER, Ph.D.
Professor and Chairman

Department of Meteorology
Naval Postgraduate School
Monterey, California

JOHN WILEY & SONS, INC.
New York London Sydney Toronto

Library of Congress Catalogue Card Number: 79-150610
ISBN 0-471-34580-6

Printed in the United States of America.

10 9 8 7 6 5 4 3 2 1

Preface

This text on numerical weather prediction was written to fill a gap in a field which has developed very rapidly in the last two decades. Its purpose is to assist students, faculty, and other scientists who presently must rely primarily on the highly condensed articles in technical journals. For the most part, the text is self-contained, although some previous knowledge of dynamic meteorology would be desirable.

There is no clear distinction between the subject of numerical weather prediction and atmospheric dynamics. As a consequence, no two people would agree exactly on which topics should be treated under the present title. Most of the material included here evolved from a one-quarter course in numerical weather prediction I taught at the Naval Postgraduate School. However, additional background material has been added in order to make the text somewhat more self-contained and hence readable by scientists in other fields. Depending on the background of the students, the material should be adequate for a one-semester course. The decision as to which topics to include was based mainly on their relevance to the prediction of large-scale weather processes. Although the specific treatment of the various processes will undoubtedly change through the years, the basic physical and mathematical principles presented will generally remain valid.

In general, no attempt has been made to provide specific references on material which is now more or less classical, though more recent contributions generally are cited. Nevertheless, I would like to acknowledge some scientists who made important contributions to the early development of the field and from whose works I have drawn for classroom presentations; namely, A. Arakawa, J. G. Charney, E. N. Lorenz, N. A. Phillips, F. G. Shuman, J. Smagorinsky, P. D. Thompson, and A. Wiin-Nielsen.

My thanks go to my colleagues at the Naval Postgraduate School, the Fleet Numerical Weather Central, and the Navy Weather Research Facility for their comments on portions of the text. I would like especially to express my deep appreciation to Professor R. T. Williams of the Postgraduate School for numerous discussions on various facets of the subject matter and for his many valuable suggestions.

Finally, I would like to express my gratitude to Dr. A. Arakawa and his research assistant, Mrs. Vivian Lamb of UCLA, who carefully reviewed the original manuscript for the publisher and provided valuable recommendations for improving the organization, readability, and accuracy of the text.

Monterey, California
April, 1971

G. J. HALTINER

Contents

List of Symbols

a	Radius of earth; constant
a_o	Clear sky albedo
a_s	Surface albedo
c	Phase velocity
c_i, c_r	Imaginary and real parts of c
c_p	Specific heat at constant pressure
c_l	Specific heat of liquid water
c_v	Specific heat at constant volume
$dS =$	$dx\, dy$
e	Vapor pressure
e_s	Saturation vapor pressure
$f =$	$2\, \Omega \sin \varphi$, coriolis parameter
g	Gravity
g_a	Gravitation
h	Horizontal; free surface height; scale factor for distance; thickness
$\mathbf{i}, \mathbf{j}, \mathbf{k}$	Unit vectors
$i =$	$\sqrt{-1}$; index
j	Index
k	Vertical wave number; constant
l	Liquid water content of air; index
m	Index; map factor
n	Index
p	Pressure; vertical coordinate
p_t	Terrain pressure
q	Specific humidity, potential vorticity
q_s	Saturated value of q
q_w	Saturated value of q at water temperature
r	Radial distance; relative humidity
$\mathbf{r}$	Radius vector
s	Scalar distance
t	Time coordinate

u	x-component of velocity
u^*	Pressure-corrected water vapor mass
v	v-component of velocity
w	z-component of velocity; precipitable water vapor
x, y, z	Space coordinates
z, Z, z_p	Pressure height
$\overline{(\ )}$	Spatial mean (horizontal, vertical or both)
A	Available potential energy; wave amplitude
C	Fractional cloud cover
C_D	Drag coefficient
C_E	Energy exchange coefficient
$C_R =$	$U - \beta/\mu^2$ Rossby wave speed
D	Horizontal divergence; vertical scale length
$E =$	c_pT, enthalpy, total potential energy
E_v	Evaporation
$F =$	V^2/gD = Froude number; friction force; radiative flux
H	Mean height of free surface or depth of the troposphere; total energy
H_L	Latent heat
H_S	Sensible heat
H_t	Terrain height
$I =$	c_vT, internal energy
J	Jacobian
$\mathbb{J}$	Finite difference Jacobian
K	Kinetic energy per unit mass; coefficient of eddy viscosity; diffusion coefficient
K_H	Horizontal curvature
L	Wavelength; latent heat; horizontal scale length
$\mathscr{L}$	Differential operator
$dM =$	$\rho\, dx\, dy\, dz = -g^{-1}\, dx\, dy\, dp$
N	Index
P	Pressure; potential energy; precipitation
P_n^m	Legendre function
Q	Diabatic heating per unit mass per unit time
Q_s	Sensible heat
Q_L	Latent heat
R	Gas constant for air

R_c	Cloud albedo
$R_i =$	σ_s/F, Richardson Number (special)
R_{ij}	Residual during relaxation process
R_0 or $R_1 =$	V/fL, Rossby number
R_v	Gas constant for water vapor
S	Surface; solar radiation
T	Temperature
T_a	Air temperature
T_c	Cloud temperature
T_g	Ground temperature
T_s	Sea temperature; saturated air temperature
U	Basic current in x-direction, velocity component in direction of map coordinate
U_T	Thermal wind
$\mathbf{V}$	Horizontal velocity; characteristic velocity, velocity component in direction of map coordinate
$\mathbf{V}_3$	Three-dimensional velocity
$\mathbf{V}_a$	Absolute velocity
$\mathbf{V}_g$	Geostrophic wind
$\mathbf{V}_T$	Thermal wind
V_χ	Divergent wind component
V_ψ	Rotational wind component
W	Characteristic vertical velocity; amplitude; weight factor
$Y_n^{\ m}$	Spherical harmonic
α	Specific volume; angle
$\beta =$	df/dy Rossby parameter
$\gamma =$	c_p/c_v; lapse rate
γ_d	Dry adiabatic lapse rate
δ	Differential; horizontal divergence; Kronecker delta, declination angle, variation
$\times$	Vector product
Δ	Finite difference
∇	Del operator
$\mathbb{V}$, ∇_t, ∇_x, etc.	Finite difference del operators
∂	Partial differentiation
ϵ	Arbitrary variable
ζ	Vertical component of vorticity; vertical coordinate
ζ_g	Geostrophic vorticity
κ	R/c_p
η	Absolute vorticity; cloud parameter

θ	Potential temperature; colatitude
θ_c	Constant temperature
θ_e	Equivalent potential temperature
$\lambda =$	$c\,\Delta t/\Delta x$; $R_1^{-2}R_i^{-1}$; longitude; parameter, radius of deformation
$\mu =$	$2\pi/L =$ wave number; diffusion coefficient
ν	Frequency; kinematic coefficient of eddy viscosity
π	3.14159 . . . ; surface pressure
ρ	Density
ρ_v	Density of water vapor
σ	Static stability parameter; vertical coordinate; Stefan-Boltzmann constant
σ_s	Static stability parameter, constant or function of pressure
$\sum$	Summation
τ	Friction force; stress; difference of stream functions
φ	Latitude
Φ	Geopotential; phase angle; amplitude
Φ_G	Surface geopotential
χ	Velocity potential function for divergent component
ψ	Stream function for rotational velocity component
$\omega =$	$dp/dt =$ vertical velocity in p-system
Ω	Angular velocity of earth

Numerical Weather Prediction

chapter one

The Governing Equations

1.1 INTRODUCTION

The phrase "numerical weather prediction" generally connotes the prediction of meteorological parameters by numerical solution of the hydrodynamic equations governing atmospheric motions. Electronic computers are usually needed for this purpose because of the enormous number of arithmetic and logical operations necessary to effect a numerical solution of even rather unsophisticated prediction models, although graphical methods have been devised for several of the simpler models.

The principal equations which are relevant to motions in the atmosphere are the following: (*a*) Newton's second law of motion; (*b*) the first law of thermodynamics; (*c*) the law of conservation of mass or continuity equation; (*d*) the equation of state; and (*e*) the conservation equation for the water substance. For a large range of scales of atmospheric motions the atmosphere may be treated as if it were a perfect gas, and this will be assumed throughout this text.

The basic principles underlying numerical weather prediction were discovered early in this century. V. Bjerknes first recognized that the fundamental equations cited in the foregoing formed a determinate system which, in principle, could be solved to forecast the subsequent state of the atmosphere from a known initial state. He recognized also that this highly nonlinear system did not possess an analytic solution and that data were wholly inadequate to determine the initial conditions. In 1921 L. F. Richardson published a monograph, "Weather Prediction by Numerical Process," which described a method for numerically integrating the governing equations. Unfortunately, his experimental results, which required several months of computation on manual calculators, were in error by several orders of magnitude. As a consequence, Richardson's monumental work was ignored for several decades. By the late 1940's the electronic computer had been invented, and in 1950 Drs. Jule Charney, R. Fjortoft and John von Neumann published the first successful numerical predictions using a simple model based on some earlier work of C. G. Rossby.

1.2 EQUATION OF MOTION

Newton's second law of motion for a "fixed" reference frame (inertial) may be expressed as

$$\frac{d_a \mathbf{V}_a}{dt} = \mathbf{M}, \tag{1.1}$$

where **M** represents the vector sum of the forces per unit mass. The subscript a denotes the values of velocity and acceleration as observed in the inertial frame. However, since these quantities are observed on earth, which is moving through space, it is desirable to express the equation of motion in terms of variables measured relative to the earth. The principal motion that must be accounted for is the rotation of the earth, while other motions such as the orbital movement around the sun, etc., may be neglected. Assuming the earth to be rotating at angular velocity $\mathbf{\Omega}$, the absolute velocity $\mathbf{V}_a$ of a particle may be written as the sum of the velocity relative to the earth $\mathbf{V}$ and the velocity due to the rotation.

$$\mathbf{V}_a = \mathbf{V} + \mathbf{\Omega} \times \mathbf{r}. \tag{1.2}$$

Here $\mathbf{r}$ represents the position vector of the particle as measured from the origin at the earth's center.

Now let $\mathbf{i}$, $\mathbf{j}$, and $\mathbf{k}$ be unit vectors along an orthogonal set of axes in the absolute frame of reference and $\mathbf{i}'$, $\mathbf{j}'$, and $\mathbf{k}'$ in the rotating frame. Then an arbitrary vector $\mathbf{A}$ may be expressed in the two frames as

$$\mathbf{A} = A_x\mathbf{i} + A_y\mathbf{j} + A_z\mathbf{k} = A'_x\mathbf{i}' + A'_y\mathbf{j}' + A'_z\mathbf{k}',$$

and its time derivative as

$$\begin{aligned}\frac{d\mathbf{A}}{dt} &= \frac{dA_x}{dt}\mathbf{i} + \frac{dA_y}{dt}\mathbf{j} + \frac{dA_z}{dt}\mathbf{k} \\ &= \frac{dA'_x}{dt}\mathbf{i}' + \frac{dA'_y}{dt}\mathbf{j}' + \frac{dA'_z}{dt}\mathbf{k}' + A'_x\frac{d\mathbf{i}'}{dt} + A'_y\frac{d\mathbf{j}'}{dt} + A'_z\frac{d\mathbf{k}'}{dt}.\end{aligned}$$

The derivatives of the unit vectors are given by $d\mathbf{i}'/dt = \mathbf{\Omega} \times \mathbf{i}'$, etc.; hence

$$\frac{d_a\mathbf{A}}{dt} = \frac{d\mathbf{A}}{dt} + \mathbf{\Omega} \times \mathbf{A}, \tag{1.3}$$

where d_a/dt represents the rate of change with respect to the fixed system and d/dt with respect to the rotating system.

Thus it has been shown that the total derivative of an arbitrary vector as measured in the fixed system is expressible as the sum of the derivative in the rotating system and the term $\mathbf{\Omega} \times \mathbf{A}$.

From this theorem it follows that

$$\frac{d_a\mathbf{V}_a}{dt} = \frac{d\mathbf{V}_a}{dt} + \mathbf{\Omega} \times \mathbf{V}_a.$$

Substituting further from (1.2) gives

$$\frac{d_a\mathbf{V}_a}{dt} = \frac{d(\mathbf{V} + \mathbf{\Omega} \times \mathbf{r})}{dt} + \mathbf{\Omega} \times (\mathbf{V} + \mathbf{\Omega} \times \mathbf{r})$$

or

$$\frac{d_a\mathbf{V}_a}{dt} = \frac{d\mathbf{V}}{dt} + 2\mathbf{\Omega} \times \mathbf{V} + \mathbf{\Omega} \times (\mathbf{\Omega} \times \mathbf{r}). \tag{1.4}$$

The principal forces (per unit mass) in atmospheric motions are the pressure force, gravitation $\mathbf{g}_a$, and the friction force $\mathbf{F}$. The subscript 3 will be used to denote the three-dimensional velocity and del operator. Introducing these forces and the expression for the absolute acceleration (1.4) into (1.1) gives the equation of relative motion

$$\frac{d\mathbf{V}_3}{dt} = -\alpha\nabla_3 p - 2\mathbf{\Omega} \times \mathbf{V}_3 + \mathbf{g} + \mathbf{F}, \tag{1.5}$$

where α is the specific volume and gravity $\mathbf{g}$ is the sum of gravitation and the centrifugal force-$\mathbf{\Omega} \times (\mathbf{\Omega} \times \mathbf{r})$. That is,

$$\mathbf{g} = \mathbf{g}_a - \mathbf{\Omega} \times (\mathbf{\Omega} \times \mathbf{r}).$$

The second term on the right of (1.5). $-2\mathbf{\Omega} \times \mathbf{V}_3$, is referred to as the *coriolis force.*

1.3 CONTINUITY EQUATION

The second fundamental law is the conservation of mass, which may be expressed in mathematical form as follows. Consider an infinitesimal volume $\delta x\, \delta y\, \delta z$ as shown in Figure 1.1. The mass inflow per unit time into the x-face is $\rho u\, \delta y\, \delta z$, where ρ is the density and u is the x component of velocity.

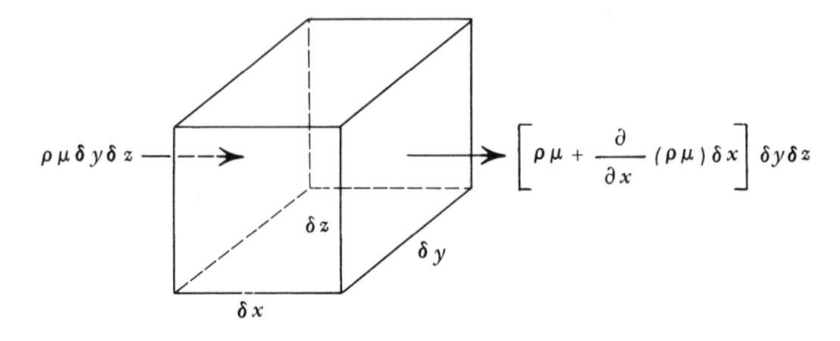

Figure 1.1. Mass conservation.

The outflow from the opposite face is $[\rho u + (\partial \rho u/\partial x)\delta x]\delta y\ \delta z$. The difference between the inflow and the outflow per unit volume is $-\partial(\rho u)/\partial x$, which represents the x contribution to the local rate of change of mass per unit volume, namely, the rate of density change. Considering all three directions, the result is

$$\frac{\partial \rho}{\partial t} = -\mathbf{\nabla}_3 \cdot (\rho \mathbf{V}_3). \tag{1.6}$$

An alternate form is obtainable by expanding the right member of (1.6), giving, with $\alpha = 1/p$,

$$\frac{1}{\alpha}\frac{d\alpha}{dt} = \mathbf{\nabla}_3 \cdot \mathbf{V}_3. \tag{1.7}$$

1.4 EQUATION OF STATE

As suggested by the early discoveries of Boyle and Charles, experience has shown that the three thermodynamic variables p, α, and T are not independent. For every substance there exists a relationship known as the equation of state of the form

$$f(p, \alpha, T) = 0.$$

For real substances this relationship is highly complex, and no simple analytical expression exists. However, there are analytic forms which are satisfactory over limited ranges of the variables. A general equation may be derived by assuming certain idealized conditions. For example, a *perfect* gas may be defined as one which obeys Boyle's and Charles' laws exactly, in which case it is readily shown that the equation of state takes the form

$$p\alpha = RT. \tag{1.8}$$

Although no perfect gas exists, (1.8) may be used to approximate real gases for many processes and will suffice for our purposes here. Although the atmosphere is, in fact, a mixture of gases, (1.8) is nevertheless satisfactory, provided the gas constant R is a weighted average of the gas constants of the individual gases constituting the atmosphere.

1.5 FIRST LAW OF THERMODYNAMICS

The first law of thermodynamics expresses the principle of conservation of energy. The latter states that the change in energy of a system is equal to the net transfer of energy across the boundaries of the system. In complete generality, this principle embraces all types of energy: potential, kinetic, thermal, radiant, magnetic, electrical, chemical, etc.; however, for application to the meteorological thermodynamics considered here, the relationship may be considerably simplified. As will be shown later, Newton's law of motion implies a balance of certain forms of mechanical energy. As a consequence, the change in the sum of the organized kinetic energy of motion and the potential energy due to the field of gravity is equal to part of the work done by the pressure and frictional forces. Thus these particular terms may be omitted from the first law of thermodynamics. The remaining work done on (or by) the gas by the pressure and frictional forces is a consequence of the expansion and deformation of the gas.

If the chemical, magnetic, and electrical effects are omitted and a nonviscous gas is considered, the first law of thermodynamics may be written in the simple form

$$Q = \frac{du}{dt} + W,$$

where Q is the rate of heat addition, W is the rate at which work is done by the gas on its surroundings by expansion, and du/dt is the rate of change of internal energy. For a perfect gas the rate of change of internal energy per unit mass is given by $c_v\, dT/dt$, where c_v is the specific heat at constant volume, while the work done per unit mass per unit time during a reversible expansion of a nonviscous gas is $p\, d\alpha/dt$. Substituting these expressions into the thermodynamical equation gives

$$Q = c_v \frac{dT}{dt} + p \frac{d\alpha}{dt}. \tag{1.9a}$$

Use of (1.8) gives the alternate form

$$Q = c_p \frac{dT}{dt} - \alpha \frac{dp}{dt}. \tag{1.9b}$$

When there are changes of phase of the water substance from vapor to liquid, (1.9*b*) takes the form of

$$Q - L\frac{dq}{dt} - c_l l\frac{dT}{dt} \doteq c_p\frac{dT}{dt} - \alpha\frac{dp}{dt}, \tag{1.10}$$

where q is the specific humidity of the water-vapor constituent, l is the liquid water per unit mass of air, L is the latent heat, and c_l is the specific heat of the liquid water.

1.6 THE COMPLETE SYSTEM OF EQUATIONS

For dry air, (1.5), (1.7), (1.8), and (1.9) comprise a *complete system* of *six scalar equations* and *six unknowns*, p, α, T, u, v, and w. The friction force $\mathbf{F}$ and diabatic heating Q are assumed to be either known functions or expressible in terms of the other variables; hence, in principle, all future states can be determined by solution of this system.

When moisture is included, some modifications are necessary in the equation of state and the first law of thermodynamics; in addition, an equation is needed to express the conservation of the water substance. For the present, only dry air will be considered; in Chapter 9 the effects of moisture will be discussed.

For large-scale motions of the air, the atmosphere may be assumed to be in hydrostatic equilibrium, i.e., the vertical acceleration may be neglected along with the vertical component of the coriolis force. This result is established in Chapter 3 by a scale analysis which consists of a systematic comparison of the various terms comprising the equations of motion. When the hydrostatic approximation is valid, (1.5) is equivalent to the pair

$$\frac{d\mathbf{V}}{dt} = -\alpha\nabla p - \mathbf{f}\times\mathbf{V} + \mathbf{F}_H, \tag{1.11}$$

$$0 = -\alpha\frac{\partial p}{\partial z} - g. \tag{1.12}$$

Here $\mathbf{V}$ is the horizontal wind velocity, ∇ is the two-dimensional del operator, $\mathbf{F}_H$ is the horizontal component of the frictional force, $\mathbf{f} = f\mathbf{k}$, and $f = 2\Omega\sin\varphi$ is the *coriolis parameter* (which is the vertical component of the earth's vorticity).

1.7 THE (x, y, p, t) COORDINATE SYSTEM

When the hydrostatic approximation is valid, it is advantageous to introduce pressure as the vertical coordinate because some of the terms in the fundamental equations then take on a simpler form. For this purpose consider arbitrary variables $\epsilon(x, y, z, t)$ and $\sigma(x, y, z, t)$, where the latter represents a new vertical coordinate.

Now let $\epsilon(x, y, \sigma, t)$ be the functional form of ϵ in the σ system and $\epsilon'(x, y, z, t)$ in the ordinary Cartesian system, then

$$\epsilon(x, y, \sigma, t) \equiv \epsilon'[x, y, z(x, y, \sigma, t), t].$$

It follows that

$$\left(\frac{\partial \epsilon}{\partial x}\right)_{y,\sigma,t} = \left(\frac{\partial \epsilon'}{\partial x}\right)_{y,z,t} + \left(\frac{\partial \epsilon'}{\partial z}\right)_{x,y,t}\left(\frac{\partial z}{\partial x}\right)_{y,\sigma,t}.$$

Similar expressions hold for y and t; whereas for the vertical direction,

$$\frac{\partial \epsilon}{\partial \sigma} = \frac{\partial \epsilon'}{\partial z}\frac{\partial z}{\partial \sigma}.$$

From the foregoing results the following relationships may easily be shown to hold:

$$\mathbf{\nabla}_\sigma \epsilon = \mathbf{\nabla}\epsilon + \frac{\partial \epsilon}{\partial z}\mathbf{\nabla}_\sigma z \qquad (1.13)^*$$

and

$$\mathbf{\nabla}_\sigma \cdot \mathbf{V} = \mathbf{\nabla} \cdot \mathbf{V} + \frac{\partial \mathbf{V}}{\partial z} \cdot \mathbf{\nabla}_\sigma z, \qquad (1.14)$$

where the subscript σ denotes the two-dimensional del operator for a constant σ surface. These relations may be used to express the hydrodynamical equations in a general σ coordinate system. For the present, the special case of $\sigma = p$ will be considered. To obtain the proper expression for the horizontal pressure force in (1.11), place $\epsilon = \sigma = p$ into (1.13), giving

$$\mathbf{\nabla}_p p = \mathbf{\nabla} p + \frac{\partial p}{\partial z}\mathbf{\nabla}_p z.$$

Now by substituting from (1.12) and noting that $\mathbf{\nabla}_p p \equiv 0$, the foregoing equation becomes simply

$$\alpha \mathbf{\nabla} p = \mathbf{\nabla}_p \Phi,$$

where $\Phi = gz$ is the geopotential. With similar manipulations the equation of continuity may be transformed into the (x, y, p, t) coordinate system. The resulting forms of the hydrodynamical equations, (1.11), (1.7), (1.9*b*), and

* Alternatively one could write

$$\nabla_\sigma \epsilon + \frac{\partial \epsilon}{\partial \sigma}\nabla \sigma = \nabla \epsilon.$$

(1.12) are:

$$\frac{d\mathbf{V}}{dt} = -\nabla_p \Phi - \mathbf{f} \times \mathbf{V} + \mathbf{F}_H, \tag{1.15}$$

$$\nabla_p \cdot \mathbf{V} + \frac{\partial \omega}{\partial p} = 0, \tag{1.16}$$

$$Q = c_p \frac{dT}{dt} - \frac{RT}{p} \omega, \tag{1.17}$$

$$\frac{\partial \Phi}{\partial p} = -\frac{RT}{p}, \tag{1.18}$$

where $\omega = dp/dt$ is the "vertical" velocity in the (x, y, p, t) coordinate system. The total derivative operator now becomes

$$\frac{d\epsilon}{dt} \equiv \frac{\partial \epsilon}{\partial t} + \mathbf{V} \cdot \nabla_p \epsilon + \omega \frac{\partial \epsilon}{\partial p} \tag{1.19}$$

for an arbitrary variable, $\epsilon(x, y, p, t)$, vector, or scalar. The foregoing system comprises five scalar equations in five unknowns, u, v, ω, T, and Φ, the specific volume α having been eliminated by means of the equation of state. Note that the continuity equation (1.16) has a simpler form than (1.7), and so does the pressure force in (1.15) as compared to (1.11).

1.8 VORTICITY AND DIVERGENCE EQUATIONS

For the purpose of dynamical analysis and numerical weather prediction it is often advantageous to replace the vector equation of horizontal motion (1.15) by two scalar equations, the *vorticity* and *divergence* equations. To obtain the former, apply the operator $\mathbf{k} \cdot \nabla_p \times$ to (1.15), giving

$$\frac{\partial \zeta}{\partial t} + \mathbf{V} \cdot \nabla_p(\zeta + f) + \omega \frac{\partial \zeta}{\partial p} = -(\zeta + f)\nabla_p \cdot \mathbf{V} + \mathbf{k} \cdot \left(\frac{\partial \mathbf{V}}{\partial p} \times \nabla_p \omega\right) + \mathbf{k} \cdot \nabla_p \times \mathbf{F}. \tag{1.20}$$

Some vector identities which are convenient for this task are the following.

$$\nabla \cdot (a\mathbf{A}) = a\nabla \cdot \mathbf{A} + \mathbf{A} \cdot \nabla a,$$

$$\nabla \times (a\mathbf{A}) = a\nabla \times \mathbf{A} - \mathbf{A} \times \nabla a,$$

$$\nabla \cdot (\mathbf{A} \times \mathbf{B}) = \mathbf{B} \cdot \nabla \times \mathbf{A} - \mathbf{A} \cdot \nabla \times \mathbf{B},$$

$$\nabla \times (\mathbf{A} \times \mathbf{B}) = \mathbf{A}\nabla \cdot \mathbf{B} - \mathbf{B}\nabla \cdot \mathbf{A} - (\mathbf{A} \cdot \nabla)\mathbf{B} + (\mathbf{B} \cdot \nabla)\mathbf{A},$$

$$\mathbf{A} \times \mathbf{B} \times \mathbf{C} = (\mathbf{A} \cdot \mathbf{C})\mathbf{B} - (\mathbf{A} \cdot \mathbf{B})\mathbf{C},$$

$$\nabla \times \nabla a \equiv 0, \nabla \cdot \nabla \times \mathbf{A} = 0,$$

$$(\nabla \cdot \nabla)\mathbf{V} = \nabla \frac{V^2}{2} + \zeta\mathbf{k} \times \mathbf{V} \qquad \text{where} \qquad \zeta = \mathbf{k} \cdot \nabla \times \mathbf{V}.$$

$$\nabla \times \nabla \times \mathbf{A} = \nabla(\nabla \cdot \mathbf{A}) - \nabla^2 \mathbf{A}.$$

The subscript p has been omitted on the del operator, as will be usually done henceforth in order to simplify the notation. The divergence equation is obtained by taking the divergence (two-dimensional, ∇_p) of (1.15), with the result

$$\frac{\partial \delta}{\partial t} + \nabla \cdot [(\mathbf{V} \cdot \nabla)\mathbf{V}] + \nabla \omega \cdot \frac{\partial \mathbf{V}}{\partial p} + \omega \frac{\partial \delta}{\partial p} = -\nabla^2 \Phi - \nabla \cdot (\mathbf{f} \times \mathbf{V}) + \nabla \cdot \mathbf{F}, \quad (1.21)$$

where

$$\delta = \nabla \cdot \mathbf{V}.$$

Equation 1.21 has been only partially expanded and is not one of the usual forms. In later chapters various approximations of (1.20) and (1.21) will be presented, which will be used for discussions of the dynamics of atmospheric motions and for numerical weather prediction. For the latter purpose it is generally desirable to transform the hydrodynamical equations into the coordinates of one of several map projections. For large-scale motions of the atmosphere the curvature of the earth is important; hence, spherical coordinates would appear to be the most logical. Nevertheless, for portraying meteorological analyses and predictions, the usual practice has been to project the spherical earth onto a plane map. In Section 1.11 the transformation of the equations of motion into map coordinates is illustrated for several commonly used projections. However, in most of the discussions to follow, the equations will not be transformed into map coordinates in order to simplify the discussion and to focus attention on other important physical or mathematical considerations, such as the dynamics, energetics, numerical methods, etc. In Chapter 3 a scale analysis of the vorticity and divergence equations is presented for the purpose of assessing the relative magnitudes of the various terms comprising these equations in order to develop consistent approximations of the equations. On the other hand, Chapter 4 deals with the integral properties of these equations with regard to vorticity and energy, which also provide guide lines for the specification of approximate forms of the vorticity, divergence, and thermodynamic equations that are energetically consistent.

1.9 SPHERICAL CURVILINEAR COORDINATES

The large-scale motions of the atmosphere are quasi-horizontal with respect to the earth's surface and, hence, are approximately in spherical surfaces with only small vertical displacements compared to those in the horizontal. It would therefore seem logical to express the equations of motion in spherical or spherical curvilinear coordinates. Although there are some disadvantages, occasional use will be made of these coordinates in this text; hence, the basic equations will be presented in that particular form now. If λ

denotes the longitude, φ, the latitude, and r the radial distance of the particle as measured from the center of the earth, then the curvilinear velocities along latitude circles, meridians, and in the local vertical are given by

$$u = r \cos \varphi \frac{d\lambda}{dt}, \qquad v = r \frac{d\varphi}{dt}, \qquad w = \frac{dr}{dt} = \frac{dz}{dt}, \tag{1.22}$$

where z is the altitude measured from mean sea level. The velocity vector is given by

$$\mathbf{V}_3 = u\mathbf{i} + v\mathbf{j} + w\mathbf{k},$$

where **i**, **j**, and **k** are unit vectors in the directions of the curvilinear coordinates x, y, and z. If the velocity vector is differentiated with respect to time to obtain acceleration, and the variation of the unit vectors along the curvilinear coordinates is taken into account, the following equations are obtained for the three coordinate directions (see Haltiner and Martin for the details).

$$\begin{gathered}
\frac{du}{dt} - \frac{uv \tan \varphi}{r} + \underbrace{\frac{uw}{r}}_{small} = -\alpha \frac{\partial p}{\partial x} + fv - \underbrace{2w\Omega \cos \varphi}_{small}, \\
\frac{dv}{dt} + \frac{u^2 \tan \varphi}{r} + \underbrace{\frac{vw}{r}}_{small} = -\alpha \frac{\partial p}{\partial y} - fu, \\
\frac{dw}{dt} - \underbrace{\frac{u^2 + v^2}{r}}_{small} = -\alpha \frac{\partial p}{\partial z} - g + \underbrace{2u\Omega \cos \varphi}_{small}.
\end{gathered} \tag{1.23}$$

For all scales of motion the terms labeled "small" can be omitted. When the hydrostatic approximation is valid, these terms must be omitted in order to yield a consistent kinetic-energy equation. Also r may then be replaced by the mean radius of the earth a, where it appears as a factor in (1.22) and (1.23). If the equations are desired in terms of λ, φ, and r, the expressions for velocity (1.22) may be substituted back into (1.23), together with the following additional replacements:

$$\delta x = r \cos \varphi \, \delta\lambda, \qquad \delta y = r \, \delta\varphi, \qquad \delta z = \delta r. \tag{1.24}$$

The continuity equation takes the following form in spherical curvilinear coordinates:

$$\frac{1}{\rho} \frac{d\rho}{dt} + \frac{\partial u}{\partial x} + \frac{\partial v}{\partial y} - \frac{v \tan \varphi}{r} + \frac{\partial w}{\partial z} + \frac{2w}{r} = 0. \tag{1.25}$$

The equation of state and the first law of thermodynamics remain unchanged in format; thus (1.8) and (1.9) obtain for the curvilinear coordinates as well.

1.10 GENERAL CURVILINEAR COORDINATES

Although there is no intention to provide a discussion of generalized curvilinear coordinates here, it is useful to include the expressions for some of the principal vector operators in general orthogonal coordinates. Consider coordinates x_1, x_2, x_3, and the corresponding scalar curvilinear measures of distance, s_1, s_2, s_3, such that

$$ds_1 = h_1\,dx_1, \qquad ds_2 = h_2\,dx_2, \qquad ds_3 = h_3\,dx_3. \tag{1.26}$$

If $\mathbf{a}_1$, $\mathbf{a}_2$, and $\mathbf{a}_3$ denote unit vectors in the x_1, x_2, and x_3 directions, the expressions for some of the common vector operators are as follows:

1. $$\operatorname{grad} A = \nabla A = \frac{\mathbf{a}_1}{h_1}\frac{\partial A}{\partial x_1} + \frac{\mathbf{a}_2}{h_2}\frac{\partial A}{\partial x_2} + \frac{\mathbf{a}_3}{h_3}\frac{\partial A}{\partial x_3}. \tag{1.27}$$

2. $$\operatorname{div}\mathbf{F} = \boldsymbol{\nabla}\cdot\mathbf{F} = \frac{1}{h_1h_2h_3} \times \left[\frac{\partial}{\partial x_1}(h_2h_3F_1) + \frac{\partial}{\partial x_2}(h_1h_3F_2) + \frac{\partial}{\partial x_3}(h_1h_2F_3)\right]. \tag{1.28}$$

3. $$\nabla^2 A = \frac{1}{h_1h_2h_3} \times \left[\frac{\partial}{\partial x_1}\left(\frac{h_2h_3}{h_1}\frac{\partial A}{\partial x_1}\right) + \frac{\partial}{\partial x_2}\left(\frac{h_1h_3}{h_2}\frac{\partial A}{\partial x_2}\right) + \frac{\partial}{\partial x_3}\left(\frac{h_1h_2}{h_3}\frac{\partial A}{\partial x_3}\right)\right]. \tag{1.29}$$

4. $$\operatorname{curl}\mathbf{F} = \frac{\mathbf{a}_1}{h_2h_3}\left[\frac{\partial}{\partial x_2}(h_3F_3) - \frac{\partial}{\partial x_3}(h_2F_2)\right] + \frac{\mathbf{a}_2}{h_3h_1}\left[\frac{\partial}{\partial x_3}(h_1F_1) - \frac{\partial}{\partial x_1}(h_3F_3)\right] + \frac{\mathbf{a}_3}{h_1h_2}\left[\frac{\partial}{\partial x_1}(h_2F_2) - \frac{\partial}{\partial x_2}(h_1F_1)\right]. \tag{1.30}$$

5. $$\begin{aligned}\frac{dA}{dt} &= \frac{\partial A}{\partial t} + \frac{dx_1}{dt}\frac{\partial A}{\partial x_1} + \frac{dx_2}{dt}\frac{\partial A}{\partial x_2} + \frac{dx_3}{dt}\frac{\partial A}{\partial x_3}\\ &= \frac{\partial A}{\partial t} + \frac{v_1}{h_1}\frac{\partial A}{\partial x_1} + \frac{v_2}{h_2}\frac{\partial A}{\partial x_2} + \frac{v_3}{h_3}\frac{\partial A}{\partial x_3},\end{aligned} \tag{1.31}$$

where $v_1 = ds_1/dt$, etc.

As a specific example of the foregoing general expressions, consider spherical coordinates such that

$$x_1 = \lambda, \qquad x_2 = \varphi, \qquad x_3 = r.$$

Then [see (1.24)],

$$ds_1 = r\cos\varphi\,d\lambda, \qquad ds_2 = r\,d\varphi, \qquad ds_3 = dr,$$

hence,

$$h_1 = r \cos \varphi, \qquad h_2 = r, \qquad h_3 = 1.$$

For $\mathbf{V} = u\mathbf{a}_1 + v\mathbf{a}_2 + w\mathbf{a}_3$,

$$\nabla \cdot \mathbf{V} = \frac{1}{r \cos \varphi} \frac{\partial u}{\partial \lambda} + \frac{1}{r} \frac{\partial v}{\partial \varphi} - \frac{v \tan \varphi}{r} + \frac{\partial w}{\partial r} + \frac{2w}{r}, \tag{1.32}$$

which appears in (1.25).

1.11 MAP PROJECTIONS

For purposes of weather analysis, prediction, and viewing data, it is often desirable to depict the meteorological parameters observed in the earth's atmosphere on charts. This requires the representation of the earth surface on a plane surface. Naturally it would be desirable to have the map as nearly like the spherical surface as possible and, hence, preserve such features as distance, direction, area, shape, etc. Map projections which preserve distance are termed *isometric*, and those which preserve the angle between two intersecting curves, *conformal*. The sphere, or a part thereof, may be projected directly on a plane or on a surface such as a cylinder or cone, which can then be cut to form a plane surface. Some commonly used maps are the polar stereographic plane projection, the Mercator cylindrical projection, and the Lambert conical projection, all of which are conformal and are approximately isometric near the so-called standard parallels. Figure 1.2 shows examples of several map projections.

When projections of the type described here are used to define grids for the purpose of numerical weather prediction, it is necessary to transform the hydrodynamical equations into the map coordinates. We may begin with the equations of motion in spherical curvilinear coordinates, namely (1.23), where x, y, and z are the distances along latitude circles, meridians, and in the local vertical. If the curvilinear velocities (1.22) are substituted into (1.23), the equations in ordinary spherical coordinates (λ, φ, r) are obtained (see Brunt, 1939):

$$\begin{aligned}
&r\ddot{\lambda} \cos \varphi + 2(\dot{\lambda} + \Omega)(\dot{r} \cos \varphi - r\dot{\varphi} \sin \varphi) = F_\lambda, \\
&r\ddot{\varphi} + 2\dot{r}\dot{\varphi} + r\dot{\lambda}(\dot{\lambda} + 2\Omega) \sin \varphi \cos \varphi = F_\varphi, \\
&\ddot{r} - r\dot{\varphi}^2 - r\dot{\lambda}(\dot{\lambda} + 2\Omega) \cos^2 \varphi = F_r.
\end{aligned} \tag{1.33}$$

The coriolis terms are included on the left side of these equations; thus F_λ, F_φ, and F_r include the components of the pressure, gravity, and friction forces in the directions of increasing λ, φ, and r respectively.

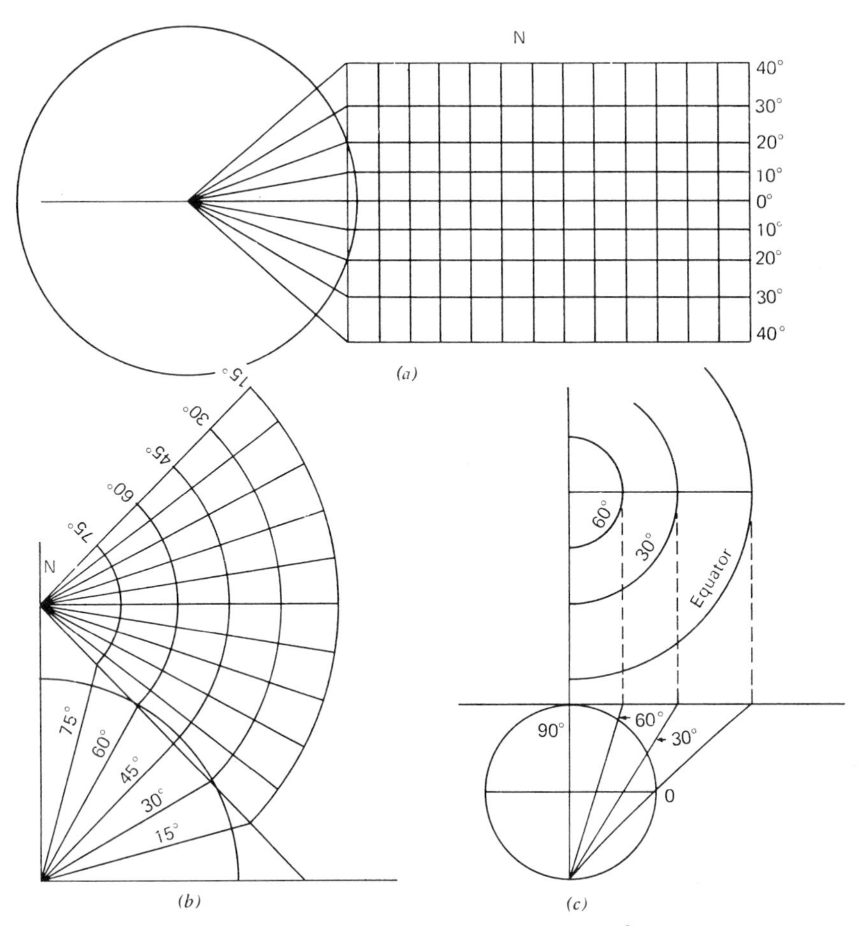

Figure 1.2. (*a*) Mercator cylindrical projection true at 22.5° latitude; (*b*) Lambert conical projection true at 30° and 60° latitude; (*c*) polar stereographic projection true at 90°.

1.11.1 Polar Stereographic Coordinates

Equations of motion

The coordinate transformation for a polar stereographic projection is given by the equations

$$x = \frac{2a \cos \varphi \cos \lambda}{1 + \sin \varphi}, \qquad y = \frac{2a \cos \varphi \sin \lambda}{1 + \sin \varphi}, \qquad r = r, \tag{1.34}$$

where r is the distance from the center of the earth and a is an arbitrary constant which will be taken as the mean radius of the earth. Differentiating the foregoing quantities leads to

$$dx = \frac{2a}{1 + \sin\varphi}(-\cos\varphi \sin\lambda\, d\lambda - \cos\lambda\, d\varphi),$$
$$dy = \frac{2a}{1 + \sin\varphi}(\cos\varphi \cos\lambda\, d\lambda - \sin\lambda\, d\varphi). \tag{1.35}$$

If the latitude and longitude differentials are replaced by the spherical curvilinear differentials dx_s and dy_s, as given by (1.24), (1.35) may be expressed in the following matrix form:

$$\begin{pmatrix} dx \\ dy \end{pmatrix} = \frac{2(a/r)}{1 + \sin\varphi} \begin{pmatrix} -\sin\lambda & -\cos\lambda \\ \cos\lambda & -\sin\lambda \end{pmatrix} \begin{pmatrix} dx_s \\ dy_s \end{pmatrix}. \tag{1.36}$$

In (1.36) the 2×2 matrix corresponds to a rotation of the spherical coordinates through an angle of $270 - \lambda$, while the coefficient $2(a/r)(1 + \sin\varphi)^{-1}$ represents a magnification of distance in the transformation from spherical to the polar stereographic coordinates. Evidently the latter are orthogonal; however the transformation is clearly not isometric as apparent from Figure 1.2.

Since the quantities dx_s, dy_s represent true distances on the sphere, the metric coefficients h_x and h_y for the polar stereographic coordinates are, in accordance with (1.26),

$$h_x = h_y = \frac{1}{m} = \frac{1 + \sin\varphi}{2(a/r)} \doteq \frac{1 + \sin\varphi}{2},$$
$$h_r = 1. \tag{1.37}$$

It also follows from (1.36) that the "map" velocities, $\dot{x} = dx/dt$, $\dot{y} = dy/dt$, are related to the spherical curvilinear velocities, $u = dx_s/dt$, $v = dy_s/dt$, by a similar equation. If the metric coefficients are now coupled with the map velocities to give the velocity components relative to the earth, U, V, in the directions of the polar stereographic coordinates, i.e.,

$$U = h_x\dot{x}, \qquad V = h_y y, \tag{1.38}$$

then (1.36) yields

$$\begin{pmatrix} U \\ V \end{pmatrix} = \begin{pmatrix} -\sin\lambda & -\cos\lambda \\ \cos\lambda & -\sin\lambda \end{pmatrix} \begin{pmatrix} u \\ v \end{pmatrix}. \tag{1.39}$$

An equation similar to (1.39) holds between the horizontal components of force in the spherical coordinates and those in the polar stereographic system.

Thus to obtain the equations of motion in the polar stereographic coordinates, merely apply the rotation operator

$$\begin{pmatrix} -\sin\lambda & -\cos\lambda \\ \cos\lambda & -\sin\lambda \end{pmatrix}. \tag{1.40}$$

to the first two equations of (1.33) and then replace λ and φ by x and y by means of (1.34), using (1.36) through (1.39) where convenient. Alternatively, (1.23) could be utilized instead of (1.33). The resulting equations of motion in polar stereographic coordinates are

$$\begin{aligned} &\frac{dU}{dt} - V\left(f - \frac{xV - yU}{2a^2}\right) - \frac{W}{a}[(1 + \sin\varphi)\Omega y - U] = -m\alpha\frac{\partial p}{\partial x}, \\ &\frac{dV}{dt} + U\left(f - \frac{xV - yU}{2a^2}\right) + \frac{W}{a}[(1 + \sin\varphi)\Omega x + V] = -m\alpha\frac{\partial p}{\partial y}, \\ &\frac{dW}{dt} - \frac{1}{a}[U^2 + V^2 + \Omega(1 + \sin\varphi)(xV - yU)] = -\alpha\frac{\partial p}{\partial z} - g. \end{aligned} \tag{1.41}$$

In these equations $W = \dot{r} = \dot{z}$, r has been replaced by a wherever it appears as a factor, $z = r - a$, and the total derivatives are, for example,

$$\frac{dU}{dt} = \frac{\partial U}{\partial t} + mU\frac{\partial U}{\partial x} + mV\frac{\partial U}{\partial y} + W\frac{\partial U}{\partial z}, \tag{1.42}$$

etc.

For synoptic scale motions of the atmosphere, the *hydrostatic approximation is valid*, and the vertical coordinate z may be replaced by p, in which case the components of the horizontal pressure force in (1.41) become

$$-m\frac{\partial \Phi}{\partial x} \quad \text{and} \quad -m\frac{\partial \Phi}{\partial y}. \tag{1.43}$$

At this point note that, if a kinetic-energy equation is formed by multiplying the three equations of (1.41) by U, V, and W respectively, and then adding them, the terms involving W in the horizontal equations would exactly cancel all of the terms on the left side of the vertical equation save dW/dt. On the other hand, if the hydrostatic equation is used for the vertical equation, namely,

$$0 = -\alpha\frac{\partial p}{\partial z} - g \quad \text{or} \quad \frac{\partial \Phi}{\partial p} = -\alpha, \tag{1.44}$$

then, when the kinetic-energy equation is formed, it will involve only $\frac{1}{2}(U^2 + V^2)$ and not include $W^2/2$. Hence to retain consistency with respect to kinetic energy, the terms involving W in the horizontal equations of (1.41) must be omitted. It should be noted here that these terms are at least one order of magnitude less than the horizontal acceleration terms. With these adjustments, the first two equations of (1.41) become

$$\frac{dU}{dt} - V\left(f - \frac{xV - yU}{2a^2}\right) = -m\frac{\partial\Phi}{\partial x}, \tag{1.45}$$

$$\frac{dV}{dt} + U\left(f - \frac{xV - yU}{2a^2}\right) = -m\frac{\partial\Phi}{\partial y}. \tag{1.46}$$

Equation of continuity

The equation of continuity (1.6) is easily transformed by utilizing (1.28) for the divergence, where in this instance $h_1 = h_2 = 1/m$ and $h_3 = 1$; hence (1.6) becomes

$$\frac{\partial\rho}{\partial t} + m^2\left[\frac{\partial(\rho U/m)}{\partial x} + \frac{\partial(\rho V/m)}{\partial y}\right] + \frac{\partial(\rho W)}{\partial z} = 0. \tag{1.47}$$

The variation of m with z has been neglected in the last term. When pressure is used as the vertical coordinate [see (1.16)], the continuity equation becomes

$$m^2\left[\frac{\partial(U/m)}{\partial x} + \frac{\partial(V/m)}{\partial y}\right] + \frac{\partial\omega}{\partial p} = 0. \tag{1.48}$$

If the last equation is integrated from $p = 0$ (where $\omega = 0$) to an arbitrary pressure level p, we obtain

$$\omega = -m^2\int_0^p\left[\frac{\partial(U/m)}{\partial x} + \frac{\partial(V/m)}{\partial y}\right]\delta p. \tag{1.49}$$

The kinematic boundary condition at the earth's surface states that the velocity normal to the earth's surface must vanish. It follows that the vertical velocity at the earth's surface is

$$W = \mathbf{V}\cdot\nabla H_t = mU\frac{\partial H_t}{\partial x} + mV\frac{\partial H_t}{\partial y}, \tag{1.50}$$

where $\mathbf{V}$ is the horizontal velocity and $H_t(x, y)$ is the height of the terrain. Expanding ω gives

$$\omega = \frac{dp}{dt} = \frac{\partial p}{dt} + W\frac{\partial p}{\partial z} + mU\frac{\partial p}{\partial x} + mV\frac{\partial p}{\partial y}. \tag{1.51}$$

This equation may be used to calculate the surface pressure change $\partial p/\partial t$, since ω at the ground (G) is available. Thus from (1.50) and (1.51),

$$\left(\frac{\partial p}{\partial t}\right)_G = \omega_G + g\rho_G m\left(U\frac{\partial H_t}{\partial x} + V\frac{\partial H_t}{\partial y}\right) - m\rho_G\left(U\frac{\partial \Phi}{\partial x} + V\frac{\partial \Phi}{\partial y}\right),$$

$$\left(\frac{\partial p}{\partial t}\right)_G = \omega_G + m\rho_G\left(U\frac{\partial(\Phi_G - \Phi)}{\partial x} + V\frac{\partial(\Phi_G - \Phi)}{\partial y}\right), \tag{1.52}$$

where $\Phi_G = gH_t$, or in terms of the surface geopotential tendency,

$$\left(\frac{\partial \Phi}{\partial t}\right)_G = (\omega/\rho)_G + m\left[U\frac{\partial(\Phi_G - \Phi)}{\partial x} + V\frac{\partial(\Phi_G - \Phi)}{\partial y}\right]. \tag{1.53}$$

First law of thermodynamics

The first law of thermodynamics may be written as

$$c_p\frac{T}{\theta}\frac{d\theta}{dt} = Q.$$

Expanding gives

$$\frac{\partial \theta}{\partial t} + \dot{x}\frac{\partial \theta}{\partial x} + \dot{y}\frac{\partial \theta}{\partial y} + w\frac{\partial \theta}{\partial z} = \frac{\theta Q}{c_p T}.$$

Substituting from (1.38) yields

$$\frac{\partial \theta}{\partial t} + mU\frac{\partial \theta}{\partial x} + mV\frac{\partial \partial}{\partial y} + W\frac{\partial \theta}{\partial z} = \frac{\theta Q}{c_p T}. \tag{1.54}$$

In the (x, y, p, t) stereographic coordinates, (1.54) becomes

$$\frac{\partial \theta}{\partial t} + mU\frac{\partial \theta}{\partial x} + mV\frac{\partial \theta}{\partial y} + \omega\frac{\partial \theta}{\partial p} = \frac{\theta Q}{c_p T}. \tag{1.55}$$

1.11.2 Mercator Coordinates

A map which has some advantageous properties near the equator is the Mercator projection which is defined by the coordinates

$$\begin{aligned} x &= a\lambda, \\ y &= -a\ln\frac{\cos\varphi}{1 + \sin\varphi}, \\ r &= r. \end{aligned} \tag{1.56}$$

Procedures similar to those previously described for polar stereographic coordinates lead to the following equations of motion for Mercator coordinates:

$$\frac{\partial U}{\partial t} + mU\frac{\partial U}{\partial x} + mV\frac{\partial U}{\partial y} + \omega\frac{\partial U}{\partial p} - V\left(f + \frac{U\tan\varphi}{a}\right) = -m\frac{\partial \Phi}{\partial x},$$

$$\frac{\partial V}{\partial t} + mU\frac{\partial V}{\partial x} + mV\frac{\partial V}{\partial y} + \omega\frac{\partial V}{\partial p} + U\left(f + \frac{U\tan\varphi}{a}\right) = -m\frac{\partial \Phi}{\partial y}, \qquad (1.57)$$

$$m = \sec\varphi. \qquad (1.58)$$

Here U and V are velocities relative to the earth.

1.11.3 Concluding Remarks

The map projections described here are suitable only for a limited portion of the earth—the polar stereographic for either hemisphere, the Mercator for perhaps a band from 40 N to 40 S. This requires the imposition of lateral boundary conditions when the hydrodynamical equations are numerically integrated. Such boundaries are obviously not present in the real atmosphere, and these artificial constraints cause errors near the boundaries which may eventually contaminate the entire forecast region. For this reason, as well as the desirability of having worldwide computerized forecasts, global integrations of the hydrodynamical equations will eventually be made on a routine basis. Experiments along this line have already been conducted. Obviously a global coordinate system is needed for this purpose. Ordinary spherical coordinates (λ, φ, r) are feasible but have the disadvantage that distance corresponding to a fixed value of $\Delta\lambda$ decreases to zero at the poles which are singular points. Sadourny, Arakawa, and Mintz (1968) have integrated the barotropic vorticity equation on a icosohedral-hexagonal grid for the sphere. Other schemes have also been tried; it suffices to say here that, in a few years, computer advances will permit global forecasts on an operational basis. A feature of such global integrations will involve "zooming in" on various areas to provide detailed forecasts of meso-scale features on a finer grid.

chapter two

Simple Types of Wave Motion in the Atmosphere

2.1 LINEARIZED EQUATIONS

In order to understand the extremely complex motions of the atmosphere, it is desirable first to isolate and analyze some simple types of motion. For this purpose consider motion only in the $x - z$ plane. Assume uniformity in the lateral direction (y) and also neglect the rotation of the earth, friction, and diabatic heating. The Newtonian momentum equations, thermodynamic equation, and the continuity equation are then expressible in the form

$$\begin{aligned} \frac{du}{dt} + \alpha\frac{\partial p}{\partial x} &= 0, \\ \frac{dw}{dt} + \alpha\frac{\partial p}{\partial z} + g &= 0, \\ \alpha\frac{dp}{dt} + p\gamma\frac{d\alpha}{dt} &= 0, \\ \alpha\nabla \cdot \mathbf{V} - \frac{d\alpha}{dt} &= 0, \end{aligned} \tag{2.1}$$

where

$$\gamma = \frac{c_p}{c_v}, \qquad \alpha = \frac{1}{\rho}.$$

Now these equations will be linearized by the so-called perturbation method (see Haltiner and Martin). For simplicity assume a constant basic current U. Next express the dependent variables as the sum of the basic or undisturbed value plus a perturbation, e.g., $u = U + u'$, and substitute these expressions into the system (2.1). After subtracting the equations for the basic flow only and neglecting the products of perturbation quantities, the resulting linear equations for the perturbation quantities are:

$$
\begin{aligned}
\frac{\partial u'}{\partial t} + U\frac{\partial u'}{\partial x} + \bar{\alpha}\frac{\partial p'}{\partial x} &= 0, \\
\delta_1\left(\frac{\partial w'}{\partial t} + U\frac{\partial w'}{\partial x}\right) + \bar{\alpha}\frac{\partial p'}{\partial z} - \frac{g\alpha'}{\bar{\alpha}} &= 0, \\
\bar{\alpha}\left(\frac{\partial p'}{\partial t} + U\frac{\partial p'}{\partial x}\right) - gw' + \bar{p}\gamma\left(\frac{\partial \alpha'}{\partial t} + U\frac{\partial \alpha'}{\partial x} + w'\frac{\partial \bar{\alpha}}{\partial z}\right) &= 0, \\
\left(\frac{\partial u'}{\partial x} + \frac{\partial w'}{\partial z}\right)\bar{\alpha} - \delta_2\left(\frac{\partial \alpha'}{\partial t} + U\frac{\partial \alpha'}{\partial x} + w'\frac{\partial \bar{\alpha}}{\partial z}\right) &= 0.
\end{aligned}
\tag{2.2}
$$

The symbols δ_1 and δ_2 are merely to identify the vertical acceleration and the compressibility terms during the subsequent analysis and will take on values of either unity or zero according to whether the terms are omitted or included. For the purposes of the following discussion, all the coefficients of the dependent variables will be treated as constants which are valid when the scale of the phenomena is small.

Now assume the perturbation quantities to be harmonic in x, z, and t with constant coefficients as follows:

$$
\begin{aligned}
u' &= Se^{i(\mu x + kz - \nu t)}, \qquad & w' &= We^{i(\mu x + kz - \nu t)}, \\
p' &= Pe^{i(\mu x + kz - \nu t)} \qquad & \alpha' &= Ae^{i(\mu x + kz - \nu t)}.
\end{aligned}
\tag{2.3}
$$

Here μ and k are wave numbers in the x and z directions and ν is the frequency. Since the equations (22) are linear, any linear combination of solutions will also be a solution, hence only a single harmonic need be considered. Substituting the functions (2.3) into (2.2) leads to a system of homogeneous algebraic equations for the amplitudes S, W, P, and A as follows:

$$
\begin{aligned}
(\mu U - \nu)S + \bar{\alpha}\mu P &= 0, \\
i\delta_1(\mu U - \nu)W + i\bar{\alpha}kP - \frac{g}{\bar{\alpha}}A &= 0, \\
\left(\bar{p}\gamma\frac{\partial \bar{\alpha}}{\partial z} - g\right)W + i\bar{\alpha}(\mu U - \nu)P + \bar{p}\gamma i(\mu U - \nu)A &= 0, \\
i\mu\bar{\alpha}S + i\bar{\alpha}kW - i\delta_2(\mu U - \nu)A - \delta_2\frac{\partial \bar{\alpha}}{\partial z}W &= 0.
\end{aligned}
\tag{2.4}
$$

The foregoing system may be written in matrix form as

$$\begin{pmatrix} (\mu U - \nu) & 0 & \bar{\alpha}\mu & 0 \\ 0 & \delta_1(\mu U - \nu) & \bar{\alpha}k & \dfrac{gi}{\bar{\alpha}} \\ 0 & \left(g - \gamma\bar{p}\dfrac{\partial\bar{\alpha}}{\partial z}\right)i & \bar{\alpha}(\mu U - \nu) & \bar{p}\gamma(\mu U - \nu) \\ \bar{\alpha}\mu & \left(\bar{\alpha}k + i\delta_2\dfrac{\partial\bar{\alpha}}{\partial z}\right) & 0 & -\delta_2(\mu U - \nu) \end{pmatrix} \begin{pmatrix} S \\ W \\ P \\ A \end{pmatrix} = 0. \tag{2.5}$$

For a nonzero solution of the amplitudes S, W, P, and A, the determinant of the coefficients must vanish. This leads to a fourth-order equation for permissible values of the frequency ν in terms of the other parameters describing the basic state and the wavelength of the perturbation.

The four roots of this equation represent four modes of oscillation of the atmosphere which, as will be seen shortly, represent sound and gravity waves. Inclusion of the coriolis force requires the other horizontal equation of motion to complete the system. When the variation of the coriolis parameter is taken into account, as will be done in Section 2.7, a fifth mode of oscillation appears in the form of large-scale quasi-horizontal waves, which are characteristic of synoptic weather charts.

2.2 VERTICAL SOUND WAVES

At this time some special cases of the system (2.5) will be considered, the first of which will be vertically propagating sound waves. For this purpose the first row and column may be omitted, and μ is identically zero. Now write

$$k = \frac{2\pi}{L}, \qquad \nu = \frac{2\pi}{T}, \qquad c = \frac{L}{T} = \frac{\nu}{k},$$

where L is the vertical wavelength, T is the period, and c is the phase speed of the waves. The ratio of the two terms comprising the element in the last row and second column of the matrix (2.5) is

$$\frac{\partial\bar{\alpha}/\partial z}{\bar{\alpha}k} \sim \frac{L}{2\pi H},$$

where H is the vertical scale of the atmosphere ($\sim$10 km). For sound waves this ratio is small; hence the shear term can be neglected here. With this simplification and the foregoing notations the determinant of the reduced system of equations becomes

$$\begin{vmatrix} -\delta_1 ck & \bar{\alpha}k & \dfrac{gi}{\bar{\alpha}} \\ i\left(g - \gamma\bar{p}\dfrac{\partial\bar{\alpha}}{\partial z}\right) & -\bar{\alpha}ck & -\bar{p}\gamma ck \\ \bar{\alpha}k & 0 & \delta_2 ck \end{vmatrix}. \tag{2.6}$$

For a nontrivial solution of the homogeneous system of equations this determinant must vanish, giving

$$\delta_1\delta_2 c^2k^2 - R\bar{T}\gamma k^2 + kig - \delta_2 ki\left(g - \gamma\bar{p}\frac{\partial\bar{\alpha}}{\partial z}\right) = 0, \tag{2.7}$$

where $\bar{p}\bar{\alpha}$ has been replaced by $R\bar{T}$ from the equation of state. If compressibility is retained in the continuity equation and vertical accelerations in the vertical motion equation, both δ_1 and δ_2 are equal to one. In this case two terms in (2.7) cancel. Moreover, for sound waves, the term involving $\partial\bar{\alpha}/\partial z$ is small compared to $R\bar{T}k^2$. Hence, as a good approximation, the phase velocities are given by

$$c \doteq \pm\left(\frac{c_pR\bar{T}}{c_v}\right)^{1/2}, \tag{2.8}$$

which is the Laplacian speed of sound and represents vertical propagation of sound waves in both directions.

On the other hand, if *either* δ_1 or δ_2 is taken to be zero, i.e., if the *incompressible* form of the continuity equation is used *or* the perturbations are assumed to be *hydrostatic*, there are *no* vertically propagating *sound waves*. However, the incompressibility assumption would be a rather drastic limitation when modeling the atmosphere for meteorological purposes; whereas the hydrostatic assumption, which also filters vertically propagating sound waves, is a commonly used approximation to determine upper air data and does not constitute a significant limitation on the investigation of large-scale atmospheric disturbances.

2.3 HORIZONTAL SOUND WAVES AND INTERNAL GRAVITY WAVES

Next consider wave propagation in the x direction only, in which case the second row and column need not be included, and $k \equiv 0$ as well. The frequency equation now becomes

$$\begin{vmatrix} (\mu U - \nu) & \bar{\alpha}\mu & 0 \\ 0 & \bar{\alpha}(\mu U - \nu) & \bar{p}\gamma(\mu U - \nu) \\ \bar{\alpha}\mu & 0 & \delta_2(\mu U - \nu) \end{vmatrix} = 0.$$

Placing $c = \nu/\mu$ and simplifying leads to

$$\delta_2(U - c)^3 - \gamma(U - c)\bar{p}\bar{\alpha} = 0.$$

Now if the compressibility term is included in the continuity equation, $\delta_2 = 1$, and the roots of the frequency equation are the trivial root, $c = U$, and again the phase velocities for sound waves,

$$c = U \pm \left(\frac{c_p R\bar{T}}{c_v}\right)^{1/2}.$$

On the other hand, if the compressibility term in the continuity equation is omitted by taking $\delta_2 = 0$, which is equivalent to assuming nondivergent flow, only the trivial solution, $c = U$, occurs. Thus the sound waves have been filtered out.

In the foregoing treatment, the vertical variation of the dependent variables was neglected as well as the vertical velocity which eliminated the vertical motion equation entirely. If the hydrostatic condition is imposed on the pressure field, the expanded system of equations can be solved to give a particular type of sound wave. To simplify the discussion, assume

$$U = 0, \qquad \frac{\partial \bar{\alpha}}{\partial z} = 0,$$

and again consider the perturbation motion to be strictly horizontal, i.e., $W \equiv 0$. However, a vertical dependency will be permitted by retaining k. The equation for horizontal motion, the thermodynamic and continuity equations, then lead to the same frequency equation (2.8) for the horizontally propagating waves. In addition, the hydrostatic equation together with the equation for horizontal motion and thermodynamic equation can be utilized to determine k, which turns out to be

$$k = \frac{-gi}{\gamma RT}.$$

In accordance with (2.3), this imaginary vertical wave number represents an exponential increase of the perturbation velocities with height by the factor $e^{gz/\gamma RT}$. This type of wave is referred to as a *Lamb wave*.

Returning now to the general case represented by (2.5) and setting the determinant equal to zero results in the frequency equation

$$\delta_1\,\delta_2(\mu U - \nu)^4 - \left[R\bar{T}\gamma(k^2 + \mu^2\delta_1) + gik(\delta_2 - 1) + \delta_2\frac{g}{\bar{\alpha}}\frac{\partial\bar{\alpha}}{\partial z}\right] \times (\mu U - \nu)^2 - g\mu^2\left(g - \gamma R\bar{T}\,\frac{1}{\bar{\alpha}}\frac{\partial\bar{\alpha}}{\partial z}\right) = 0.$$

To simplify this equation, neglect the somewhat smaller terms involving g in the quadratic term and also express the last term as a function of potential temperature. The result is

$$\delta_1 \delta_2(\mu U - \nu)^4 - [R\bar{T}\gamma(k^2 + \mu^2 \delta_1)](\mu U - \nu)^2 + \frac{gR\bar{T}\mu^2\gamma}{\bar{\theta}} \frac{\partial \bar{\theta}}{\partial z} \tag{2.9}$$

The four roots of this equation correspond to a pair of sound waves and a pair of internal gravity waves. Inspection shows that the gravity waves may be removed by omitting the last term involving g. For simplicity also neglect the basic current and place $\delta_2 = \delta_1 = 1$, in which case (2.9) reduces to the well-known sound-speed formula

$$c^2 = \gamma R\bar{T},$$

where

$$c = \nu(k^2 + \mu^2)^{-1/2}.$$

It is evident that either the incompressibility or the hydrostatic assumption will eliminate sound waves which have a vertical component to the phase velocity.

To isolate the gravity waves, take $\delta_2 = 0$ and place $\delta_1 = 1$. Taking $U = 0$ gives the following frequencies (relative to the basic current)

$$\nu^2 = \frac{\mu^2}{k^2 + \mu^2} \frac{g}{\bar{\theta}} \frac{\partial \bar{\theta}}{\partial z}. \tag{2.10}$$

Substituting for ν from the foregoing gives the phase velocities

$$c = \pm \frac{\mu}{\mu^2 + k^2} \left(\frac{g}{\bar{\theta}} \frac{\partial \bar{\theta}}{\partial z}\right)^{1/2}. \tag{2.10a}$$

These gravitational oscillations are stable if the lapse rate is subadiabatic, i.e., if $\partial\bar{\theta}/\partial z > 0$. An amplified disturbance occurs if the lapse rate is superadiabatic, as might be expected, i.e., if $\partial\bar{\theta}/\partial z < 0$.

If the depth of the disturbance is large compared to the horizontal scale, $\mu^2 > k^2$, and from (2.10),

$$\nu \doteq \pm \left(\frac{g}{\bar{\theta}} \frac{\partial \bar{\theta}}{\partial z}\right)^{1/2}. \tag{2.10b}$$

This is the Brunt-Vaisala frequency for essentially vertical oscillations. If the lapse rate is superadiabatic, ν is imaginary and the perturbation amplifies exponentially with time.

Now if the horizontal scale is much larger than the vertical scale, the hydrostatic approximation is valid and δ_1 may be taken to be zero in (2.9). In this case,

$$\nu^2 = \frac{\mu^2}{k^2} \frac{g}{\bar{\theta}} \frac{\partial \bar{\theta}}{\partial z},$$

which also follows from (2.10) with $k^2 > \mu^2$.

Furthermore, for strictly horizontal propagation, $\nu = c\mu$, and the above result becomes

$$c = \pm \frac{1}{k}\left(\frac{g}{\bar{\theta}}\frac{\partial \bar{\theta}}{\partial z}\right)^{1/2}, \tag{2.11}$$

which is the phase velocity for horizontally propagating internal gravity waves. This result will be derived later in slightly different form from another point of view.

2.4 SURFACE GRAVITY WAVES

The next case of horizontally propagating waves to be considered will permit both horizontal and vertical motions; however, the atmosphere will be assumed to be incompressible, i.e., $\alpha' \equiv 0$, and a slightly different procedure will be used.

Since the fluid is incompressible, an upper surface will exist which will be permitted to be free. The linearized equations of motion (2.2) are (dropping primes):

$$\frac{\partial u}{\partial t} + U\frac{\partial u}{\partial x} + \frac{1}{\bar{\rho}}\frac{\partial p}{\partial x} = 0, \tag{2.12a}$$

$$\delta\left(\frac{\partial w}{\partial t} + U\frac{\partial w}{\partial x}\right) + \frac{1}{\bar{\rho}}\frac{\partial p}{\partial z} = 0, \tag{2.12b}$$

$$\frac{\partial u}{\partial x} + \frac{\partial w}{\partial z} = 0. \tag{2.12c}$$

The hydrostatic equation applies to the undisturbed flow

$$\frac{\partial \bar{p}}{\partial z} = -g\bar{\rho}.$$

Integrating this equation from $z = 0$ to the top of the undisturbed fluid H gives

$$g\bar{\rho}H = p_0. \tag{2.13}$$

Next assume the perturbation quantities to be of the harmonic form

$$\begin{aligned} u &= \psi(z)e^{i\mu(x-ct)}, \\ w &= \Phi(z)e^{i\mu(x-ct)}, \\ \frac{p}{\bar{\rho}} &= P(z)e^{i\mu(x-ct)}. \end{aligned} \tag{2.14}$$

Substituting (2.14) into (2.12) and simplifying leads to

$$\begin{aligned} (U - c)\psi(z) + P(z) &= 0, \\ i\mu\delta(U - c)\Phi(z) + P'(z) &= 0, \\ i\mu\psi(z) + \Phi'(z) &= 0. \end{aligned} \tag{2.15}$$

Eliminating $P(z)$ from the first two equations of (2.15) and then further elimination of $\psi(z)$ between the resulting equation and the last equation of (2.15) gives

$$\Phi''(z) - \mu^2\delta\Phi(z) = 0. \tag{2.16}$$

Consider two cases: $\delta = 1$, in which *vertical accelerations* are *permitted* and $\delta = 0$, where the *perturbations are hydrostatic*. The solutions are respectively,*

$$\Phi(z) = a_1 e^{\mu z} + a_2 e^{-\mu z}, \qquad \delta = 1,$$

$$\Phi(z) = a_1' z + a_2', \qquad \delta = 0,$$

where the a's are arbitrary constants to be determined by the boundary conditions. At the lower boundary, which is assumed to be horizontal, the vertical velocity vanishes. Hence $a_1 = -a_2 \equiv a$, when vertical accelerations are present ($\delta = 1$) and

$$\Phi(z) = a(e^{\mu z} - e^{-\mu z}), \qquad \delta = 1. \tag{2.17}$$

Expanding the exponential terms as power series leads to the approximation

$$\Phi(z) = 2a\mu z + \cdots.$$

In the hydrostatic case ($\delta = 0$), $a_2' = 0$, and placing $a_1' = a$ gives

$$\Phi(z) = az. \tag{2.18}$$

The second boundary condition is that the total pressure of a surface particle (which must remain at the boundary) remains unchanged. Hence

$$d(\bar{p} + p)/dt = 0$$

at the free surface. This may be approximated by linearizing and applying this condition at $z = H$. Thus,

$$\frac{\partial p}{\partial t} + U\frac{\partial p}{\partial x} + w\frac{\partial \bar{p}}{\partial z} = 0 \qquad \text{at} \qquad z = H. \tag{2.19}$$

* Note that this exponential solution is of the form (2.3) and hence have could been obtained as a particular case of (2.5).

Utilizing the solutions (2.17) and (2.18) and the system (2.15) gives the following results:

CASE 1. Nonhydrostatic ($\delta = 1$):

$$\begin{aligned} \psi(z) &= ia(e^{\mu z} + e^{-\mu z}), \\ P(z) &= -ia(U - c)(e^{\mu z} + e^{-\mu z}). \end{aligned} \tag{2.20}$$

Substituting (2.17) and (2.20) into (2.19) and simplifying yields the roots of the frequency equation

$$c = U \pm \left(\frac{gL}{2\pi} \tanh \frac{2\pi H}{L}\right)^{1/2}. \tag{2.21}$$

CASE 2. Hydrostatic ($\delta = 0$):

$$\begin{aligned} \psi(z) &= \frac{a'i}{\mu}, \\ P(z) &= -\frac{ia'(U - c)}{\mu}. \end{aligned} \tag{2.22}$$

Substituting (2·18) and (2.22) into (2.19) and simplifying gives

$$c = U \pm \sqrt{gH}. \tag{2.23}$$

Waves traveling with the phase velocity given by (2.23) are generally referred to as "shallow-water" or "long" waves. When the ratio H/L in (2.21) is relatively large (about 0.5 is sufficient), the phase velocity is approximately

$$c \doteq U \pm \sqrt{\frac{gL}{2\pi}}.$$

The waves are then called "deep-water" waves, and the fluid particle trajectories are nearly circular. On the other hand, for small values of H/L (≤ 0.04), (2.21) reduces to (2.23), and the particle trajectories are very elongated ellipses and thus nearly horizontal lines. If one assumes a homogeneous atmosphere, the hydrostatic equation gives $gH = p_0/\rho = RT$, and (2.23) may be written

$$c = U \pm \sqrt{RT}. \tag{2.24}$$

Thus the speed of "long" gravity waves is nearly the speed of sound waves given by (2.8).

The phase velocity for the so-called long gravity waves given by (2.23) was obtained when vertical accelerations were omitted, i.e., when the perturbations are hydrostatic. This result could have been obtained in a somewhat more direct manner as follows. Assuming hydrostatic equilibrium, it follows immediately that for any point in the fluid, $g\bar{\rho}(h - z) = p$, where h is the height of the free surface. Placing $h = H + h'$ and $p = \bar{p} + p'$ leads to $g\bar{\rho}h' = p'$, and thus to (dropping primes)

$$g\frac{\partial h}{\partial x} = \frac{1}{\bar{\rho}}\frac{\partial p}{\partial x},$$

etc.

Thus the first equation in (2.12) may be written as

$$\frac{\partial u}{\partial t} + U\frac{\partial u}{\partial x} + g\frac{\partial h}{\partial x} = 0, \tag{2.25}$$

where u and h are now the perturbation quantities.

A second equation in u and h is obtainable by integrating the continuity equation (2.12c) in the vertical. However, this mass conservation equation may be easily derived directly in terms of h as follows (see Figure 2.1):

$$\frac{\partial(\bar{\rho}h)}{\partial t} = -\frac{\partial(\bar{\rho}uh)}{\partial x}.$$

Linearizing and simplifying, yields

$$\frac{\partial h}{\partial t} + U\frac{\partial h}{\partial x} + H\frac{\partial u}{\partial x} = 0, \tag{2.26}$$

where h and u are now perturbation quantities. Since h and hence u are not functions of z, they may be taken to be of the form

$$u = u_0 e^{i\mu(x-ct)}, \qquad h = h_0 e^{i\mu(x-ct)},$$

where u_0 and h_0 are constants. Substituting the expressions into (2.25) and (2.26) and setting the determinant of the homogeneous system to zero again gives the phase velocity for shallow-water waves,

$$c = U \pm \sqrt{gH},$$

which is identical to (2.23), as might have been expected.

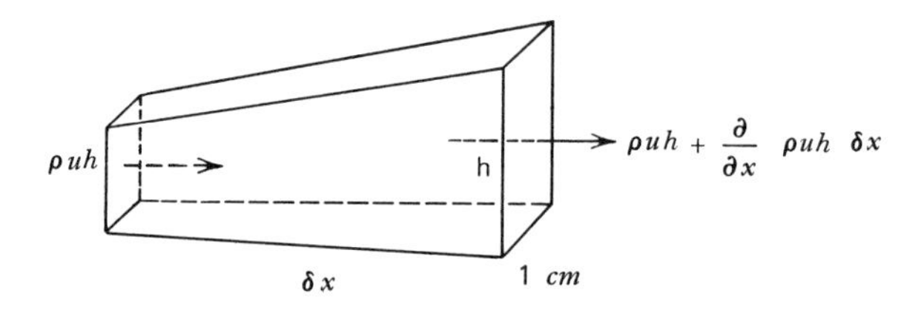

Figure 2.1. Illustrating mass continuity.

The gravity waves just described are usually termed *external*, since their maximum amplitude is at the boundary of the fluid. There are also *internal* gravity waves that may develop on interfaces of density or velocity discontinuity. The propagation speeds of such *internal* gravity waves may be quite different from those mentioned earlier. For example, in the simple case of two semi-infinite layers separated by a zero-order discontinuity of both density and velocity, the wave speed is given by (see Haltiner and Martin)

$$c = \frac{\rho U + \rho' U'}{\rho + \rho'} \pm \left[\frac{gL(\rho - \rho')}{2\pi(\rho + \rho')} - \frac{\rho\rho'(U - U')^2}{(\rho + \rho')^2}\right]^{1/2},$$

where the primes denote the parameters for the upper layer. To obtain this result the perturbations were assumed to vanish at $\pm\infty$. The phase velocity of the "deep-water" waves discussed earlier may be obtained as a special case of the foregoing formula by placing ρ' equal to zero. Other special cases include nonshearing waves, $U = U'$, and pure shearing waves ($\rho = \rho'$). Note that, if the quantity under the radical sign becomes negative, the phase velocity is complex and unstable waves occur.

As discussed in Section 2.3, internal gravity waves are possible without density or velocity discontinuities. As another approach, consider the linearized form of (1.15) through (1.18), assuming a basic current $U(p)$ and perturbations which are of infinite lateral extent, and neglecting rotation and friction

$$\frac{\partial u}{\partial t} + U\frac{\partial u}{\partial x} + \frac{\partial \Phi}{\partial x} + \omega\frac{\partial U}{\partial p} = 0,$$

$$\frac{\partial}{\partial t}\frac{\partial \Phi}{\partial p} + U\frac{\partial}{\partial x}\frac{\partial \Phi}{\partial p} + \sigma\omega = 0,$$

$$\frac{\partial u}{\partial x} + \frac{\partial \omega}{\partial p} = 0, \qquad \sigma = -\frac{\alpha}{\theta}\frac{\partial \theta}{\partial p},$$

where σ will be taken as a constant.

Next take perturbations of the form

$$u = Ae^{i\mu[x+(kp/\mu)-ct]},$$

$$\Phi = Be^{i\mu[x+(kp/\mu)-ct]},$$

$$\omega = We^{i\mu[x+(kp/\mu)-ct]}.$$

For simplicity, the vertical shear may be neglected, i.e., take U to be a constant. Treating A, B, and W as constants and substituting the perturbation quantities into the foregoing differential equations leads to a system of linear equations which may be expressed in matrix form as follows:

$$\begin{pmatrix} (U-c)\mu & \mu & 0 \\ 0 & -\mu k(U-c) & \sigma \\ \mu & 0 & k \end{pmatrix} \begin{pmatrix} A \\ B \\ W \end{pmatrix} = 0.$$

For a nontrivial solution the determinant of this matrix must vanish, giving the frequency equation

$$(U-c)^2 - \frac{\sigma}{k^2} = 0,$$

or

$$c = U \pm \left(\frac{\sigma}{k^2}\right)^{1/2}, \tag{2.27}$$

which is similar to (2.11).

For a positive static stability parameter, the waves are stable and propagate in both directions relative to the basic current at speed of $\pm(\sigma/k^2)^{1/2}$.

When the atmosphere is statically unstable ($\sigma < 0$), (2.27) shows that c has an imaginary part; hence the perturbations are dynamically unstable and, moreover, are stationary relative to the basic current U. The exponential growth rate with time is proportional to $c_i\mu = (\mu/k)\sqrt{|\sigma|}$, which implies a continually increasing rate of growth as the horizontal wavelength $L = 2\pi/\mu$ decreases to zero. The latter implication is an error introduced by the hydrostatic approximation, and a more complete treatment [see (2.10*a*)] gives

$$c_i\mu = \frac{\mu^2}{\mu^2 + k^2}\sqrt{|\sigma|},$$

which shows an asymptotic approach of the growth rate to $\sqrt{|\sigma|}$ as the horizontal wavelength L approaches zero.

2.5 INERTIAL GRAVITY WAVES

In the previous examples of wave motion, the rotation of the earth was neglected; however, it is well known that this factor is very important in large-scale weather phenomena. As an extension of the previous analysis of gravity waves, the rotational effect will be included. Consider a basic constant zonal current U in geostrophic balance, namely,

$$U = -\frac{g}{f}\frac{\partial H}{\partial y}, \tag{2.28}$$

where H is the depth of a fluid of constant density as before. If the perturbations u, v, and h as well as f are taken to be independent of y, the following system of equations results:

$$
\begin{aligned}
\frac{\partial u}{\partial t} + U\frac{\partial u}{\partial x} - fv + g\frac{\partial h}{\partial x} &= 0,\\
\frac{\partial v}{\partial t} + U\frac{\partial v}{\partial x} + fu + g\frac{\partial h}{\partial y} &= 0, \qquad (2.29)\\
\frac{\partial h}{\partial t} + U\frac{\partial h}{\partial x} + H\frac{\partial u}{\partial x} + v\frac{\partial H}{\partial y} &= 0.
\end{aligned}
$$

Treating the coefficient H as a constant and assuming harmonic perturbations of the form $u_0e^{i\mu(x-ct)}$, etc., requires the following determinant to vanish for nontrivial solutions for u_0, v_0, h_0.

$$
\begin{vmatrix}
\mu(U-c)i & -f & g\mu i\\
f & \mu(U-c)i & 0\\
H\mu i & -\dfrac{fU}{g} & \mu(U-c)i
\end{vmatrix} = 0. \qquad (2.30)
$$

Expanding this determinant leads to a cubic frequency equation

$$
(U-c)^3 - gH(U-c) + \frac{cf^2}{\mu^2} = 0. \qquad (2.31)
$$

Approximate roots for this cubic may be found by first assuming that at least one of the phase velocities is considerably slower than that of gravity waves, i.e., $gH \gg (U-c)^2$. In this event the first term in (2.31) may be neglected in comparison to the second, giving

$$
c \doteq \frac{U}{1 + f^2/(\mu^2 gH)}. \qquad (2.32)
$$

For typical values of f, μ, and H for the atmosphere, $f^2/(\mu^2 gH) \sim (L/20)^2$, with L in thousands of km; hence $c \doteq U$, justifying the initial assumption.

The other two roots, which may be expected to represent gravity waves, may be obtained by assuming $c \gg U$, in which case Uf^2/μ^2 may be added to the last term of (2.31) with little error, giving a particularly simple form, namely,

$$
(U-c)^3 - gH(U-c) - \frac{f^2}{\mu^2}(U-c) \doteq 0.
$$

Solving for c gives $c \doteq U$ as above, and

$$
c \doteq U \pm \left(gH + \frac{f^2}{\mu^2}\right)^{1/2}, \qquad (2.33)
$$

which are the phase velocities for *inertial-gravity* waves. As shown above, f^2/μ^2 is generally small compared to gH.

2.6 INERTIAL OSCILLATIONS

It is of interest now to inquire whether horizontal oscillations are possible in the atmosphere when the pressure fluctuations are negligible. For this purpose, neglect the spatial variation of the perturbations and omit the basic current U as well. The resulting equations are simply

$$\frac{\partial u}{\partial t} - fv = 0,$$
$$\frac{\partial v}{\partial t} + fu = 0. \tag{2.34}$$

Next, analogously to the previous procedure, place

$$u = u_0 e^{-i\nu t}, \qquad v = v_0 e^{-i\nu t}, \tag{2.35}$$

which leads to

$$\begin{vmatrix} -i\nu & -f \\ f & -i\nu \end{vmatrix} = 0,$$

and gives a frequency

$$\nu = \pm f. \tag{2.36}$$

The period of an inertial oscillation is $2\pi/\nu$, as seen from (2.35). Substituting for ν from (2.36) gives 12 hr/sin ϕ, which is about 17 hr at 45° latitude. Thus the frequency of inertial oscillations in middle latitudes is considerably higher than for ordinary synoptic waves which have periods of several days to a week. Either of the equations in (2.34) relates u_0 and v_0 for each of the roots $\pm f$ of (2.36), giving

$$v_{0+} = -iu_{0+}, \qquad v_{0-} = iu_{0-}.$$

Hence the complete solution of the system (2.34) is

$$u = u_{0+}e^{-ift} + u_{0-}e^{ift},$$
$$v = -iu_{0+}e^{-ift} + iu_{0-}e^{ift}. \tag{2.37}$$

From (2.34) it is easily seen that

$$\frac{\partial(u^2 + v^2)}{\partial t} = 0, \tag{2.38}$$

which may also be verified from the real parts of (2.37).

Some further insight into the character of inertial oscillations may be gained by consideration of the associated particle trajectories. In natural coordinates, the equations of motion in the absence of a pressure force are (see Haltiner and Martin)

$$\dot{V} = 0 \qquad \text{and} \qquad K_H V^2 = -fV.$$

It follows immediately that the speed is constant, as deduced in (2.38), and the trajectory is anticyclonic with a radius of curvature

$$K_H^{-1} = R_H = -\frac{V}{f}. \tag{2.39}$$

For small to moderate values of wind speed in middle to high latitudes R_H is nearly constant; the trajectory is nearly circular and referred to as an *inertial circle*. The period is obtained by dividing the circumference $2\pi R_H$ by the velocity V giving $2\pi/f$, as before. The particle speed is constant and the velocity vector rotates through a full 360° during one cycle. Hence the hodograph of the inertial velocity vector has the appearance of spokes of a wheel.

A simple extension of this concept, though somewhat artificial, is the superposition of an inertial oscillation $\mathbf{V}'$ on a constant wind current $\mathbf{V}_g$ in geostrophic balance. The hodograph is shown in the adjacent figure where the total wind $\mathbf{V}$ is the sum of the geostrophic and inertial wind. The latter is shown for an entire cycle; however only one value of $\mathbf{V}$ is illustrated.

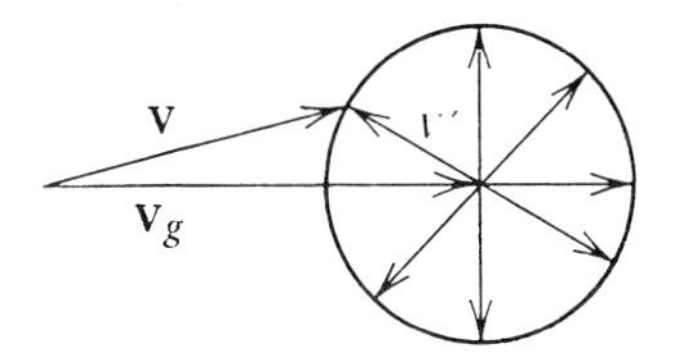

It is apparent that in this case a particle would experience a sinusoidal path and tangential accelerations associated with the cross isobar flow.

The artificiality of this analysis is that the pressure is assumed to be constant whereas, in nature, adjustment in the pressure field would be brought about by gravity waves generated by the imbalance between the mass and wind fields. The adjustment problem has been studied by Rossby, Cahn, Bolin, and others, and has become important recently in relation to initial conditions for integration of the Newtonian equations. Nevertheless there are a number of instances observed in nature where the inertial effects similar to those described previously are evident. For example, the observed paths of constant-pressure balloons show loops and cusps that are suggestive of an inertial influence. Another phenomenon that has been explained to some degree by an inertial oscillation is the boundary-layer jet. Here a diurnal variation in the eddy viscosity can result in a significant nocturnal reduction in the friction force at about 1,000 to 2,000 ft due to the temporal variation of the static stability. The decrease in the friction force, in turn, will set up an ageostrophic wind component which then experiences an inertial oscillation, modified by friction, of course. A classical example of inertial motion was observed in the Baltic Sea by Kullenberg and Gustafson after a squall provided a sudden impulse. Subsequent current measurements clearly showed an inertial oscillation superimposed on a general northward drift.

The solution of the nonlinear equations of motion in the absence of a pressure force may be easily found as follows:

$$\frac{du}{dt} = fv, \qquad \frac{dv}{dt} = -fu. \tag{2.40}$$

Multiplying the second equation by $i = \sqrt{-1}$ and adding gives

$$\frac{d(u + iv)}{dt} + if(u + iv) = 0. \tag{2.41}$$

The solution of this equation for a constant f is

$$V = V_0 e^{-ift}, \tag{2.42}$$

where $V = u + iv$ and V_0 is the initial value. The period of the oscillation is seen to be $2\pi/f$, as obtained earlier.

Although certain phenomena in the atmosphere are explainable, at least in part, by these inertial effects, they do not account for the behavior of the large-scale atmospheric waves observed on synoptic weather charts. The essential physical basis for this phenomenon as discovered by C. G. Rossby, is the variability of the coriolis parameter. In the following section a model which permits both gravity and synoptic waves will be analyzed.

2.7 ROSSBY GRAVITY WAVES

Continuing the investigation of the effects of the earth's rotation on atmospheric wave motions, consider again a frictionless homogeneous incompressible fluid with a horizontal lower boundary and an upper free surface. Assuming the pressure is determined hydrostatically, the equations of motion are expressible in the form

$$\frac{\partial u}{\partial t} + u\frac{\partial u}{\partial x} + v\frac{\partial u}{\partial y} - fv + g\frac{\partial h}{\partial x} = 0, \tag{2.43a}$$

$$\frac{\partial v}{\partial t} + u\frac{\partial v}{\partial x} + v\frac{\partial v}{\partial y} + fu + g\frac{\partial h}{\partial y} = 0. \tag{2.43b}$$

Because of the hydrostatic assumption and constant density, the horizontal pressure force is independent of height. By assuming that the velocity field is initially independent of height, it will remain so; thus the vertical advection terms have been omitted in (2.43). With the incompressibility assumption, the continuity equation is linear:

$$\frac{\partial u}{\partial x} + \frac{\partial v}{\partial y} + \frac{\partial w}{\partial z} = 0. \tag{2.44}$$

Integrating (2.44) with respect to z gives

$$\left(\frac{\partial u}{\partial x} + \frac{\partial v}{\partial y}\right)h + w_h - w_0 = 0. \tag{2.45}$$

In accordance with the kinematic boundary condition, w must vanish at the lower boundary, i.e., $w_0 = 0$. On the other hand, the vertical velocity $w = dz/dt$ at the upper boundary represents the rate at which the free surface is rising. Thus $w_h = dh/dt$, and (2.45) becomes

$$h\left(\frac{\partial u}{\partial x} + \frac{\partial v}{\partial y}\right) = -\frac{dh}{dt} = -\left(\frac{\partial h}{\partial t} + u\frac{\partial h}{\partial x} + v\frac{\partial h}{\partial y}\right), \tag{2.46}$$

which is the same form as obtained in Section 2.3, but derived here in a somewhat different manner. Equations 2.43*a*, 2.43*b*, and 2.46 constitute a system of three equations in three unknowns, u, v, and h. An essentially equivalent system may be obtained by substituting the vorticity equation for one of the equations of motion, namely, (2.43*b*). The vorticity equation for this model is obtained by differentiating (2.43*b*) with respect to x and (2.43*a*) with respect to y and forming the difference, with the result

$$\frac{\partial \zeta}{\partial t} + u\frac{\partial \zeta}{\partial x} + v\frac{\partial \zeta}{\partial y} + \beta v = -(\zeta + f)\left(\frac{\partial u}{\partial x} + \frac{\partial v}{\partial y}\right) - (\zeta + f)\frac{1}{h}\frac{dh}{dt}, \tag{2.47a}$$

which can also be written as

$$\frac{d}{dt}\left(\frac{\zeta + f}{h}\right) = 0. \tag{2.47b}$$

This equation shows that the quantity $(\zeta + f)/h$, called the *potential vorticity*, is conserved. The purpose of the vorticity equation is to introduce the latitudinal variation of the coriolis parameter β, otherwise f would have to be treated as a function of y in (2.43), which would complicate the analysis somewhat. To obtain analytical solutions for the system (2.43*a*), (2.46), and (2.47*a*), the equations will be linearized by the perturbation method. Assuming the zonal current U and the perturbations to be independent of y, the linearized versions of (2.43*a*), (2.46) and (2.47*a*) are

$$\delta\left(\frac{\partial}{\partial t} + U\frac{\partial}{\partial x}\right)u - fv + g\frac{\partial h}{\partial x} = 0, \tag{2.48a}$$

$$\left(\frac{\partial}{\partial t} + U\frac{\partial}{\partial x}\right)h + \frac{\partial H}{\partial y}v + H\frac{\partial u}{\partial x} = 0, \tag{2.48b}$$

$$\left(\frac{\partial}{\partial t} + U\frac{\partial}{\partial x}\right)\frac{\partial v}{\partial x} + \beta v + f\frac{\partial u}{\partial x} = 0. \tag{2.48c}$$

Here H is the mean depth of the fluid, $\partial v/\partial x$ is the vorticity, and u, v, and h now represent the perturbation quantities. The coefficient δ of the linear differential operator in the first equation will identify the source of this term to be the momentum equation. In the undisturbed state (2.43*b*) reduces to

$$fU = -g\frac{\partial H}{\partial y}, \tag{2.49}$$

which may be substituted in (2.48*b*) for $\partial H/\partial y$.

Next harmonic solutions will be assumed for u, v, and h, as done previously, namely,

$$\begin{aligned} u &= u_0 e^{i\mu(x-ct)}, \\ v &= v_0 e^{i\mu(x-ct)}, \\ h &= h_0 e^{i\mu(x-ct)}, \qquad \mu = \frac{2\pi}{L}. \end{aligned} \tag{2.50}$$

To simplify the analysis, f and β will be assumed constant in (2.48), in which case u_0, v_0, and h_0 also may be assumed to be constants. Substituting the expressions (2.50) into (2.48) leads to the algebraic system

$$\begin{aligned} \delta(U - c)i\mu u_0 - fv_0 + gi\mu h_0 &= 0, \\ Hi\mu u_0 - \frac{fU}{g}v_0 + (U - c)i\mu h_0 &= 0, \\ fi\mu u_0 + [\beta - \mu^2(U - c)]v_0 &= 0. \end{aligned} \tag{2.51}$$

A nonzero solution of this homogeneous system for the amplitudes u_0, v_0, h_0 exists only if the determinant of the coefficients is zero, namely,

$$\begin{vmatrix} \delta(U - c) & -f & g \\ f & \beta - \mu^2(U - c) & 0 \\ H & \dfrac{-fU}{g} & U - c \end{vmatrix} = 0.$$

The common factor $i^2\mu^2$ has been omitted. Expanding the determinant leads to

$$[\beta - \mu^2(U - c)][gH - \delta(U - c)(U - c)] - f^2[(U - c) - U] = 0. \quad (2.52)$$

Equation 2.52 determines permissible values of the phase speed c as a function of certain parameters characterizing the mean flow and the superimposed wave. Approximate roots may be found in a manner similar to the treatment of (2.31), i.e., first assume a phase velocity much less than that of gravity waves,

$$(U - c)^2 \ll gH. \quad (2.53)$$

Then (2.52) becomes approximately

$$[\beta - \mu^2(U - c)]gH - f^2[(U - c) - U] \doteq 0.$$

Solving for $(c - U)$ gives the result

$$c \doteq U - \frac{\beta + f^2U/gH}{\mu^2 + f^2/gH}. \quad (2.54)$$

This result is obtainable directly from (2.47b) by linearizing and making the geostrophic approximation for ζ and v. Returning to (2.54), it is seen that, for values of H that correspond in a reasonable sense to the atmosphere and for wavelengths typical of middle latitudes, the second term of (2.54) is quite small so that the tentative assumption (2.53) is justified.

To approximate the remaining roots, tentatively assume that the wave speed greatly exceeds the speed C_R (called the Rossby wave speed).

$$|c| \gg |C_R| \equiv \left| U - \frac{\beta}{\mu^2} \right|.$$

Since C_R and U are of the same order of magnitude,

$$|c| \gg |U|, \qquad |c - U| \gg U.$$

With these approximations, the β term may be omitted in (2.52), as well as the last U in the last term, leaving

$$-\mu^2[gH - \delta(U - c)(U - c)] - f^2 \doteq 0$$

or

$$(U - c)(U - c) \doteq gH + \frac{f^2}{\mu^2}, \qquad \delta \equiv 1. \quad (2.55)$$

Thus $|U - c|$ slightly exceeds $\sqrt{gH}$, the speed of "shallow-water" or "long" gravity waves, which have a velocity nearly the Newtonian speed of sound for a homogeneous atmosphere. This justifies the tentative approximations used to obtain the two roots represented by (2.55), namely,

$$c \doteq U \pm \sqrt{gH + \frac{f^2}{\mu^2}}. \tag{2.56}$$

The solutions corresponding to these two roots are the shallow-water, inertial-gravity waves obtained earlier in Section 2.4, (2.33).

It is apparent from (2.52) and (2.55) that these high-speed gravity waves owe their presence to the term $\delta(U - c)(U - c)$ in the frequency equation. If it is desired to exclude these waves at the onset, steps must be taken to prevent this term from appearing in the frequency equation. This could be accomplished by eliminating the operator giving rise to either $\delta(U - c)$ or $(U - c)$. In the former case, reference to (2.48*a*), from which the operator $\delta(U - c)$ originated, shows that elimination of this operator is equivalent to utilization of the geostrophic value for the meridional velocity component, a well-founded approximation, i.e.,

$$fv = g\frac{\partial h}{\partial x}. \tag{2.57}$$

Thus (2.57) together with (2.48*b*) and (2.48*c*) constitute a complete system which includes the synoptic waves of great importance to the meteorologist, but excludes the high-speed gravity waves. The former are essentially Rossby waves modified somewhat by the influence of gravity introduced through the presence of vertical motions. Care must be taken not merely to introduce the geostrophic approximation for the velocity in both "primitive" equations of (2.43), leaving only the continuity equation (2.46) for prediction purposes since the geostrophic approximation is not suitable for estimating velocity divergence. Nevertheless in nonlinear models the geostrophic approximation may be used for both the u and v velocity components in the *vorticity equation after the horizontal velocity divergence has been replaced by substitution from the continuity equation.* This avoids the pitfall of substituting geostrophic wind divergence for horizontal wind divergence. In a more general treatment of the filtering problem, P. D. Thompson (1956) has shown that a necessary and sufficient condition for filtering gravity waves is the vanishing of the total derivative of horizontal divergence in the divergence equation.

To isolate pure Rossby waves ($c = C_R$), the gravity waves may be excluded by assuming that the motion is strictly horizontal as well as incompressible. In this event the horizontal velocity divergence $\partial u/\partial x$ vanishes in the vorticity equation (2.48*c*), and with the geostrophic approximation the vorticity equation involves but one variable h. Substitution of the harmonic form (2.50) into (2.48*c*) then yields the phase speed for pure Rossby waves, namely,

$$c = U - \frac{\beta}{\mu^2}. \tag{2.58}$$

Haurwitz waves (see Haltiner and Martin) are similar to Rossby waves but of finite lateral extent; the phase speed in this case is

$$c = U - \frac{\beta/\mu^2}{1 + L^2/d^2}, \tag{2.59}$$

where d is the width of the disturbance. Note further that the hydrostatic approximation removes deep-water gravity waves. In all of the foregoing models, sound waves have been excluded by the assumption of incompressibility; however the hydrostatic approximation is sufficient to remove the sound waves as discussed earlier except for the Lamb wave.

In the model for combined synoptic-gravity waves discussed here, the initial assumptions imply no vertical nor lateral variation of the velocity components, neither basic nor the perturbation velocities (2.50). As a result the solution of the differential system for harmonic waves immediately reduced to the algebraic system (2.51) from which the frequency equation was easily obtained.

More complex models in which the perturbation quantities vary laterally and/or vertically are generally more difficult to solve and the "eigenvalue" problem associated with the determination of the necessary and sufficient conditions for dynamic instability is often troublesome indeed.

In this chapter, with one minor exception, only simple linear systems have been examined; however these are adequate to provide general information on several kinds of wavelike motion which are possible in the atmosphere. Nevertheless, for prediction purposes the nonlinear equations must be used simply because natural processes are usually highly nonlinear. Analytical solutions of nonlinear systems are possible only in rather rare instances; hence numerical solutions must be sought. Even here the problem is by no means simple, for numerical methods have their limitations and pitfalls. In Chapter 5 numerical solutions of an elementary prediction equation will be obtained and compared to the analytical solutions in order to illustrate the basic concepts and some of the difficulties involved.

2.8 GEOSTROPHIC ADJUSTMENT PROCESS

A problem which has recently received considerable attention in one form or another is the mutual adjustment of the mass and velocity fields when an initial imbalance exists. Its present interest lies principally in the need for the proper determination of initial conditions of mass and velocity for numerical integration of the primitive equations of motion. C. G. Rossby first posed the adjustment problem in the form of an initial narrow zonal current imposed on a fluid of infinite width with zero pressure gradient and then considered the subsequent mass and velocity changes. A. Cahn gave an analytical treatment of this problem which has been investigated further by Bolin, Obukhov, Blumen, and more recently with some numerical integrations by Winninghoff (Ph.D. Thesis, UCLA), which are described below. Consider a one-layer barotropic fluid of depth H with these initial conditions imposed at $t = 0$,

$$\begin{aligned} u &= u_0, & |y| &\leq a, \\ u_0 &= 0, & |y| &> a, \\ v_0 &= 0, & (\nabla h)_0 &= 0. \end{aligned} \tag{2.60}$$

Since the lateral transport of mass is of principal interest here, variations in the x direction will be neglected. The resulting linear equations are, following Winninghoff,

$$\frac{\partial u}{\partial t} - fv = 0, \tag{2.61}$$

$$\frac{\partial v}{\partial t} + fu + g\frac{\partial h}{\partial y} = 0, \tag{2.62}$$

and

$$\frac{\partial h}{\partial t} + H\frac{\partial v}{\partial y} = 0. \tag{2.63}$$

Now differentiate (2.62) with respect to time and (2.63) with respect to y, giving

$$\frac{\partial^2 v}{\partial t^2} + f\frac{\partial u}{\partial t} = -g\frac{\partial^2 h}{\partial y\,\partial t}, \tag{2.64}$$

and

$$\frac{\partial^2 h}{\partial y\,\partial t} = -H\frac{\partial^2 v}{\partial y^2}. \tag{2.65}$$

Substituting from (2.61) and (2.65) into (2.64) yields an equation for v:

$$\frac{\partial^2 v}{\partial t^2} + f^2 v - gH\frac{\partial^2 v}{\partial y^2} = 0. \tag{2.66}$$

If wave-type solutions of the form $Ae^{i(ky-\nu t)}$ are sought, the inertial-gravity waves propagate with the frequency [(2.56) with $c = \nu/k$],

$$\nu = \pm\sqrt{f^2 + gHk^2}, \tag{2.67}$$

where k is the wave number in the y direction. However, for an analysis of the adjustment problem, the process of dispersion must be included. For this purpose the solution may be represented by a Fourier integral

$$v(y, t) = \frac{1}{2\pi} \int_{-\infty}^{\infty} \tilde{v}(k, t) e^{iky}\, dk, \tag{2.68}$$

where

$$\tilde{v}(k, t) = \int_{-\infty}^{\infty} v(y, t) e^{-iky}\, dy. \tag{2.69}$$

The integrals, (2.68) and (2.69), are said to be the Fourier transforms of one another. Next the differential equation (2.66) may be transformed into the (k, t) space by multiplying by e^{-iky} and integrating with respect to y from $-\infty$ to $+\infty$. The first two terms are easily seen to be $\partial^2 \tilde{v}(k, t)/\partial t^2$ and $f^2 \tilde{v}(k, t)$, while the third term $\partial^2 v/\partial y^2$ transforms to $-k^2 \tilde{v}(k, t)$. Thus the transformed equation becomes

$$\frac{\partial^2 \tilde{v}(k, t)}{\partial t^2} + (f^2 + gHk^2)\tilde{v}(k, t) = 0. \tag{2.70}$$

Equation 2.70 has the general solution

$$\tilde{v}(k, t) = A(k) \cos \nu t + B(k) \sin \nu t, \tag{2.71}$$

where ν is defined earlier in (2.67). To determine $A(k)$, place $t = 0$ in (2.71) and (2.69), giving

$$A(k) = \frac{1}{2\pi} \int_{-\infty}^{\infty} v(y, 0) e^{-iky}\, dy. \tag{2.72}$$

Furthermore, from (2.71) and (2.69) it may be seen that

$$\frac{\partial \tilde{v}(k, t)}{\partial t} = -\nu A(k) \sin \nu t + \nu B(k) \cos \nu t$$

and

$$\frac{\partial \tilde{v}(k, 0)}{\partial t} = \nu B(k) = \int_{-\infty}^{\infty} \frac{\partial v(y, 0)}{\partial t} e^{-iky}\, dy.$$

Thus

$$B(k) = \frac{1}{\nu} \int_{\infty}^{\infty} e^{-iky} \frac{\partial v(y, 0)}{\partial t}\, dy. \tag{2.73}$$

It follows from (2.62), using the initial conditions of (2.60), that

$$\frac{\partial v(y, 0)}{\partial t} = -fu_0, \qquad y \leq |a|.$$

From (2.72), $A(k) \equiv 0$; and from (2.73)

$$B(k) = \frac{1}{\nu} \int_{-\infty}^{\infty} -fe^{-iky} u_0\, dy = \frac{fu_0 e^{-iky}}{\nu ik} \bigg|_{-a}^{a},$$

or

$$B(k) = -\frac{2fu_0}{k\nu}\sin ka.$$

Substituting these values of $A(k)$ and $B(k)$ into (2.71) gives

$$\tilde{v}(k, t) = -\frac{2fu_0}{k\nu}\sin ka \sin \nu t, \tag{2.74}$$

and, furthermore, from (2.68),

$$v(y, t) = \frac{-2fu_0}{2\pi}\int_{-\infty}^{\infty}\frac{\sin ka}{k}\frac{\sin \nu t}{\nu}e^{iky}\,dk. \tag{2.75}$$

Since the initial distribution is symmetric with respect to y, the foregoing solution may be written as the cosine integral

$$v(y, t) = \frac{-2fu_0}{\pi}\int_{0}^{\infty}\frac{\sin ka}{k}\frac{\sin \nu t}{\nu}\cos ky\,dk,$$

with ν given by (2.67).

Some interesting results may be obtained with typical values of the various parameters; e.g., $u_0 = 10$ m/sec, $f = 10^{-4}$ sec^{-1}, $H = 10^3$ m, and the radius of deformation, λ,

$$\lambda \equiv \left(\frac{gH}{f^2}\right)^{\frac{1}{2}} = 1{,}000 \text{ km}.$$

The comparatively small value of H is chosen to reduce the maximum group velocity $\sqrt{gH}$ of the external (surface) gravity waves to a value closer to that of internal gravity waves in the atmosphere, which would mainly effect the mass redistribution in an actual situation of imbalance.

Figures 2.2, 2.3, and 2.4, after Winninghoff, which are similar to Cahn's results, show the surface height as a function of time at $y = -a$, the surface height as a function of y after 80 hr, and the lateral velocity as a function of time at $y = 0$.

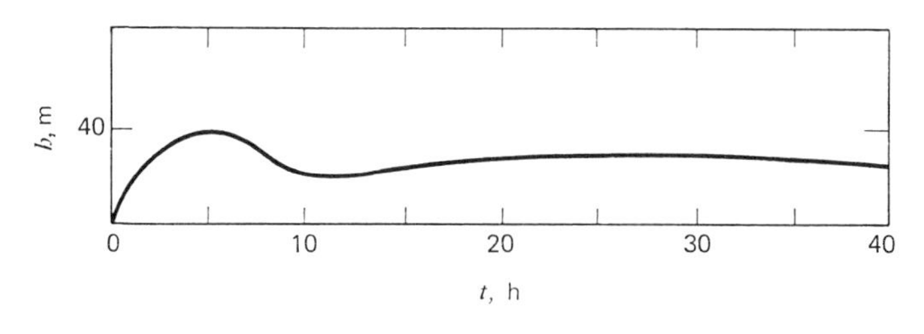

Figure 2.2. Surface height as a function of time at $y = -a = -500$ km.

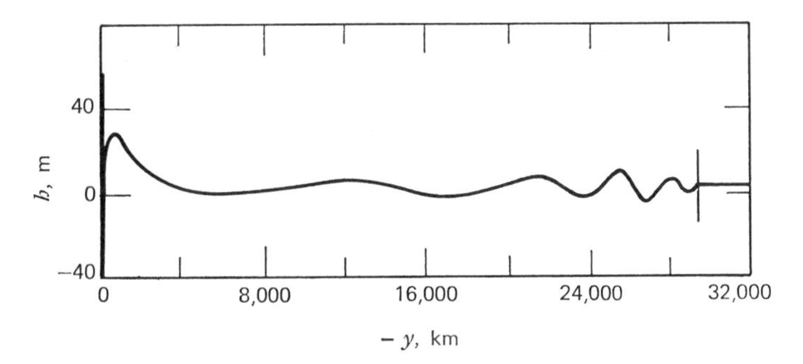

Figure 2.3. Surface height as a function of y after 80 hr; $a = 500$ km.

Note an initial rise of the surface at point $y = -a$ in Figure 2.2 as mass is transported rapidly to the right of the center of the current, $y = 0$. The surface then oscillates as inertial gravity waves propagate laterally across the current with the oscillations diminishing with time as the current tends toward a geostrophic balance. The vertical line in Figure 2.3 at slightly less than 30,000 km shows the present extent of the region of influence which propagates at the maximum group velocity $G = \sqrt{gH}$. More generally,

$$G = c - L\frac{dc}{dL}.$$

Figure 2.4 shows the variation of the lateral velocity with time at the middle of the initial current $y = 0$. The initial displacement of the mass to the right is reflected in the negative velocities in the first few hours.

Inertial-gravity motions in the atmosphere are small in amplitude, since they tend to be damped by energy dispersion in the spectrum of wavelengths normally present. As a result, the total energy of these motions tends to be dispersed into a larger and larger region and is eventually dissipated by

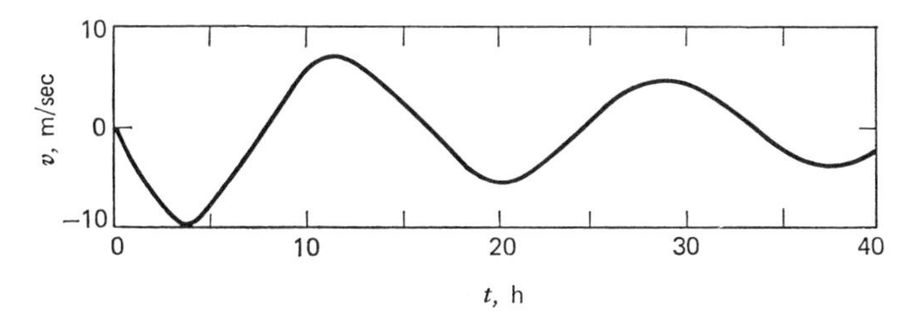

Figure 2.4. Lateral velocity as a function of time at $y = 0$; $a = 2{,}000$ km.

friction. With respect to synoptic disturbances, an important function of the *inertial gravity* waves is *to adjust imbalances between the pressure and wind fields* which then tend toward a quasi-geostrophic state, a process usually referred to as *geostrophic adjustment.*

The precise nature of this process depends critically on the scale L. If the scale is large compared to the radius of deformation, $L > \lambda$, the adjustment is effected primarily through the wind field changes; whereas for smaller scales, $L < \lambda$, the mass field changes more rapidly.

In the preceding problem a zonal velocity field was imposed instantaneously on a previously stationary fluid in equilibrium. The coriolis force then generated laterally propagating inertial-gravity waves which redistributed the mass and energy. An analogous problem consists of suddenly imposing a laterally varying mass distribution on a stationary fluid in equilibrium and then determining the character of the evolving velocity field and the redistribution of mass.

The basic equations for this problem are identical to the previous set, (2.61) through (2.63), but here it is convenient to obtain a differential equation for the surface height rather than velocity, after which the preceding method may be followed. In this instance, however, the final form of the surface may be obtained indirectly by utilizing the potential vorticity equation for the model, which is easily shown to be

$$\frac{\partial q}{\partial t} = 0,$$

where

$$q = \frac{\partial v}{\partial x} - \frac{\partial u}{\partial y} - \frac{fh}{H}. \tag{2.76}$$

When the condition of equality for the initial and final values of potential vorticity is imposed, together with the assumption that the final state is geostrophic, it is fairly straightforward to obtain the expressions for the limiting distributions of height and velocity. In the case where the initial height distribution is given by $h = h_0$, $|y| \leqq a$, and $h_0 = 0$, $|y| > a$, the resulting fields after an infinite time are symmetrical about $y = 0$ with westerly winds to the north and easterly winds to the south. Since the potential vorticity must remain zero for $|y| > a$, the increase in h there must be offset by lateral shear in accordance with (2.76).

There are two important problems (e.g., see Arakawa, 1970) that arise in connection with seeking the numerical solution of the primitive equations of motion, (1.15) through (1.18), for the purpose of weather prediction. The first is to properly simulate the geostrophic adjustment of the atmosphere to its characteristic quasi-geostrophic, quasi-nondivergent state, for example, with respect to the initial conditions where there may be erroneous imbalances between the mass and velocity fields because of sparse and inaccurate data. Secondly, the numerical integration scheme must correctly predict the large-scale flow of the atmosphere and, thus, the transformations between the potential and kinetic energies.

chapter three
Scale Analysis

3.1 INTRODUCTION

A scale analysis, which will be carried out in this chapter, provides a systematic method of comparing the magnitudes of the various terms comprising the hydrodynamical equations governing atmospheric motions. This theory, together with energy considerations, permits the design of consistent dynamical-mathematical models for dynamical analysis and numerical weather prediction. J. Charney introduced this technique to the study of large scale meteorological dynamics in 1948; further developments have been added since that time, especially by Burger (1958) and Green (1960).

The physical variables are assumed to have characteristic values and scale in space and time as follows:

L = characteristic horizontal scale (roughly a quarter wave-length of disturbances);
D = characteristic vertical scale;
H = scale height of the atmosphere (usually the troposphere);
V, W = characteristic horizontal and vertical velocities;
L/V = characteristic period ($\sim$advective period). For example, for typical synoptic waves $L \sim 1{,}000$ km, $H \sim 10$ km, $V \sim 10$ m/sec or sometimes more.

It is further assumed that the approximate magnitudes of the derivatives are as follows:

$$\frac{\partial v}{\partial x} \sim \frac{\partial u}{\partial x} \sim \frac{V}{L}, \text{ etc.,} \qquad \frac{\partial u}{\partial z} \sim \frac{V}{D}, \qquad \frac{\partial u}{\partial t} \sim \frac{V}{L/V} = \frac{V^2}{L}.$$

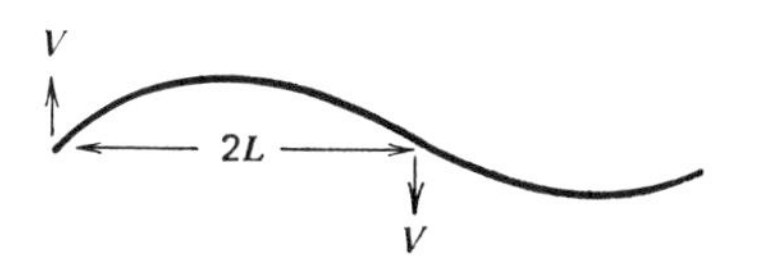

Figure 3.1. Illustrating characteristic length.

The equations of frictionless motion to be analyzed are

$$
\begin{aligned}
&\frac{d\mathbf{V}_3}{dt} + 2\boldsymbol{\Omega} \times \mathbf{V}_3 = -\alpha \boldsymbol{\nabla}_3 p + \mathbf{g}, \\
&\frac{1}{\rho}\frac{d\rho}{dt} = -\boldsymbol{\nabla}_3 \cdot \mathbf{V}_3, \\
&\frac{c_p T}{\theta}\frac{d\theta}{dt} = Q, \qquad \theta = T\left(\frac{P}{p}\right)^{R/c_p} \\
&p\alpha = RT, \qquad \alpha = \frac{1}{\rho},
\end{aligned}
\tag{3.1}
$$

where the subscript 3 denotes three-dimensional space.

3.2 THERMODYNAMIC VARIABLES

Assume the atmosphere to be of constant temperature $\bar{\theta}$ and integrate the hydrostatic equation $dp = -g\rho\, dz$. The result is

$$p = Pe^{-gz/R\bar{\theta}} = Pe^{-z/H},$$

where

$$H = \frac{R\bar{\theta}}{g}, \tag{3.2}$$

defines the scale height to be the level where the atmospheric pressure is equal to P/e, i.e., about 37% of its sea-level value.

Next write the thermodynamic variables in dimensionless form, as perturbations from a reference (or standard) state which depends on z only, as follows:

$$
\begin{aligned}
p &= p_s(z) + \delta p = P(p_s' + \delta p'), \\
T &= T_s(z) + \delta T = \bar{\theta}(T_s' + \delta T'), \\
\theta &= \theta_s(z) + \delta\theta = \bar{\theta}(\theta_s' + \delta\theta'), \\
\rho &= \rho_s(z) + \delta\rho = \frac{P}{R\bar{\theta}}(\rho_s' + \delta\rho') = \frac{P}{gH}(\rho_s' + \delta\rho'), \\
\alpha &= \alpha_s(z) + \delta\alpha = \frac{R\bar{\theta}}{P}(\alpha_s' + \delta\alpha') = \frac{gH}{P}(\alpha_s' + \delta\alpha').
\end{aligned}
\tag{3.3}
$$

The dimensionless quantities T'_s, p'_s, ρ'_s, α'_s, θ'_s, are clearly near unity whereas $\delta T'$, $\delta p'$, $\delta\rho'$, $\delta\alpha'$ and $\delta\theta'$ may be expected to be significantly less than unity. It is readily shown that the dimensionless variables for the standard atmosphere obey the following

$$\begin{aligned} \rho'_s &= \frac{1}{\alpha'_s}, & p'_s &= \rho'_s T'_s, \\ \theta'_s &= T'_s p_s'^{-R/c_p}, & H\frac{\partial p'_s}{\partial z} &= -\rho'_s. \end{aligned} \tag{3.4}$$

If the vertical coordinate is scaled by means of the height H, i.e., $\partial A_s/\partial z = H^{-1}\,\partial A_s/\partial z'$, the hydrostatic relation becomes

$$\frac{\partial p'_s}{\partial z'} = -\rho'_s \qquad \text{or} \qquad \frac{\partial \ln p'_s}{\partial z'} = -\frac{1}{T'_s}. \tag{3.5}$$

In general, the thermodynamic variables representing the reference state of the atmosphere will be scaled vertically with the parameter H, whereas the dynamic variables $\delta p'$, $\delta T'$, etc., are scaled according to the characteristic scale length, which for the vertical direction is D. Introducing the dimensionless variables into the equation of state gives

$$P(p'_s + \delta p')\,\frac{R\bar{\theta}}{P}(\alpha'_s + \delta\alpha') = R\bar{\theta}(T'_s + \delta T').$$

Neglecting the product of perturbations and using (3.4) gives

$$p'_s\,\delta\alpha' + \alpha'_s\,\delta p' = \delta T',$$

and

$$\frac{\delta\alpha'}{\alpha'_s} + \frac{\delta p'}{p'_s} = \frac{\delta T'}{T'_s}. \tag{3.6}$$

3.3 CONTINUITY EQUATION

In spherical curvilinear coordinates the equation of continuity is

$$\begin{array}{cccccccc} \dfrac{d\ln\rho}{dt} & + \dfrac{\partial u}{\partial x} & + \dfrac{\partial v}{\partial y} & - \dfrac{v\tan\varphi}{r} & + \dfrac{\partial w}{\partial z} & + \dfrac{2w}{r} & = 0, \\ \dfrac{V}{L} & \dfrac{V}{L} & \dfrac{V}{L} & \dfrac{V}{a} & \dfrac{W}{D} & \dfrac{2W}{a} & \sim 0. \end{array} \tag{3.7}$$

(The expression below each term gives its magnitude.) Since the radial distance from the center of the earth, r, is approximately equal to the earth's radius a,

$$\frac{2w}{r} \sim \frac{2W}{a} \ll \frac{W}{D}.$$

Also

$$\frac{1}{\rho}\frac{d\rho}{dt} \sim \frac{1}{\rho}\frac{\Delta\rho}{L/V} \leq \frac{V}{L}.$$

It follows at once that the last term in (3.7) may be neglected. Moreover, the first three terms in (3.7) are all of order V/L or less; hence

$$\frac{W}{D} \leqslant \frac{V}{L} \quad \text{or} \quad W \leqslant \frac{D}{L} V. \tag{3.8}$$

For large-scale motions, $D/L \ll 1$; therefore

$$W \ll V. \tag{3.9}$$

Empirical evidence indicates that the terms $\partial u/\partial x$ and $\partial v/\partial y$ tend to cancel one another and that their sum is an order of magnitude less than each individually; therefore the inequality in (3.8) holds, namely,

$$W < \frac{D}{L} V. \tag{3.10}$$

This will also be demonstrated later by a scale analysis of the thermodynamic equation.

3.4 HORIZONTAL EQUATION OF MOTION

Next the horizontal equation of motion will be considered. Obviously the coriolis term $2w\Omega \cos \varphi$ may be neglected in comparison to the contribution due to the horizontal motion, namely, $2V\Omega \sin \varphi$, by virtue of (3.9); hence the horizontal equation of motion may be written

$$\frac{d\mathbf{V}}{dt} + f\mathbf{k} \times \mathbf{V} = -\alpha\nabla p$$

or

$$\frac{\partial \mathbf{V}}{\partial t} + (\mathbf{V}\cdot\nabla)\mathbf{V} + w\frac{\partial \mathbf{V}}{\partial z} + f\mathbf{k} \times \mathbf{V} = -\alpha\nabla p. \tag{3.11}$$

$$\frac{V^2}{L} \qquad \frac{V^2}{L} \qquad W\frac{V}{D} \qquad fV.$$

The magnitude of each of the terms on the left side of (3.11) is given directly below.

Now $\bar{f} = 2\Omega \sin \varphi \sim 10^{-4}$ sec^{-1}, whereas for large-scale motions, $V/L \sim 10^{-5}$. Consequently the horizontal acceleration is one order of magnitude less than the coriolis force. It follows that for this scale of motions the pressure force must be the same order of magnitude as the coriolis force.

The ratio of the acceleration to the coriolis force is referred to as the Rossby number R_0,

$$R_0 = \frac{V}{\bar{f}L}, \tag{3.12}$$

which, for typical large-scale flow, has a magnitude of about 10^{-1}. It is apparent that a sufficient condition for the validity of the geostrophic approximation is that $R_0 \leqslant 0.1$; for then the acceleration is an order of magnitude less than both the coriolis and pressure forces which are in approximate balance.

More generally, (3.11) may be written in the dimensionless form by using (3.3), other characteristic values, and the Rossby number as follows:

$$R_0 \frac{d\mathbf{V}'}{dt'} + \frac{f}{\bar{f}} \mathbf{k} \times \mathbf{V}' = -\frac{H}{D}\frac{R_0}{F} \alpha' \boldsymbol{\nabla}' \, \delta p', \tag{3.13}$$

where $\bar{f}$ is the mean latitude and

$$t = \frac{L}{V} t', \qquad \mathbf{V} = V\mathbf{V}', \qquad \boldsymbol{\nabla} = L^{-1}\boldsymbol{\nabla}', \qquad \alpha = \frac{gH}{P}\alpha', \tag{3.14}$$

and

$$F = \frac{V^2}{gD} \qquad \text{(Froude number).}$$

It is evident from (3.13), since $V' \sim 1 \sim \alpha'$, that

$$\begin{aligned} \delta p' &\sim DF/HR_0 \qquad \text{for} \qquad R_0 < 1, \\ \delta p' &\sim DF/H \qquad \text{for} \qquad R_0 \geqslant 1, \end{aligned} \tag{3.15}$$

or

$$\delta p' \sim DF/HR_1,$$

where

$$\begin{aligned} R_1 &= R_0 \qquad \text{for} \qquad R_0 < 1, \\ R_1 &= 1 \qquad \text{for} \qquad R_0 \geqslant 1. \end{aligned}$$

Hence a small value of R_0 or a large F tends to be associated with a large pressure perturbation.

3.5 VERTICAL EQUATION OF MOTION

The vertical equation of motion may be written in spherical coordinates as (see Chapter 1):

$$\frac{dw}{dt} - \frac{u^2 + v^2}{r} - u2\Omega \cos \varphi + g + \alpha \frac{\partial p}{\partial z} = 0.$$

Considering motions such that $V \sim 10$ m/sec, $f \sim 10^{-4}$ sec^{-1}, it is apparent that the second and third terms are at least four orders of magnitude less than the force of gravity; hence the foregoing equation may be immediately reduced to

$$\frac{dw}{dt} + g + \alpha \frac{\partial p}{\partial z} \doteq 0. \tag{3.16}$$

Introducing dimensionless variables leads to

$$\frac{VW}{L}\frac{dw'}{dt'} + g + \frac{gH}{P}(\alpha_s' + \delta\alpha')\frac{P}{H}\frac{\partial p_s'}{\partial z'} + \frac{gH}{P}(\alpha_s' + \delta\alpha')\frac{P}{D}\frac{\partial\, \delta p'}{\partial z'} = 0,$$

which reduces to

$$\frac{VW}{L}\frac{dw'}{dt'} - g\frac{\delta\alpha'}{\alpha_s'} + \frac{gH}{D}(\alpha_s' + \delta\alpha')\frac{\partial\, \delta p'}{\partial z'} = 0. \tag{3.17}$$

The first term has magnitude VW/L, while the last term, by (3.15), has magnitude V^2/DR_1. Hence the ratio of the acceleration to the vertical perturbation pressure force is

$$\frac{(VW/L)HR_1}{(gH/D)DF} \sim \frac{WDR_1}{VL} \lessapprox \frac{D^2R_1}{L^2} \lessapprox \frac{D^2}{L^2},$$

which shows that the condition $D^2/L^2 \ll 1$ is a sufficient condition for the *hydrostatic approximation.* This latter is clearly valid for the typical large-scale synoptic motions because D is very small compared to L and an advective time scale prevails; however it is not valid, for example, when studying convection where $D \sim L$. With the hydrostatic approximation (3.17) reduces to

$$\begin{aligned} \frac{\delta\alpha'}{\alpha_s'} &= \frac{H}{D}(\alpha_s' + \delta\alpha')\frac{\partial\, \delta p'}{\partial z'}, \\ \frac{\delta\alpha'}{\alpha_s'} &\sim \frac{H}{D}\frac{DF}{HR_1} = \frac{F}{R_1} \sim \frac{\delta\rho'}{\rho_s'}. \end{aligned} \tag{3.18}$$

Using this relationship and (3.15) in (3.6) gives

$$\frac{\delta T'}{T_s'} \sim \frac{F}{R_1} + \frac{DF}{HR_1} \sim \frac{F}{R_1}. \tag{3.19}$$

The plus signs as used here obviously have no significance but merely delineate the various terms. Similarly, the introduction of dimensionless quantities into the definition of potential temperature leads to

$$\frac{\delta\theta'}{\theta_s'} = \frac{\delta T'}{T_s'} - \frac{R}{c_p}\frac{\delta p'}{p_s'},$$

hence

$$\frac{\delta\theta'}{\theta_s'} \sim \frac{F}{R_1} + \frac{R}{c_p}\frac{DF}{HR_1} \sim \frac{F}{R_1}. \tag{3.20}$$

3.6 FIRST LAW OF THERMODYNAMICS

The adiabatic thermodynamic equation may be written in the form

$$\frac{\partial \ln \theta}{\partial t} + \mathbf{V}\cdot\nabla \ln \theta + w\frac{\partial \ln \theta}{\partial z} = 0.$$

The several terms are expressible as

$$\left(\frac{\partial}{\partial t} + \mathbf{V}\cdot\nabla\right)\ln\left[\theta_s\left(1 + \frac{\delta\theta}{\theta_s}\right)\right] = \left(\frac{\partial}{\partial t} + \mathbf{V}\cdot\nabla\right)\ln\left(1 + \frac{\delta\theta'}{\theta_s'}\right) \lessdot \frac{V}{L}\frac{F}{R_1}$$

$$\begin{aligned} w\frac{\partial}{\partial z}\ln\left[\theta_s\left(1 + \frac{\delta\theta'}{\theta_s'}\right)\right] &\doteq w\frac{1}{\theta_s}\frac{\partial\theta_s}{\partial z} + w\frac{\partial}{\partial z}\ln\left(1 + \frac{\delta\theta'}{\theta_s'}\right) \\ &\doteq w\frac{\partial \ln \theta_s}{\partial z} + w\frac{\partial}{\partial z}\frac{\delta\theta'}{\theta_s'} \\ &\sim \frac{W\sigma_s}{D} + \frac{WF}{DR_1}, \end{aligned}$$

where

$$\sigma_s \equiv D\frac{\partial \ln \theta_s}{\partial z}. \tag{3.21}$$

is a static stability parameter based on a standard atmosphere and has a value of about 0.1. It follows that

$$\frac{VF}{LR_1} + \frac{W\sigma_s}{D} + \frac{WF}{DR_1} \sim 0. \tag{3.22}$$

Now for the synoptic scale motions the Froude number has a value of $F \sim 10^{-3}$; hence $W\sigma_s/D \gg WF/DR_1$. Moreover the terms on which (3.22) is based must add to zero; therefore the first two terms must be of the same order of magnitude but generally of opposite sign. It follows that

$$\frac{VF}{LR_1} \sim \frac{W\sigma_s}{D} \quad \text{or} \quad W \sim \frac{DVF}{L\sigma_s R_1}, \tag{3.23}$$

which provides a somewhat more precise estimate of vertical velocity than given by equation (3.9). This result shows that *vertical velocity increases* with the *square* of the *horizontal wind*, with *decreasing static stability* σ_s, and with *increasing latitude* for $R_1 = R_0$. Moreover $F/\sigma_s R_1 < 1$ for large-scale flow; hence $W/D < V/L$ as indicated in (3.10).

3.7 HORIZONTAL ACCELERATION

The acceleration of the horizontal wind may be expanded to

$$\begin{aligned} \frac{d\mathbf{V}}{dt} &= \left(\frac{\partial}{\partial t} + \mathbf{V}\cdot\nabla\right)\mathbf{V} + w\frac{\partial \mathbf{V}}{\partial z}, \\ \frac{V^2}{L} &\sim \frac{V^2}{L} + \frac{WV}{D} \sim \frac{V^2}{L} + \frac{V^2F}{L\sigma_s R_1}. \end{aligned} \tag{3.24}$$

For large-scale motions $F/\sigma_s R_1 \sim 10^{-1}$; hence the vertical advection of momentum is usually about one order of magnitude less than the horizontal advection of momentum for large-scale flow.

3.8 CONTINUITY EQUATION

We shall now return to the continuity equation (3.7) for more precise approximations. With the equation of state, potential temperature may be expressed in the form

$$\ln \theta = -\ln \rho + \frac{c_v}{c_p}\ln p + \text{const.}$$

Hence

$$\frac{\partial \ln \rho}{\partial z} = \frac{c_v}{c_p}\frac{\partial \ln p}{\partial z} - \frac{\partial \ln \theta}{\partial z} = -\frac{c_v}{c_p}\frac{g}{RT} - \frac{\partial \ln \theta}{\partial z}.$$

Substituting this result into the continuity equation gives

$$\left(\frac{\partial}{\partial t} + \mathbf{V}\cdot\nabla\right)\ln \rho - \frac{c_v}{c_p}\frac{gw}{RT} - w\frac{\partial \ln \theta}{\partial z} + \nabla\cdot\mathbf{V} + \frac{\partial w}{\partial z} = 0.$$

The magnitudes of the various terms are as follows:

$$(a) \qquad \left(\frac{\partial}{\partial t} + \mathbf{V} \cdot \nabla\right) \ln \rho \sim \frac{V}{L} \frac{F}{R_1},$$

$$(b) \qquad \frac{c_v}{c_p} \frac{gw}{RT} \sim \frac{c_v}{c_p} \frac{W}{H} \sim \frac{c_v}{c_p} \frac{D}{H\sigma_s} \frac{F}{R_1} \frac{V}{L},$$

$$(c) \qquad w \frac{\partial \ln \theta}{\partial z} \sim \frac{W\sigma_s}{D} \sim \frac{F}{R_1} \frac{V}{L}, \tag{3.25}$$

$$(d) \qquad \nabla \cdot \mathbf{V} \leqslant \frac{V}{L},$$

$$(e) \qquad \frac{\partial w}{\partial z} \sim \frac{W}{D} \sim \frac{F}{\sigma_s R_1} \frac{V}{L}.$$

Since $F/\sigma R_1 \sim 10^{-1}$ for large-scale disturbances, it follows that

$$\nabla \cdot \mathbf{V} < \frac{V}{L}, \tag{3.26}$$

which establishes that $\partial u/\partial x$ and $\partial v/\partial y$, each of order V/L, tend to be of opposite sign in general and $\delta = \partial u/\partial x + \partial v/\partial y$ is of the order

$$\delta \sim 10^{-1} \frac{V}{L} \sim 10^{-6} \sec^{-1}. \tag{3.27}$$

For *deep disturbances* $D \sim H$; also $\sigma_s < 1$. Then term (*b*) exceeds terms (*a*) and (*c*), so that the continuity equation may be approximated as

$$\nabla \cdot \mathbf{V} + \frac{\partial w}{\partial z} - \frac{c_v}{c_p} \frac{gw}{RT} \doteq 0 \qquad \text{(deep disturbances)}, \tag{3.28}$$

or, for simplicity, including both terms involving the vertical advection of density,

$$\nabla \cdot \mathbf{V} + \frac{\partial w}{\partial z} + w \frac{\partial \ln \rho}{\partial z} \doteq 0. \tag{3.29}$$

When the disturbances are shallow, $D/H < 1$. The last term in (3.29) can also be dropped and only (*d*) and (*e*) need be retained. The continuity equation then reduces to the incompressible form; i.e., for $(c_v D/c_p H) < 1$, $\sigma_s < 1$, we have

$$\nabla \cdot \mathbf{V} + \frac{\partial w}{\partial z} \doteq 0 \qquad \text{(shallow disturbances)}. \tag{3.30}$$

This latter case is referred to as *quasi-Boussinesq* or *quasi-incompressible*. However, in treating convection, the compressibility must be retained in calculating the buoyancy term for the vertical momentum equation.

As a final simplification of (3.30), the last term, $\partial w/\partial z$, is sometimes omitted, which is justified if $F/\sigma_s R_1$ is at least two orders of magnitude less than unity. Then even if the two major terms comprising $\nabla \cdot \mathbf{V} = \partial u/\partial x + \partial v/\partial y - (v/a)\tan\varphi$, namely, $\partial u/\partial x$ and $\partial v/\partial y$ are of the same order of magnitude and opposite sign (making $\nabla \cdot \mathbf{V}$ generally one order of magnitude less than either individually), the term $\partial w/\partial z$ can be omitted, reducing (3.30) to

$$\nabla \cdot \mathbf{V} = 0 \qquad \text{(quasi nondivergent).} \tag{3.31}$$

3.9 PRESSURE COORDINATES

When the hydrostatic approximation is valid, it is convenient to utilize (x, y, p, t) coordinates. The equations of motion become [see (1.15) through (1.18) and omit the p subscript on ∇]

$$\begin{gathered}
\frac{\partial \mathbf{V}}{\partial t} + (\mathbf{V} \cdot \nabla)\mathbf{V} + \omega \frac{\partial \mathbf{V}}{\partial p} + f\mathbf{k} \times \mathbf{V} = -\nabla \Phi, \\
\frac{\partial \theta}{\partial t} + \mathbf{V} \cdot \nabla \theta + \omega \frac{\partial \theta}{\partial p} = 0, \\
\nabla \cdot \mathbf{V} + \frac{\partial \omega}{\partial p} = 0, \\
\frac{\partial \Phi}{\partial p} = -\alpha = \frac{-RT}{p},
\end{gathered} \tag{3.32}$$

where

$$\omega \equiv \frac{dp}{dt} = \frac{\partial p}{\partial t} + \mathbf{V} \cdot \nabla p + w \frac{\partial p}{\partial z} = \left(\frac{\partial}{\partial t} + \mathbf{V} \cdot \nabla\right) p - g\rho w,$$

$$\omega \sim \frac{VP}{L}\left(\frac{\partial}{\partial t'} + \mathbf{V}' \cdot \nabla'\right) \delta p' - \frac{WP}{H}(\rho_s' + \delta\rho').$$

By (3.23) and (3.15)

$$W \sim \frac{DVF}{L\sigma_s R_1} = \frac{DFf}{\sigma_s}, \qquad \delta p' \sim \frac{DF}{HR_1}.$$

Therefore,

$$\omega \sim \frac{VPDF}{LHR_1} + \frac{DVFP}{L\sigma_s HR_1} \sim \frac{D}{H}\frac{V}{L}\frac{P}{\sigma_s}\frac{F}{R_1}.$$

The second term above obviously dominates, suggesting the last expression as a scaling factor for ω:

$$\omega = \frac{DVPF}{HL\sigma_s R_1}\,\omega' \equiv \frac{DVP}{HLR_i R_1}\,\omega', \tag{3.33}$$

where

$$R_i\,(\text{Richardson number}) = \frac{g\partial \ln \theta/\partial z}{(\partial V/\partial z)^2} \sim \frac{gD\sigma_s}{V^2} = \frac{\sigma_s}{F}. \tag{3.34}$$

For $R_1 = R_0 < 1$, $\omega \sim 10^{-3}$ mb/sec. The foregoing result indicates ω (like W) increases with V^2 and latitude, and with decreasing static stability.

The scaling of the geopotential, $\Phi = gz$ is consistent with

$$\Phi = gH\Phi'. \tag{3.35}$$

With these scale factors the equation of motion may be written as

$$\frac{V^2}{L}\left(\frac{\partial}{\partial t'} + \mathbf{V}' \cdot \nabla'\right)\mathbf{V}' + \frac{V}{P}\frac{FPV}{L\sigma_s R_1}\,\omega'\,\frac{\partial \mathbf{V}'}{\partial p'}$$
$$+ fV\mathbf{k} \times \mathbf{V}' + \frac{gH}{L}\,\nabla'(\Phi' + \delta\Phi') = 0$$

where $\partial p = (D/H)P\,\partial p'$. Dividing by $V\bar{f}$ and simplifying leads to

$$R_1\left(\frac{\partial}{\partial t'} + \mathbf{V}' \cdot \nabla'\right)\mathbf{V}' + \frac{F}{\sigma_s}\,\omega'\,\frac{\partial \mathbf{V}'}{\partial p'} + \frac{f}{\bar{f}}\mathbf{k} \times \mathbf{V}' = \frac{-gH}{LV\bar{f}}\,\nabla'\,\delta\Phi'. \tag{3.36}$$

For the purpose of comparing the terms comprising the equations, the Rossby number may be introduced into the second and last terms as follows:

$$R_1\left(\frac{\partial}{\partial t} + \mathbf{V}' \cdot \nabla'\right)\mathbf{V}' + \frac{R_1^{\,2}}{R_i R_1^{\,2}}\,\omega'\,\frac{\partial \mathbf{V}'}{\partial p'} + \frac{f}{\bar{f}}\mathbf{k} \times \mathbf{V}' + \frac{HR_1}{DF}\nabla'\,\delta\Phi' = 0. \tag{3.37}$$

Here $R_i R_1^2$ is of the order of unity for synoptic waves ($R_1 \sim 10^{-1}$, $\sigma_s \sim 10^{-1}$, $F \sim 10^{-3}$, $R_i \sim 10^2$).

Comparing the pressure and coriolis forces in (3.37) with $R_1 \sim 0.1$ shows that

$$\delta\Phi' \sim DF/HR_1.$$

Hence

$$\Phi'' \equiv \frac{HR_1}{DF}\,\delta\Phi' \sim 1. \tag{3.38}$$

In a similar fashion, the thermodynamic, continuity, and hydrostatic equations may be normalized as follows:

$$\left(\frac{\partial}{\partial t'} + \mathbf{V}' \cdot \nabla'\right)\theta'' - \omega'\left(\alpha_s'\theta_s' - \frac{1}{R_i R_1}\frac{\partial \theta''}{\partial p'}\right) = 0, \tag{3.39}$$

$$\nabla' \cdot \mathbf{V}' + \frac{R_1}{R_i R_1^2}\frac{\partial \omega'}{\partial p'} = 0, \tag{3.40}$$

$$\alpha'' \equiv \frac{R_1\,\delta\alpha'}{F} \qquad \theta'' \equiv \frac{R_1\,\delta\theta'}{F},$$

$$\frac{\partial \Phi''}{\partial p'} + \alpha'' = 0. \tag{3.41}$$

These forms bring out the relative importance of the various terms comprising the equations and also permit the establishment of approximations which are consistent with respect to the entire system of hydrodynamical equations.

3.10 DIVERGENT WIND

A theorem by Helmholtz permits the partition of the wind into *rotational* and *divergent* parts, namely,

$$\mathbf{V} = \mathbf{V}_\psi + \mathbf{V}_\chi \qquad \mathbf{V}_\psi = \mathbf{k} \times \nabla\psi, \qquad \mathbf{V}_\chi = \nabla\chi.$$

Then the continuity equation may be written as

$$\delta = \nabla \cdot \mathbf{V}_\chi = -\frac{\partial \omega}{\partial p}.$$

Scaling this equation gives

$$\frac{V_\chi}{L} \sim \frac{DVPF}{(DP/H)HL\sigma_s R_1} = \frac{VF}{L\sigma_s R_1} = \frac{V}{R_i R_1 L}$$

and

$$V_\chi \sim \frac{FV}{\sigma_s R_1}, \qquad V_\chi \sim \frac{1}{R_i R_1} V,$$

and

$$V_\psi \sim V. \tag{3.42}$$

Hence the divergent part of the wind is an order of magnitude less than the rotational part of the wind in synoptic scale motions. Equation 3.40 can now be rewritten as

$$\nabla' \cdot \mathbf{V}_\chi' + \frac{\partial \omega'}{\partial p'} = 0$$

where

$$V_\chi = \frac{V}{R_i R_1} V'_\chi.$$

Note from (3.42) that the horizontal velocity divergence has magnitude V/R_iR_1L which for synoptic scale disturbance is about 10^{-6} sec^{-1} although peak values may exceed this by a factor of almost ten. It is of interest to mention here that the divergence of the traditional geostrophic wind, $\nabla \cdot (f^{-1}\mathbf{k} \times \nabla\Phi) = -\beta v_g/f$, has the same general magnitude as the actual divergence, but may be of the opposite sign.

3.11 VORTICITY EQUATION

Next the vorticity equation will be considered on the basis of scale considerations utilizing the previous results concerning the magnitudes of the various terms.

$$\frac{\partial \zeta}{\partial t} + (\mathbf{V}_\psi + \mathbf{V}_\chi) \cdot \nabla\zeta + \beta(v_\psi + v_\chi) + \omega\frac{\partial \zeta}{\partial p} + (\zeta + f)\,\delta + \mathbf{k} \cdot \nabla\omega \times \frac{\partial(\mathbf{V}_\psi + \mathbf{V}_\chi)}{\partial p} = 0, \qquad (3.43)$$

$$\frac{V^2}{L^2} + \frac{V^2}{L_2}\left(1 + \frac{1}{R_iR_1}\right) + (2\Omega\cos\varphi)\frac{V}{a}\left(1 + \frac{1}{R_iR_1}\right) + \frac{V^2}{R_iR_1L^2} + \left(\frac{V}{L} + 10^{-4}\right)\frac{1}{R_iR_1}\frac{V}{L} + \frac{1}{R_iR_1}\frac{V^2}{L^2}\left(1 + \frac{1}{R_iR_1}\right) = 0.$$

Here the order of magnitude of the terms involving ω has been taken from previous estimates of vertical velocity with $f \sim 10^{-4}$ sec^{-1} $\sim 2\Omega\cos\varphi$, $V/L \sim 10^{-5}$, and $(R_iR_1)^{-1} \sim 10^{-1}$ for medium to large scale motions; and $V/a \lessdot V/L$. Comparing this result with (3.42) also shows that for these medium- to large-scale motions the vorticity is an order of magnitude greater than the horizontal velocity divergence. It is evident that the term $\mathbf{k} \cdot \nabla\omega \times \partial\mathbf{V}_\chi/\partial p$ is the least important; hence, as a first approximation, the vorticity equation may be reduced to

$$\frac{\partial \zeta}{\partial t} + \mathbf{V} \cdot \nabla(\zeta + f) + \omega\frac{\partial \zeta}{\partial p} + (\zeta + f)\nabla \cdot \mathbf{V}_\chi + \mathbf{k} \cdot \nabla\omega \times \frac{\partial \mathbf{V}_\psi}{\partial p} = 0. \qquad (3.43a)$$

As a second approximation, terms of order $V^2/L^2R_iR_1$ may be omitted, leaving

$$\frac{\partial \zeta}{\partial t} + \mathbf{V}_\psi \cdot \nabla(\zeta + f) + \mathbf{V}_\chi \cdot \nabla f + f\nabla \cdot \mathbf{V}_\chi = 0. \qquad (3.44a)$$

Next, as a still further approximation, advection of the coriolis parameter with the divergent wind may be dropped since it is somewhat smaller than the divergence term, at least for typical synoptic wavelengths, which leaves

$$\frac{\partial \zeta}{\partial t} + \mathbf{V}_\psi \cdot \nabla(\zeta + f) + f \nabla \cdot \mathbf{V}_\chi = 0.$$

However to maintain the integral constraint on global vorticity, a mean value of f [see (4.9)] must be used in the divergence term, giving the quasi-geostrophic form:

$$\frac{\partial \zeta}{\partial t} + \mathbf{V}_\psi \cdot \nabla(\zeta + f) = \bar{f} \frac{\partial \omega}{\partial p}. \tag{3.44b}$$

Here $\nabla \cdot \mathbf{V}_\chi$ has been replaced by $-\partial\omega/\partial p$ from the continuity equation. N. Phillips (1963) justified the use of cartesian coordinates (β-plane) with the geostrophic approximation.

As a final approximation the divergence term in (3.44*b*) is omitted, which yields the simple *barotropic vorticity equation*

$$\frac{\partial \zeta}{\partial t} + \mathbf{V}_\psi \cdot \nabla(\zeta + f) = 0 \qquad \text{(barotropic)}. \tag{3.45}$$

It should be noted that for planetary scale motions (say, wave numbers 1, 2, and 3), i.e., for very large values of L, the term βv in the vorticity equation dominates, giving rise to rapid retrogression of troughs and ridges [see (6.10)], a phenomenon not found in nature. Hence the quasi-geostrophic approximations are unsuitable, particularly as prediction equations for ultra-long waves, and the complete equation must be retained. Retention of the divergence term $f\delta$ is helpful in this regard, as discussed later in Chapter 6. Nevertheless, for $L \sim a$, retention of only the largest terms of (3.43*a*) leads to the diagnostic equation, $\mathbf{V} \cdot \nabla f = -f \nabla \cdot \mathbf{V}$, which is satisfied by the traditional geostrophic wind.

The preceding set of approximations was based on a Rossby number $R_0 < 1$, corresponding to synoptic scale disturbances in middle latitudes. For low latitudes, $R_0 \gtrsim 1$ and $R_1 = 1$; hence, according to (3.33) and (3.42), the vertical velocity and divergent wind are given by

$$\omega \sim \frac{DVP}{HLR_i} \qquad \text{and} \qquad V_\chi \sim \frac{V}{R_i},$$

which are about one-tenth of their middle latitude values. Thus synoptic scale motions in the tropics are more nearly horizontal and quasi-nondivergent, and there is little vertical coupling, at least under frictionless, adiabatic conditions with advective time scales. With respect to the vorticity equation, the approximations leading to (3.45) are even more accurate than for the middle latitude case, as shown by Charney (1963).

Other possibilities also exist with respect to values for V, L, and σ, for example, stronger winds as in the jet stream, a shorter wavelength and smaller static stability, which lead to different values of F, R_1, and R_i.

3.12 DIVERGENCE EQUATION

Next, a similar analysis will be carried out for the divergence equation (1.21), which may be written in the form

$$\frac{\partial \delta}{\partial t} + \mathbf{V} \cdot \nabla \delta + \omega \frac{\partial \delta}{\partial p} + \delta^2 + \nabla \omega \cdot \frac{\partial \mathbf{V}}{\partial p} - f\zeta + \beta u - 2J(u, v) + \nabla^2 \Phi = 0,$$

$$\frac{1}{R_i R_1} \frac{V^2}{L^2} + \frac{1}{R_i R_1} \frac{V^2}{L^2} + \frac{1}{(R_i R_1)^2} \frac{V^2}{L^2} + \frac{1}{(R_i R_1)^2} \frac{V^2}{L^2} + \frac{1}{R_i R_1} \frac{V^2}{L^2}$$

$$+ 10^{-4} \frac{V}{L} + 10^{-4} \frac{V}{a} + \frac{V^2}{L^2} + \frac{1}{R_1} \frac{V^2}{L^2} = 0. \quad (3.46)$$

The term βu is a simplified version of $(\mathbf{k} \times \mathbf{V}) \cdot \nabla f$, with the x-axis toward the east. The largest terms clearly are $f\zeta$ and $\nabla^2 \Phi$ followed by βu and $J(u, v)$; hence the following equation is an internally consistent approximation of the divergence equation for synoptic scale motions:

$$\nabla^2 \Phi - f\zeta + (\mathbf{k} \times \mathbf{V}) \cdot \nabla f - 2J(u, v) = 0. \quad (3.47a)$$

Noting now that the divergent portion of the wind is an order of magnitude less than the rotational part, the previous equation requires the use of the rotational wind component, $\mathbf{V} = \mathbf{k} \times \nabla \psi$, in the last two terms, with the result

$$\nabla^2 \Phi - f \nabla^2 \psi - \nabla \psi \cdot \nabla f + 2\left(\frac{\partial^2 \psi}{\partial x\, \partial y}\right)^2$$

$$- 2 \frac{\partial^2 \psi}{\partial x^2} \frac{\partial^2 \psi}{\partial y^2} = 0 \qquad \text{(balance equation)}. \quad (3.47b)$$

The use of this approximation implies a continual balance between the rotational wind component and the geopotential field.

Although not immediately apparent from (3.43) and (3.46), but as shown by Lorenz (1960) and Arakawa (1960), the vorticity equation which is energetically consistent with the balance equation is (3.43*a*).

Next omitting the last two nonlinear terms in (3.47*b*) gives the so-called *linear balance equation* namely,

$$\nabla \cdot (f \nabla \psi) - \nabla^2 \Phi = 0, \quad (3.48a)$$

which is energetically consistent with (3.44*a*). An equation very similar to the linear balance equation gives Shuman's semigeostrophic stream function ψ_s:

$$f\nabla^2\psi_s + f^{-1}\nabla f \cdot \nabla\Phi = \nabla^2\Phi.$$

Finally a constant value of the coriolis parameter in (3.48*a*) gives the quasi-geostrophic approximation

$$\bar{f}\nabla^2\psi - \nabla^2\Phi = 0, \tag{3.48b}$$

which forms a consistent pair with the geostrophic vorticity equation (3.44*b*). The reader is referred to Chapter 4 for a further discussion of the integral constraints on the hydrodynamic equations with respect to vorticity and energy.

3.13 THERMODYNAMIC EQUATION

Although the thermodynamic equation has already been treated, it is convenient to examine it in the manner applied to the vorticity and divergence equations.

$$\frac{\partial \ln \theta}{\partial t} + \mathbf{V} \cdot \nabla \ln \theta + \omega \frac{\partial \ln \theta}{\partial p} = 0,$$

$$\frac{\partial \ln \theta}{\partial t} + (\mathbf{V}_\psi + \mathbf{V}_\chi) \cdot \nabla \ln \theta + \omega \frac{\partial \ln \theta_s}{\partial p} + \omega \frac{\partial}{\partial p} \ln \left(1 + \frac{\delta\theta'}{\theta'_s}\right) = 0. \tag{3.49}$$

From previous estimates [e.g. (3.20), (3.21), and (3.42)], the orders of magnitudes are

$$\frac{V}{L}\frac{F}{R_1} + \frac{VF}{LR_1} + \frac{VF}{LR_1}\frac{F}{\sigma_s R_1} + \frac{V}{L}\frac{\sigma_s}{R_i R_1} + \frac{V}{L}\frac{1}{R_i R_1}\frac{F}{R_1}$$

or

$$\frac{V}{L}\frac{F}{R_1} + \frac{VF}{LR_1} + \frac{VF}{LR_1}\frac{1}{R_i R_1} + \frac{V}{L}\frac{F}{R_1} + \frac{VF}{LR_1}\frac{1}{R_i R_1}.$$

These magnitudes show that, for synoptic scale motions, advection with the divergent wind and the vertical advection involving the static stability deviation from the standard value are one order of magnitude less than the other terms. Hence, on the basis of this scale analysis, an internally consistent approximation to the thermodynamic equation is the quasi-geostrophic form

$$\frac{\partial\theta}{\partial t} + \mathbf{V}_\psi \cdot \nabla\theta + \sigma_s\omega = 0, \tag{3.50}$$

where σ_s is a function of the vertical coordinate only. On the other hand, if the static stability parameter is allowed to vary laterally and temporally as well as vertically, then advection of temperature with the divergent wind must be included for consistency.

3.14 THE ω EQUATION

Next the scale analysis will be applied to an equation for the vertical velocity ω. As an illustration of the procedure, the ω equation will be derived for the linear balanced system consisting of (3.44*a*), (3.48*a*), and (3.49). If (3.48*a*) is differentiated with respect to time, the result may be expressed as

$$f\frac{\partial \zeta}{\partial t} - \nabla^2 \frac{\partial \Phi}{\partial t} + \mathbf{\nabla} f \cdot \mathbf{\nabla} \frac{\partial \psi}{\partial t} = 0.$$

Eliminating $\partial \zeta / \partial t$ between the previous equation and the vorticity equation (3.44*a*) gives

$$\nabla^2 \frac{\partial \Phi}{\partial t} - \mathbf{\nabla} f \cdot \mathbf{\nabla} \frac{\partial \psi}{\partial t} + f\mathbf{V}_\psi \cdot \mathbf{\nabla}(\zeta + f) + f\mathbf{V}_\chi \cdot \mathbf{\nabla} f - f^2 \frac{\partial \omega}{\partial p} = 0. \quad (3.51)$$

Next replace θ in the thermal equation (3.49) by

$$\theta = T\left(\frac{p_0}{p}\right)^{R/c_p} = -\frac{p}{R}\left(\frac{p_0}{p}\right)^{R/c_p}\frac{\partial \Phi}{\partial p},$$

yielding

$$\frac{\partial}{\partial t}\frac{\partial \Phi}{\partial p} + \mathbf{V} \cdot \mathbf{\nabla} \frac{\partial \Phi}{\partial p} + \sigma\omega = 0 \quad (3.52)$$

where

$$\sigma = -\alpha \frac{\partial \ln \theta}{\partial p}.$$

Finally, differentiate (3.51) with respect to p, take the Laplacian of (3.52), and eliminate $\nabla^2 \partial^2 \Phi / \partial p\, \partial t$ between the two equations, giving

$$\begin{aligned}\nabla^2(\sigma\omega) + f^2 \frac{\partial^2 \omega}{\partial p^2} &= f\frac{\partial}{\partial p}[\mathbf{V}_\psi \cdot \mathbf{\nabla}(\zeta + f)] \\ &\quad - \nabla^2\left(\mathbf{V} \cdot \mathbf{\nabla} \frac{\partial \Phi}{\partial p}\right) + f\mathbf{\nabla} f \cdot \frac{\partial \mathbf{V}_\chi}{\partial p} - \mathbf{\nabla} f \cdot \mathbf{\nabla} \frac{\partial^2 \psi}{\partial p\, \partial t}. \quad (3.53)\end{aligned}$$

This is the diagnostic equation for the vertical velocity ω which is appropriate to the system (3.44*a*), (3.48*a*), and (3.49). The last term involving the time derivative makes it awkward to obtain ω directly, and the equation must be solved through an iterative process together with the vorticity equation.

However, as will now be shown, this term is of smaller magnitude than most of the others and is generally omitted in actual forecasting practice.

The order of magnitude of the various terms comprising (3.53) may be determined with a scale analysis, using the results obtained earlier, such as (3.14), (3.21), (3.33), (3.34), (3.38), and (3.42), as follows:

$$\frac{1}{L^2}\frac{(gH/P)^2\sigma_s}{gD}\frac{DVPF}{HL\sigma_s R_1} + f^2\left(\frac{H}{DP}\right)^2\frac{DVPF}{HLR_1\sigma_s} \approx f\left(\frac{V}{L}\right)^2\frac{1}{(D/H)P}$$
$$+ \frac{1}{L^2}\frac{V}{L}\frac{gHFD}{(D/H)PR_1H} + \frac{f\beta}{(D/H)P}\frac{V}{R_iR_1} + \frac{V}{L}\frac{\beta V}{(D/H)P}.$$

Replacing F by V^2/gD and simplifying leads to

$$\begin{array}{cccccc} \dfrac{V^3}{L^3R_1} + & \dfrac{Vf^2}{LR_1R_i} \approx & \dfrac{V^2f}{L^2} + & \dfrac{V^3}{L^3R_1} + & \dfrac{Vf\beta}{R_1R_i} + & \dfrac{V^2\beta}{L}, \\ 10^{-14} & 10^{-14} \approx & 10^{-14} & 10^{-14} & 10^{-15} & 10^{-15}. \end{array}$$

The order of magnitude of each term for a synoptic scale disturbance is given below the term in cgs units. The last two terms are of smaller magnitude and result from the inclusion of the advection of the coriolis parameter by the divergent wind in the vorticity equation. No effort was made here to isolate the thermal advection by the divergent wind in the second term on the right side of (3.53), which is clearly an order of magnitude less than thermal advection with the rotational wind. Similarly, terms due to the horizontal variation of the static stability are of lesser magnitude.

In the quasi-geostrophic system, comprised of (3.44*b*) and (3.48*b*), the ω equation reduces to

$$\sigma_s\nabla^2\omega + \bar{f}^2\frac{\partial^2\omega}{\partial p^2} = \bar{f}\frac{\partial}{\partial p}[\mathbf{V}_\psi \cdot \nabla(\zeta + f)] - \nabla^2\left(\mathbf{V}_\psi \cdot \nabla\frac{\partial\Phi}{\partial p}\right) \qquad (3.54)$$

where σ_s is at most a function of pressure. The corresponding thermodynamic equation is (3.50).

3.15 PERTURBATION EXPANSION

A final topic which will be discussed very briefly is a method of dynamical analysis referred to as a *perturbation expansion* which permits a determination of the physical variables by solution of a succession of linear systems, each contributing further to the solution of the initial nonlinear system of hydrodynamic equations by the addition of another term of an infinite series.

To permit assessment of the influence of the variable coriolis parameter, the latter will be expanded into a series involving the Rossby number.

$$
\begin{aligned}
f &= f_0 + \left(\frac{\partial f}{\partial y}\right)_0 (y - y_0) + \left(\frac{\partial^2 f}{\partial y^2}\right)_0 \frac{(y - y_0)^2}{2!} + \dots, \\
\frac{1}{f_0}\left(\frac{\partial f}{\partial y}\right)_0 &= \frac{\beta_0}{f_0} \sim \frac{1}{a}, \qquad (y - y_0) \sim L \\
\frac{1}{f_0}\left(\frac{\partial^2 f}{\partial y^2}\right)_0 &= \frac{1}{f_0}\left(\frac{d\beta}{dy}\right)_0 \sim -\frac{1}{a^2} \\
\frac{f}{f_0} &= 1 + \frac{a\beta_0}{f_0}(y' - y_0')\frac{L}{a} + \frac{a^2}{f_0}\left(\frac{d\beta}{dy}\right)_0 \frac{(y' - y_0')^2}{2!}\frac{L^2}{a^2} \dots
\end{aligned}
\tag{3.55}
$$

where $y' = y/L$. The coefficients of the terms L/a, L^2/a^2, etc., clearly have magnitudes of order unity. On the other hand, the ratio L/a is less than unity for small- to medium-scale synoptic waves. This ratio can be written in terms of the Rossby number as follows:

$$
\frac{L}{a} = \frac{L^2 f_0}{aV}\frac{V}{f_0 L}.
$$

When this is substituted into (3.55), the result is

$$
\frac{f}{f_0} = 1 + C_0 R_0 + C_1 R_0^2 \dots, \tag{3.56}
$$

where the coefficients C_0, C_1, etc., have magnitudes of roughly one for small- to medium-scale waves. Substitution of (3.56) into (3.37), with the use of (3.38) and $\lambda = R_i^{-1} R_0^{-2}$, yields

$$
\begin{aligned}
R_0\left(\frac{\partial}{\partial t'} + \mathbf{V}' \cdot \nabla'\right)\mathbf{V}' + \lambda R_0^2 \omega' \frac{\partial \mathbf{V}'}{\partial p'} \\
+ (1 + C_0 R_0 + C_1 R_0^2 + \dots)\mathbf{k} \times \mathbf{V}' = -\nabla' \Phi''
\end{aligned}
\tag{3.57}
$$

Next the variables $\mathbf{V}'$, ω', Φ'', and α'' are expressed as perturbation series in terms of the Rossby number R_0:

$$
\begin{aligned}
\mathbf{V}' &= \mathbf{V}_0' + R_0 \mathbf{V}_1' + R_0^2 \mathbf{V}_2' + \dots, \\
\lambda R_0 \omega' &= \omega_0' + R_0 \omega_1' + R_0^2 \omega_2' + \dots, \\
\Phi'' &= \Phi_0'' + R_0 \Phi_1'' + R_0^2 \Phi_2'' + \dots, \\
\alpha'' &= \alpha_0'' + R_0 \alpha_1'' + R_0^2 \alpha_2'' + \dots.
\end{aligned}
\tag{3.58}
$$

The expansion for ω' contains the factor λR_0 because of the way the vertical velocity appears in (3.57), (3.39), and (3.40).

Now substitute the series (3.58) into (3.57), (3.39), (3.40), and (3.41) and equate the coefficients of $R_0{}^0$. The following dynamically consistent hydrostatic-geostrophic system results:

$$
\begin{aligned}
\mathbf{k} \times \mathbf{V}_0' + \nabla'\Phi_0'' &= 0, \qquad \text{(zero order approximation)} \\
\omega_0' &\equiv 0, \\
\nabla' \cdot \mathbf{V}_0' &= 0, \\
\frac{\partial \Phi_0''}{\partial p'} + \alpha_0'' &= 0.
\end{aligned}
\tag{3.59}
$$

These equations represent a balanced hydrostatic-geostrophic, nondivergent flow.

Equating the terms which are linear with respect to R_0 gives the *first-order* approximation, namely,

$$
\begin{aligned}
&(a) \qquad \frac{\partial \mathbf{V}_0'}{\partial t} + (\mathbf{V}_0' \cdot \nabla')\mathbf{V}_0' + \mathbf{k} \times \mathbf{V}_1' + C_0 \mathbf{k} \times \mathbf{V}_0' + \nabla'\Phi_1'' = 0 \\
&(b) \qquad \frac{\partial \alpha_0''}{\partial t} + \mathbf{V}_0' \cdot \nabla'\alpha_0'' - \omega_1'\sigma_s' = 0, \\
&(c) \qquad \nabla' \cdot \mathbf{V}_1' + \frac{\partial \omega_1'}{\partial p'} = 0, \qquad \text{(first-order approximation)} \\
&\qquad\qquad \frac{\partial \Phi_1''}{\partial p'} + \alpha_1'' = 0,
\end{aligned}
\tag{3.60}
$$

which is the quasi-geostrophic system. The parameter C_0 involves β, the latitudinal variation of the coriolis parameter.

The next higher order approximation is obtainable by equating terms of order $R_0{}^2$, and so forth. The zero approximation is determinable from any-one of the variables $\mathbf{V}_0'$, Φ_0'', or α_0''. Now take the curl of (3.60*a*) and substitute for the divergence from (3.60*c*); this gives the quasi-geostrophic vorticity equation, an approximation therefore correct to terms of order R_0. The next approximation, i.e., terms of order $R_0{}^2$ is similar to the balanced system (3.43*a*) and (3.47*a*). Perturbation expansions sometimes can be developed to yield a sequence of linear systems which, when solved successively, provide estimates of nongeostrophic components.

chapter four

Integral Relations Regarding Vorticity and Energy

4.1. INTEGRAL THEOREMS

In this chapter, equations for the production of kinetic, potential, and internal energy will be derived, including expressions for the transformation between the various forms of energy. These relationships will help to explain how meteorological waves intensify and weaken. In addition, certain global features of the vorticity, divergence, and energy equations will be investigated, which will suggest constraints that should be heeded when designing numerical integration schemes for weather prediction.

For these purposes several integral theorems will be utilized on various occasions, namely, the divergence theorem in two or three dimensions, commonly referred to as Gauss' or Green's theorem, and Stokes' theorem involving the curl of a vector for transforming a surface integral to a line integral.

$$\int_v \nabla \cdot \mathbf{A}\, dv = \int_s \mathbf{A} \cdot \mathbf{n}\, ds,$$

$$\int_s \nabla \cdot \mathbf{A}\, ds = \int_L \mathbf{A} \cdot \mathbf{n}\, dL,$$

$$\int_s \mathbf{n} \cdot \nabla \times \mathbf{A}\, ds = \int_L \mathbf{A} \cdot d\mathbf{L}.$$

4.2 VORTICITY

The vorticity equation is usually written in (x, y, p, t) coordinates as

$$\frac{\partial \zeta}{\partial t} + \mathbf{V} \cdot \nabla \eta + \omega \frac{\partial \zeta}{\partial p} + \eta \nabla \cdot \mathbf{V} + \mathbf{k} \cdot \nabla \omega \times \frac{\partial \mathbf{V}}{\partial p} = -g\mathbf{k} \cdot \nabla \times \frac{\partial \boldsymbol{\tau}}{\partial p}, \quad (4.1)$$

but is also expressible in the form

$$\frac{\partial \zeta}{\partial t} = -\nabla \cdot \left(\eta \mathbf{V} + \omega \frac{\partial \mathbf{V}}{\partial p} \times \mathbf{k} - g\mathbf{k} \times \frac{\partial \boldsymbol{\tau}}{\partial p}\right), \quad (4.2)$$

where η is the absolute vorticity $(\zeta + f)$. In the latter equation the first term on the right comprises the horizontal advection of vorticity and the divergence term, while the second term includes the vertical advection of vorticity and the twisting term.

Next integrate (4.2) over the mass M:

$$\int \frac{\partial \zeta}{\partial t} dM = -\int \nabla \cdot \left(\eta \mathbf{V} + \omega \frac{\partial \mathbf{V}}{\partial p} \times \mathbf{k} - g\mathbf{k} \times \frac{\partial \boldsymbol{\tau}}{\partial p}\right) dM, \quad (4.3)$$

where

$$dM = \rho \, dx \, dy \, dz = -g^{-1} \, ds \, dp.$$

The two-dimensional surface integration may be transformed into a line integral by using the divergence theorem, with the result

$$\frac{\partial}{\partial t} \int \zeta \, dM = g^{-1} \iint_L \left[\eta \mathbf{V} + \left(\omega \frac{\partial \mathbf{V}}{\partial p} + g \frac{\partial \boldsymbol{\tau}}{\partial p}\right) \times \mathbf{k}\right] \cdot \mathbf{n} \, dL \, dp, \quad (4.4)$$

where $\mathbf{n}$ is a unit vector normal to the curve L enclosing the surface S. The left side of the equation is a measure of the mean generation of vorticity over the mass M. The right side of (4.4) will vanish if the component normal to L is zero or if S is a closed surface. It follows that the *mean generation of vorticity* over a *global pressure surface* is zero. Since this property holds for the entire vorticity equation, then any approximate form should possess the property for consistency and to avoid the fictitious mean generation of vorticity.

The right side of (4.3) can obviously be separated into several integrals— one involving the combined horizontal advection of vorticity and velocity divergence, the second, the vertical advection of vorticity and the twisting term, and finally, *the friction term*. It is apparent that the omission of one or several of these integrals will not alter the integral property,

$$\frac{\partial}{\partial t}\int \zeta \, dM = 0, \tag{4.5}$$

which holds true over a closed surface for the complete vorticity equation. Hence the *omission of the friction term in the vorticity equation is an acceptable approximation* in this sense. Similarly, the *omission of both the vertical advection of vorticity and the twisting term* yields an *approximation to the complete vorticity equation that fulfills the integral constraint.* On the other hand, omission or modification of one of these terms may result in a vorticity equation for which (4.5) no longer holds.

To examine the suitability of various approximations of the vorticity equation with regard to this integral constraint, it is advantageous to express the wind velocity vector as the sum of its rotational and divergent parts in accordance with Helmholtz' theorem

$$\begin{aligned} \mathbf{V} &= \mathbf{k} \times \nabla\psi + \nabla\chi = \mathbf{V}_\psi + \mathbf{V}_\chi, \\ \zeta &= \mathbf{k} \cdot \nabla \times \mathbf{V} = \nabla^2\psi \qquad \delta = \nabla \cdot \mathbf{V} = \nabla^2\chi, \end{aligned} \tag{4.6}$$

where ψ is the stream function and χ a velocity potential.

It is evident that the integral constraint (4.5) will not be violated if only the rotational part of the wind is utilized in the twisting term of (4.1). Thus an acceptable approximation to the vorticity equation (4.1) is

$$\frac{\partial\zeta}{\partial t} + \mathbf{V} \cdot \nabla\eta + \omega\frac{\partial\zeta}{\partial p} + \eta\nabla \cdot \mathbf{V} + \mathbf{k} \cdot \nabla\omega \times \frac{\partial \mathbf{V}_\psi}{\partial p} = 0, \tag{4.7}$$

where the friction term has also been omitted. This approximation is also consistent with the scale analysis of Chapter 3. Obviously the integral constraint (4.5) would be fulfilled if the vertical advection and twisting terms of (4.7) were both dropped; however, in this case, part of the divergence term must also be dropped to maintain consistency with respect to scale analysis. To examine the horizontal advection and divergence terms in more detail, they may be expanded as follows:

$$\begin{aligned} \mathbf{V} \cdot \nabla(\zeta + f) + (\zeta + f)\nabla \cdot \mathbf{V} &= \mathbf{V}_\psi \cdot \nabla(\zeta + f) + \mathbf{V}_\chi \cdot \nabla f \\ + f\nabla \cdot \mathbf{V}_\chi + \mathbf{V}_\chi \cdot \nabla\zeta + \zeta\nabla \cdot \mathbf{V}_\chi &= \nabla \cdot (\zeta + f)\mathbf{V}_\psi + \nabla \cdot f\mathbf{V}_\chi + \nabla \cdot \zeta\mathbf{V}_\chi. \end{aligned}$$

It is now obvious that any one or several of the three terms in the last equality may be omitted without violating the integral constraint. In keeping with the scale analysis of Chapter 3, the last term $\nabla \cdot \zeta\mathbf{V}_\chi$ should be omitted if the twisting and vertical advection terms of (4.7) are omitted, which gives the consistent approximation

$$\frac{\partial \zeta}{\partial t} + \mathbf{V}_\psi \cdot \nabla(\zeta + f) + \mathbf{V}_\chi \cdot \nabla f + f\nabla \cdot \mathbf{V}_\chi = 0. \tag{4.8}$$

The scale analysis also showed that $\mathbf{V}_\chi \cdot \nabla f$ tends to be somewhat smaller than the remaining terms of (4.8); however, omitting this term would violate the integral constraint unless $f\nabla \cdot \mathbf{V}_\chi$ is also dropped or altered in some appropriate fashion. If both terms were omitted, (4.8) would reduce to the simplified barotropic vorticity equation which has no mechanism for the conversion of potential to kinetic energy, as will be shown later in this chapter. To permit baroclinic development, it is necessary to retain a divergence term. This may be achieved by omitting the term $\mathbf{V}_\chi \cdot \nabla f$ and using a mean value for f in the divergence term, $f\nabla \cdot \mathbf{V}_\chi$, which gives the *quasi-geostrophic* vorticity equation

$$\frac{\partial \zeta}{\partial t} + \mathbf{V}_\psi \cdot \nabla(\zeta + f) = -\bar{f}\nabla \cdot \mathbf{V}_\chi = \bar{f}\frac{\partial \omega}{\partial p}. \tag{4.9}$$

The last form follows from the continuity equation. Equation 4.9 clearly fulfills the integral constraint (4.5), since $\mathbf{V}_\psi \cdot \nabla(\zeta + f) = \nabla \cdot (\zeta + f)\mathbf{V}_\psi$ and $\bar{f}\nabla \cdot \mathbf{V}_\chi = \nabla \cdot \bar{f}\mathbf{V}_\chi$, and both vanish when integrated over a closed surface.

The earliest numerical prediction models were barotropic forms utilizing the traditional geostrophic wind

$$\mathbf{V}_g = f^{-1}\mathbf{k} \times \nabla\Phi.$$

If the divergence term is dropped in (4.9) and the geostrophic wind utilized, the result is

$$\frac{\partial \zeta_g}{\partial t} = -\mathbf{V}_g \cdot \nabla(\zeta_g + f). \tag{4.10}$$

Since $\nabla \cdot \mathbf{V}_g \neq 0$, the right side of (4.10) will not vanish when integrated over a closed surface and, hence, there will be a fictitious source of vorticity in the mean. This is believed to have been responsible for the "spurious anticyclogenesis" that characterized early 500-mb numerical forecasts. This error essentially disappeared when $\mathbf{V}_g$ was replaced by a nondivergent form of the wind.

Several such nondivergent winds were apparently successful in this regard. A very simple method devised by F. Shuman consisted of using only the rotational part of the geostrophic wind which can be obtained by calculating the geostrophic vorticity from a known geopotential field and equating it to the Laplacian of the geostrophic stream function, say ψ_s. Thus

$$\nabla^2\psi_s = \mathbf{k} \cdot \nabla \times (f^{-1}\mathbf{k} \times \nabla\Phi) = f^{-1}\nabla^2\Phi - f^{-2}\nabla f \cdot \nabla\Phi,$$

which was shown earlier below (3.48*a*). Another choice involved the solution of the balance equation (3.47*b*), which is considerably more troublesome to solve than Shuman's equation.

The balanced winds also give better barotropic 500-mb forecasts than the geostrophic wind. It should be mentioned, however, that a minor violation of an integral constraint or scale consideration need not necessarily lead to poorer forecasts, especially in short-range prediction. Such violations are more likely to lead to serious errors after many time steps. From a purely pragmatic view, the linear balance equation often appears to be a good compromise between the desirable simplicity of the geostrophic forms and the more time-consuming solution of the complete balance equation.

4.3 ENERGY RELATIONS

The horizontal equation of motion in (x, y, p, t) coordinates is

$$\frac{\partial \mathbf{V}}{\partial t} + (\mathbf{V} \cdot \nabla)\mathbf{V} + \omega \frac{\partial \mathbf{V}}{\partial p} = -\nabla\Phi - f\mathbf{k} \times \mathbf{V} - g\frac{\partial \boldsymbol{\tau}}{\partial p}, \tag{4.11}$$

where Φ is the geopotential. Taking the dot product of this equation with $\mathbf{V}$ gives

$$\frac{\partial K}{\partial t} + \mathbf{V} \cdot \nabla K + \omega \frac{\partial K}{\partial p} = -\mathbf{V} \cdot \nabla\Phi - g\mathbf{V} \cdot \frac{\partial \boldsymbol{\tau}}{\partial p}. \tag{4.12}$$

Here K is the kinetic energy per unit mass

$$K = \tfrac{1}{2}V^2$$

Multiplying the continuity equation by K and adding it to (4.12) yields a convenient form for global integration

$$\frac{\partial K}{\partial t} + \nabla \cdot K\mathbf{V} + \frac{\partial}{\partial p} K\omega = -\mathbf{V} \cdot \nabla\Phi - g\mathbf{V} \cdot \frac{\partial \boldsymbol{\tau}}{\partial p}. \tag{4.13}$$

If the previous equation is now integrated over a mass of air for which the flux of kinetic energy across the boundaries is zero, it may be seen that kinetic-energy production takes place in the mean as a result of work done by the pressure force.

$$\frac{\partial}{\partial t}\int K\,dM = -\int \mathbf{V} \cdot \nabla\Phi\,dM + \bar{F}. \tag{4.14}$$

Here the friction term is denoted by $\bar{F}$ and ω has been assumed zero at $p = 0$ and $p = p_0$ which prevents any vertical hydrodynamic flux of energy out of the system. Note that if $\mathbf{V}$ is strictly nondivergent, then $\mathbf{V} \cdot \nabla\Phi = \nabla \cdot (\Phi\mathbf{V})$, and the right side of (4.14) will vanish when the normal wind component vanishes at the boundary or the surface is closed.

Another form of the kinetic-energy equation is of interest here. Multiplying (1.16) by Φ, adding it to the left side of equation (4.13) gives

$$\frac{\partial K}{\partial t} + \nabla \cdot K\mathbf{V} + \Phi\nabla \cdot \mathbf{V} + \frac{\partial(K\omega)}{\partial p} + \Phi\frac{\partial \omega}{\partial p} = -\mathbf{V} \cdot \nabla\Phi + F$$

or

$$\frac{\partial K}{\partial t} + \nabla \cdot (K\mathbf{V} + \Phi\mathbf{V}) + \frac{\partial}{\partial p}(K\omega + \Phi\omega) = \omega\frac{\partial \Phi}{\partial p} + F. \tag{4.15}$$

Next, integration over the entire atmosphere together with replacement of $\partial\Phi/\partial p$ by $-RT/p = -\alpha$ yields the following alternate expression for the increase in mean kinetic energy:

$$\frac{\partial}{\partial t}\int K\,dM = -R\int \frac{\omega T}{p}\,dM + \bar{F} = -\int \omega\alpha\,dM + \bar{F}. \tag{4.16}$$

According to (4.16) the kinetic-energy change depends on the correlation between ω and T. When warm air is ascending and cold air descending in the mean, $\overline{(\omega T)} < 0$, and there will be an increase in mean kinetic energy apparently at the expense of the potential energy. On the other hand, when the correlation between ω and T is positive, the mean kinetic energy will decrease.

Next an equation for the potential energy will be derived. The potential energy per unit horizontal area for a column extending to the top of the atmosphere is

$$P = \int_0^\infty gz\rho\,dz = \frac{1}{g}\int_0^{p_0}\Phi\,dp = \frac{1}{g}(\Phi p)|_0^{p_0} - \frac{1}{g}\int p\,d\Phi.$$

When p_0 is the sea-level pressure, the result is

$$P = -\frac{1}{g}\int p\,d\Phi = \frac{1}{g}\int_0^{p_0} RT\,dp.$$

Similarly the internal energy in such a column is given by

$$I = \int_0^\infty c_vT\rho\,dz = \frac{1}{g}\int_0^{p_0} c_vT\,dp.$$

Combining the internal and potential energies yields

$$I + P = \frac{1}{g}\int_0^{p_0}(c_v + R)T\,dp = \frac{1}{g}\int_0^{p_0} E\,dp, \tag{4.17}$$

where $E = c_pT$ is the *enthalpy*.

The combination $I + P$ is usually referred to as the *total potential energy* or simply the *potential energy*. When the earth's surface is not at sea level, the term $(\Phi p/g)_G$ at the ground must be retained in the potential energy.

Consider next the diabatic thermodynamic equation,

$$\frac{\partial T}{\partial t} + \mathbf{V} \cdot \nabla T + \omega\left(\frac{\partial T}{\partial p} - \frac{RT}{pc_p}\right) = \frac{Q}{c_p}. \tag{4.18}$$

Multiplying (4.18) by c_p, (1.16) by c_pT, and adding leads to the result

$$\frac{\partial E}{\partial t} + \nabla \cdot (E\mathbf{V}) + \frac{\partial}{\partial p} E\omega = \alpha\omega + Q. \tag{4.19}$$

Comparison of (4.15) and (4.19) shows that the production of kinetic energy takes place at the expense of potential energy and vice versa, since the terms $\omega\alpha$ and $\omega\, \partial\Phi/\partial p$ are equal except for sign (with the hydrostatic approximation). Adding the two equations gives

$$\frac{\partial(K + E)}{\partial t} + \nabla \cdot [(K + E + \Phi)\mathbf{V}] + \frac{\partial}{\partial p}[(K + E + \Phi)\omega] = Q + F. \tag{4.20}$$

Integration over the entire atmosphere with application of the divergence theorem and the boundary conditions $\omega = 0$ at $p = p_0$ and $p = 0$ yields

$$\frac{\partial}{\partial t}\int (K + E)\, dM = \bar{Q} + \bar{F}. \tag{4.21a}$$

Thus the sum of the kinetic energy and potential energy remains constant in the mean for adiabatic, frictionless flow, i.e.,

$$\bar{K} + \bar{E} = \text{const.} \tag{4.21b}$$

From (4.19) it is evident that the mean temperature (and hence the potential internal energy) will decrease if there is a negative correlation between ω and T, i.e., if in the mean relatively warm air is ascending and cold air is descending.

Now consider the case when only the nondivergent wind V_ψ is used for the advection of temperature in (4.18). Then the term $\mathbf{V}_\psi \cdot \nabla T$ can be written directly as $\nabla \cdot (T\mathbf{V}_\psi)$; and when the continuity equation is added, as was done to obtain (4.19), the result is

$$\frac{\partial T}{\partial t} + \nabla \cdot T\mathbf{V}_\psi + \frac{\partial}{\partial p} T\omega = \frac{RT\omega}{c_p p} - T\nabla \cdot \mathbf{V}. \tag{4.22}$$

Integration of (4.22) over the entire atmosphere would lead to a *fictitious mean generation of enthalpy* since the term $\int -T\nabla \cdot \mathbf{V}\, dM$ does not vanish in general. To consider this further, return to (4.18) and approximate $\mathbf{V}$ by $\mathbf{V}_\psi$.

Then $\mathbf{V}_\psi \cdot \nabla T = \nabla \cdot T\mathbf{V}_\psi$, and (4.18) can be written (with $Q = 0$),

$$\frac{\partial T}{\partial t} + \nabla \cdot T\mathbf{V}_\psi = \frac{RT\omega}{pc_p} - \omega \frac{\partial T}{\partial p}. \tag{4.23}$$

As this equation stands it would not give the correct value of $\partial \bar{T}/\partial t$ when integrated over the entire atmosphere because the last term does not vanish. However, if $\partial T/\partial p$ is treated as a function of p alone, then the last term does vanish, i.e.,

$$\int \omega \frac{\partial T}{\partial p} \, dS = \frac{\partial T}{\partial p} \int \omega \, dS = 0,$$

because the mean value of ω over an isobaric surface is zero. This is easily seen by first integrating the continuity equation with respect to pressure, then integrating over a global isobaric surface, and finally applying the divergence theorem as follows:

$$\omega = -\int_0^p \nabla \cdot \mathbf{V} \, dp,$$

$$\int \omega \, dS = -\int_0^p \int_S \nabla \cdot \mathbf{V} \, dS \, dp = 0.$$

The integral $\int_S \nabla \cdot \mathbf{V} \, dS$ vanishes in accordance with the divergence theorem, provided no mountains are encountered. Thus with $\partial T/\partial p$ as a function of p alone utilization of (4.23) would give the correct value of mean enthalpy change over the entire atmosphere, even though advection of temperature with the divergent wind is omitted. Nevertheless there is an inconsistency here on the basis of scale considerations since the horizontal variations of $\partial T/\partial p$ are neglected while those of RT/pc_p are permitted, yet both are of the same order of magnitude. If the horizontal variations of both are neglected, (4.23) may be written in the form

$$\frac{\partial T}{\partial t} + \mathbf{V}_\psi \cdot \nabla T - \sigma_p \omega = 0, \tag{4.24}$$

where

$$\sigma_p = \frac{RT}{pc_p} - \frac{\partial T}{\partial p},$$

is a static stability parameter which is permitted to be at most a function of p. This is the proper form of the adiabatic thermodynamic equation for quasi-geostrophic models. Global integration of (4.24) gives

$$M \frac{\partial \bar{T}}{\partial t} = \int \sigma_p \omega \, dM. \tag{4.25}$$

The right side vanishes when σ_p is a function of p or a constant, since the mean vertical velocity over an isobaric surface is zero. Thus the *mean global temperature is conserved.* While this is not the correct answer, namely (4.21), at least the result is reasonable for prediction purposes. On the other hand, if the full variation of σ were allowed in (4.25), there would be a steady increase in mean temperature because σ and ω are by and large positively correlated; i.e., rising motion ($\omega < 0$) is generally associated with a smaller static stability and sinking motion ($\omega > 0$) with a larger σ. Such a steady increase in mean temperature would not be desirable.

Now since T is conserved by the quasi-geostrophic thermodynamic equation (4.24), the mean total potential energy is conserved and it would appear that there could be no conversion of potential to kinetic energy. However, as shown by E. N. Lorenz, conversion is possible, as will be seen in the next section.

4.4 AVAILABLE POTENTIAL ENERGY

As is often the case in nature, the mere existence of potential energy does not automatically guarantee its availability for conversion to other forms of energy. To clarify this situation with respect to atmospheric processes, E. N. Lorenz (1957) defined the *available potential energy* as the difference between the total potential energy and the minimum achieved by an adiabatic rearrangement of the temperature field which yields a stable, horizontal stratification of the potential temperature field. With a horizontal stratification the p, T, and θ surfaces would coincide; and potential energy, though present, would be unavailable for conversion. It has been shown earlier that the total potential energy in a column of air of unit cross section is given by the integral

$$P + I = c_p g^{-1} \int_0^{p_0} T \, dp.$$

Introducing the potential temperature, $\theta = Tp^{-\kappa} p_{00}{}^{\kappa}$, and integrating by parts, yields (except for a constant)

$$P + I = \frac{(1+\kappa)^{-1} c_p}{g p_{00}{}^{\kappa}} \int_0^{\infty} p^{1+\kappa} \, d\theta,$$

where

$$\kappa = \frac{R}{c_p} \qquad \text{and} \qquad p_{00} = 1{,}000 \text{ mb}.$$

Then the average potential energy over an area, denoted by a bar, is

$$\overline{P + I} = \frac{(1+\kappa)^{-1} c_p}{g p_{00}{}^{\kappa}} \int_0^{\infty} \overline{p^{1+\kappa}} \, d\theta.$$

After adiabatic arrangement the pressure over a given potential temperature surface would equal the average of the initial pressure distribution, say $\bar{p}$. Thus the *average available potential energy* $\bar{A}$ is given by

$$\bar{A} = \frac{c_p(1+\kappa)^{-1}}{g p_{00}^{\kappa}} \int_0^{\infty} (\overline{p^{1+\kappa}} - \bar{p}^{1+\kappa})\, d\theta.$$

Now let $p = \bar{p} + p'$, then

$$p^{1+\kappa} = (\bar{p} + p')^{1+\kappa} = \bar{p}^{1+\kappa} + \frac{(1+\kappa)\bar{p}^{\kappa} p'}{1!} + \frac{\kappa(1+\kappa)\bar{p}^{\kappa-1} p'^2}{2!} + \dots$$

or

$$p^{1+\kappa} - \bar{p}^{1+\kappa} = \bar{p}^{1+\kappa}\left[(1+\kappa)\frac{p'}{\bar{p}} + \kappa\frac{1+\kappa}{2}\left(\frac{p'}{\bar{p}}\right)^2 + \frac{(1-\kappa)\kappa(1+\kappa)}{6}\left(\frac{p'}{\bar{p}}\right)^3 + \dots\right].$$

Taking the space average of the above expression and substituting it into the previous integral yields

$$\bar{A} \doteq \tfrac{1}{2}\kappa c_p g^{-1} p_{00}^{-\kappa} \int_0^{\infty} \bar{p}^{1+\kappa} \overline{\left(\frac{p'}{\bar{p}}\right)^2}\, d\theta.$$

Thus the average available potential energy depends on the variance of pressure over the isentropic surfaces which, in turn, is closely related to the variance of θ (or T) over an isobaric surface. If $\bar{\theta}$ and $\bar{T}$ are average values on a $\bar{p}$ surface, then approximately

$$p \doteq \bar{p}(\theta(p)) \qquad \text{and} \qquad p' \doteq \bar{p}(\theta - \theta') - \bar{p}(\theta) \doteq -\theta' \frac{\partial \bar{p}}{\partial \theta}.$$

Hence

$$\overline{\left(\frac{p'}{\bar{p}}\right)^2} \doteq \frac{1}{\bar{p}^2}\overline{\left(\theta' \frac{\partial \bar{p}}{\partial \theta}\right)^2} = \frac{1}{\bar{p}^2}\overline{\theta'^2}\left(\frac{\partial \bar{\theta}}{\partial p}\right)^{-2},$$

and

$$\bar{A} \doteq \frac{\kappa c_p}{2 g p_{00}^{\kappa}} \int_0^{\infty} \frac{\bar{p}^{\kappa}}{\bar{p}} \overline{\left(\theta' \frac{\partial \bar{p}}{\partial \theta}\right)^2}\, d\theta$$

$$\doteq \frac{+\kappa}{2\gamma_d p_{00}^{\kappa}} \int_{p_0}^{0} \bar{p}^{(\kappa-1)} \bar{\theta}^2 \overline{\left(\frac{\theta'}{\bar{\theta}}\right)^2} \left(\frac{\partial \bar{\theta}}{\partial p}\right)^{-1} dp.$$

Also,

$$\frac{\gamma_d}{\kappa}\frac{p}{\bar{\theta}}\frac{\partial \bar{\theta}}{\partial p} = -(\gamma_d - \bar{\gamma}), \qquad \gamma = -\frac{\partial T}{\partial z},$$

and

$$\frac{\theta'}{\theta} \doteq \frac{T'}{T}.$$

Thus

$$\bar{A} \doteq \frac{1}{2}\int_0^{p_0} \frac{\bar{T}}{\gamma_d - \bar{\gamma}} \overline{\left(\frac{T'}{\bar{T}}\right)^2}\, dp. \tag{4.26}$$

For the typical values, $\gamma = \frac{2}{3}\gamma_d$ and $\overline{T'^2} = (15°)^2$,

$$\bar{A}(\bar{P} + \bar{I})^{-1} \sim \tfrac{1}{200}.$$

Hence less than 1% of the total potential energy is available for conversion to kinetic energy.

The average kinetic energy per unit area of a vertical column is

$$\bar{K} = \frac{1}{2g}\int_0^{p_0} \overline{V^2}\, dp.$$

The average total potential energy may be written

$$\bar{E} = \bar{P} + \bar{I} = \frac{c_v}{gR}\int_0^{p_0} \overline{c^2}\, dp,$$

where $c^2 = c_p RT/c_v$. Since the ratio of the wind speed to the speed of sound is $V/c \sim 1/20$ it follows that

$$\frac{\bar{K}}{\bar{P} + \bar{I}} \sim \frac{1}{2{,}000}, \qquad \text{and} \qquad \frac{\bar{K}}{\bar{A}} \sim \frac{1}{10}.$$

Thus the average kinetic energy is an order of magnitude less than the average available potential energy. If kinetic energy is not released, it is not because of a lack of available potential energy.

The extremely small magnitude of the kinetic energy compared to the total potential energy suggests that the use of a conservation law which merely guarantees the conservation of the sum of the potential and kinetic energies during numerical integration may not be a sufficiently effective constraint on the solution for the velocity field (Arakawa, 1970). Note that an erroneous conversion of only 0.01% of the total potential energy to kinetic energy could result in a false increase of kinetic energy by 20%, a completely unacceptable error. On the other hand, the magnitude of the available potential energy is more nearly that of the kinetic energy; hence the imposition of appropriate constraints on these two quantities, with respect to the numerical integration scheme, would appear to have more promise. In any event, there is no generation of kinetic energy by advective processes; hence any suitable numerical treatment of the nonlinear advective terms must avoid the fictitious production of kinetic energy. The nonlinear advective processes do bring about energy transformations between existing scales which will be discussed in Chapter 12.

Next the manner in which A may change will be considered. Multiplying the equation of continuity and the thermodynamic equation by the potential temperature θ gives the following results:

$$\theta \nabla \cdot \mathbf{V} + \theta \frac{\partial \omega}{\partial p} = 0,$$

$$\theta \left(\frac{\partial \theta}{\partial t} + \mathbf{V} \cdot \nabla \theta + \omega \frac{\partial \theta}{\partial p} = \frac{1}{c_p} \frac{\bar{\theta}}{\bar{T}} Q \right),$$

$$\frac{\partial}{\partial t}\left(\frac{\theta^2}{2}\right) + \mathbf{V} \cdot \nabla \frac{\theta^2}{2} + \theta\omega \frac{\partial \theta}{\partial p} = \frac{1}{c_p} \frac{\bar{\theta}}{\bar{T}} \theta Q$$

$$\frac{\partial}{\partial t}\left(\frac{\theta^2}{2}\right) + \nabla \cdot \left(\mathbf{V} \frac{\theta^2}{2}\right) - \frac{\theta^2}{2} \nabla \cdot \mathbf{V} + \theta\omega \frac{\partial \theta}{\partial p} = \frac{1}{c_p} \frac{\bar{\theta}}{\bar{T}} \theta Q.$$

Substituting $-\partial\omega/\partial p$ for $\nabla \cdot \mathbf{V}$ gives

$$\frac{\partial}{\partial t}\left(\frac{\theta^2}{2}\right) + \nabla \cdot \left(\mathbf{V} \frac{\theta^2}{2}\right) + \frac{\partial}{\partial p}\left(\frac{\omega\theta^2}{2}\right) = \frac{1}{c_p} \frac{\bar{\theta}}{\bar{T}} \theta Q.$$

Next, taking the space average of the above equation with $\theta = \bar{\theta} + \theta'$, etc., leads to

$$\frac{1}{2} \frac{\partial \bar{\theta}^2}{\partial t} + \frac{1}{2} \frac{\partial \overline{\theta'^2}}{\partial t} + \frac{\partial}{\partial p}\left(\overline{\omega\theta'}\bar{\theta} + \frac{\overline{\omega\theta'^2}}{2}\right) = \frac{\bar{\theta}}{c_p \bar{T}} (\bar{\theta}\bar{Q} + \overline{\theta' Q'}). \qquad (4.27)$$

The space average vertical velocity over an isobaric surface is zero, i.e., $\bar{\omega} = 0$, which has been used previously. We now desire to eliminate $\partial\bar{\theta}^2/\partial t$ in the foregoing equation. Adding the continuity equation to the thermodynamic equation gives

$$\frac{\partial \theta}{\partial t} + \nabla \cdot (\mathbf{V}\theta) + \frac{\partial(\omega\theta)}{\partial p} = \frac{1}{c_p} \frac{\bar{\theta}}{\bar{T}} Q.$$

The space average is

$$\frac{\partial \bar{\theta}}{\partial t} + \frac{\partial \overline{(\omega\theta)}}{\partial p} = \frac{1}{c_p} \frac{\bar{\theta}}{\bar{T}} \bar{Q}.$$

and after multiplication with $\bar{\theta}$ we obtain

$$\frac{1}{2} \frac{\partial \bar{\theta}^2}{\partial t} + \bar{\theta} \frac{\partial \overline{\omega\theta'}}{\partial p} = \frac{1}{c_p} \frac{\bar{\theta}}{\bar{T}} \bar{\theta}\bar{Q}. \qquad (4.28)$$

Subtracting (4.28) from (4.27) yields, after simplification,

$$\frac{1}{2}\frac{\partial\overline{\theta'^2}}{\partial t} + \overline{\omega\theta'}\frac{\partial\bar{\theta}}{\partial p} + \frac{\partial}{\partial p}\frac{\overline{\omega\theta'^2}}{2} = \frac{1}{c_p}\frac{\bar{\theta}}{\bar{T}}\overline{\theta'Q'}. \tag{4.29}$$

The third term on the left is a "triple correlation" which is generally negligible, and will be omitted.

Partial differentiation of (4.26) with respect to time gives

$$\frac{\partial\bar{A}}{\partial t} \doteq \frac{1}{2}\int_0^{p_0} \frac{\bar{T}}{\gamma_d - \bar{\gamma}}\frac{\partial}{\partial t}\overline{\left(\frac{\theta'}{\bar{\theta}}\right)^2}\,dp, \tag{4.30}$$

where T'/T has been replaced by θ'/θ and the local variation of the coefficient involving lapse rate has been neglected. Substituting from (4.29) into (4.30) and substitution for $\partial\bar{\theta}/\partial p$ leads to

$$\frac{\partial\bar{A}}{\partial t} = -C + G, \tag{4.31}$$

where

$$C = -\frac{R}{g}\int_0^{p_0} p^{-1}\overline{T'\omega}\,dp = -R\int\frac{T'\omega}{p}\,dM,$$

$$G = \frac{1}{g}\int_0^{p_0} \frac{\gamma_d}{\bar{T}(\gamma_d - \bar{\gamma})}\overline{T'Q'}\,dp.$$

If the temperature in (4.16) is replaced by $\bar{T} + T'$, the kinetic-energy equation, with the friction term added, may be written in the form

$$\frac{\partial\bar{K}}{\partial t} = C - D,$$

$$C = -R\int\frac{T'\omega}{p}\,dM; \qquad D = -\int \mathbf{V}\cdot\mathbf{F}\,dM. \tag{4.32}$$

Here $-\mathbf{F}$ is the friction force per unit mass.

Under adiabatic conditions the function G in (4.31) vanishes; whereas if the atmosphere is frictionless, the dissipation function D in (4.32) is zero. If both are zero, the sum of the available potential and kinetic energies is conserved, i.e.,

$$\bar{K} + \bar{A} = \text{const.} \tag{4.33}$$

This result is similar to (4.21*b*). Strictly speaking, from (4.31) and (4.32) the proper inference is that (4.33) is only approximately true due to the approximations used in arriving at (4.31). However, it has been shown earlier that the total potential energy plus the kinetic energy is conserved in the mean for frictionless, adiabatic flow. Since the unavailable potential energy is automatically conserved if there is no diabatic heating, it follows that (4.33) is exactly true.

If we now return to the quasi-geostrophic thermodynamic energy equation (4.24), substitute $\bar{T} + T'$ for T, multiply by T', and integrate over an entire pressure surface, the result is

$$\frac{1}{2}\frac{\overline{\partial T'^2}}{\partial t} = \sigma\overline{T\omega}\,.$$

Dividing both sides by the factor $(\gamma_d - \bar{\gamma})\bar{T}$ and integrating vertically yields (4.31) for adiabatic flow. Thus the quasi-geostrophic thermodynamic equation provides for the transformation of available potential energy into kinetic energy.

From its definition it is seen that the existence of available potential energy depends on departures of the density stratification from the horizontal, which, in turn, gives rise to horizontal pressure gradients and geostrophic winds. Since the wind is nearly geostrophic, it follows that *large kinetic energy is associated with large available potential energy*, and vice versa. Thus large increases in kinetic energy should accompany large increases in available potential energy. However, the energy transformation equations (4.31) and (4.32) show that under adiabatic conditions an increase in kinetic energy takes place at the expense of available potential energy. Consequently, when both forms of energy increase simultaneously, diabatic effects must be creating available potential energy, as, for example, from summer to winter. Paradoxically, this seasonal change is accompanied by a decrease in the total potential energy.

4.5 VORTICITY AND DIVERGENCE EQUATIONS

The kinetic-energy relations discussed earlier in Section 4.3 were based on the Newtonian equation of motion, usually referred to as the primitive equation. However, many numerical prediction models have utilized the vorticity and divergence equations in one form or another. It is therefore desirable to become familiar with the technique for determining kinetic-energy relationships from the vorticity and/or divergence equations. Following Arakawa (1960) and Lorenz (1960), the former may be written as

$$\frac{\partial \zeta}{\partial t} + (\mathbf{V}_\psi + \mathbf{V}_\chi)\cdot\nabla\zeta + (\mathbf{V}_\psi + \mathbf{V}_\chi)\cdot\nabla f + (f+\zeta)\nabla\cdot(\mathbf{V}_\psi + \mathbf{V}_\chi)$$
$$+\,\omega\frac{\partial \zeta}{\partial p} + \mathbf{k}\cdot\nabla\omega \times \left(\frac{\partial \mathbf{V}_\psi}{\partial p} + \frac{\partial \mathbf{V}_\chi}{\partial p}\right) = 0. \qquad (4.34)$$

Here the horizontal wind has been expressed as the sum of the rotational and divergent components. Multiplying this equation by ψ and rearranging the terms yields

$$\psi\left[\left(\frac{\partial\zeta}{\partial t}+\mathbf{V}_\psi\cdot\nabla\zeta\right)+\mathbf{V}_\psi\cdot\nabla f+\nabla\cdot(f\mathbf{V}_\chi)+\nabla\cdot\left(\zeta\mathbf{V}_\chi+\omega\frac{\partial\nabla\psi}{\partial p}\right)\right.$$
$$\qquad(a)\qquad\qquad(b)\qquad\qquad(c)\qquad\qquad(d)$$

$$\left.+\nabla\cdot\left(\frac{\partial\mathbf{V}_\chi}{\partial p}\times\omega\mathbf{k}\right)\right]=0. \quad (4.35a)$$
$$(e)$$

These terms may be written in the form of a divergence plus other terms as follows:

$$\begin{aligned}
&(b) && \nabla\cdot(f\psi\mathbf{V}_\psi) && \lessapprox\frac{V^3}{LR_1},\\
&(a) && -\frac{\partial}{\partial t}(\tfrac{1}{2}V_\psi^2)+\nabla\cdot\left[\psi\left(\frac{\partial\nabla\psi}{\partial t}+\zeta\mathbf{V}_\psi\right)\right] && \sim\frac{V^3}{L}\\
&(c) && -f\mathbf{V}_\chi\cdot\nabla\psi+\nabla\cdot(f\psi\mathbf{V}_\chi) && \sim\frac{\lambda V^3}{L} \qquad (4.35b)\\
&(d) && -\zeta\mathbf{V}_\chi\cdot\nabla\psi-\omega\frac{\partial}{\partial p}(\tfrac{1}{2}V_\psi^2)+\nabla\cdot\left(\psi\zeta\mathbf{V}_\chi+\psi\omega\frac{\partial\nabla\psi}{\partial p}\right) && \sim\frac{\lambda R_1V^3}{L},\\
&(e) && -\left(\frac{\partial\mathbf{V}_\chi}{\partial p}\times\omega\mathbf{k}\right)\cdot\nabla\psi+\nabla\cdot\left[\psi\left(\frac{\partial\mathbf{V}_\chi}{\partial p}\times\omega\mathbf{k}\right)\right] && \lessapprox\frac{(\lambda R_1)^2V^3}{L}.
\end{aligned}$$

The quantity to the right of each group of terms represents its order of magnitude as discussed in Chapter 3.† If the complete vorticity equation is integrated globally and the divergence theorem is applied, the result is

$$\int\left[\frac{\partial}{\partial t}\frac{V_\psi^2}{2}+f\mathbf{V}_\chi\cdot\nabla\psi+\zeta\mathbf{V}_\chi\cdot\nabla\psi\right.$$
$$\quad(a)\qquad\quad(c)\qquad\qquad(d)$$

$$\left.+\omega\frac{\partial}{\partial p}\frac{V_\psi^2}{2}+\left(\frac{\partial\mathbf{V}_\chi}{\partial p}\times\omega\mathbf{k}\right)\cdot\nabla\psi\right]dM=0. \qquad (4.36)$$
$$(e)$$

Had only the terms denoted by (*a*) and/or (*b*) been retained in (4.35), then only term (*a*) in (4.36) would appear. Under these circumstances the mean kinetic energy over an isobaric surface of the rotational (or nondivergent) wind would be conserved, which is characteristic of the barotropic model. Retention of the terms denoted by (*a*), (*b*), and (*c*) results in the next term in (4.36) being also retained in the global mean, and so forth. It should be

† $R_1 = V/fL$ is the Rossby number; $R_i = \sigma_s/F$, the Richardson number; $F = V^2/gD$, the Froude number; σ_s is a static stability parameter; and $\lambda = R_1^{-2}R_i^{-1}$.

mentioned here that the various sets of terms in (4.35*b*) have been grouped in order of generally decreasing magnitude as described in more detail in Chapter 3.

The divergence equation may be treated in a similar manner. To obtain this equation, operate on the equation of motion with $\chi\nabla$.

$$\chi\nabla\cdot\left\{\left[\underset{(1)}{\frac{\partial(\mathbf{V}_\psi+\mathbf{V}_\chi)}{\partial t}}+\underset{(2)}{(\mathbf{V}_\psi+\mathbf{V}_\chi)\cdot\nabla(\mathbf{V}_\psi+\mathbf{V}_\chi)}+\underset{(3)}{f\mathbf{k}\times(\mathbf{V}_\psi+\mathbf{V}_\chi)}\right]\right.$$

$$\left.+\underset{(4)}{\omega\left(\frac{\partial\mathbf{V}_\psi}{\partial p}+\frac{\partial\mathbf{V}_\chi}{\partial p}\right)}+\underset{(5)}{\nabla\Phi}\right\}=0. \quad (4.37a)$$

The various terms may now be expressed in alternate forms as follows:

$$(5)\quad \chi\nabla^2\Phi=\nabla\cdot(\chi\nabla\Phi)-\mathbf{V}_\chi\cdot\nabla\Phi=\nabla\cdot(\chi\nabla\Phi-\Phi\mathbf{V}_\chi)-\frac{\partial}{\partial p}\,\omega\Phi-\omega\alpha \qquad \leqslant\frac{\lambda V^3}{L}.$$

(The continuity equation has been used to obtain the last form.)

$$(3a)\quad \chi\nabla\cdot(f\mathbf{k}\times\mathbf{V}_\psi)=-\chi\nabla\cdot(f\nabla\psi)=-\nabla\cdot(f\chi\nabla\psi)+f\mathbf{V}_\chi\cdot\nabla\psi \qquad \leqslant\frac{\lambda V^3}{L},$$

$$(2a)\quad \chi\nabla\cdot(\mathbf{V}_\psi\cdot\nabla\mathbf{V}_\psi)=\nabla\cdot[\chi(\mathbf{V}_\psi\cdot\nabla\mathbf{V}_\psi)]+\mathbf{V}_\chi\cdot(\zeta\nabla\psi-\nabla\tfrac{1}{2}V_\psi^2) \qquad \leqslant\frac{\lambda R_1V^3}{L}$$

$$(3b)\quad \chi\nabla\cdot(f\mathbf{k}\times\mathbf{V}_\chi)=\nabla\cdot(\chi f\mathbf{k}\times\mathbf{V}_\chi) \qquad \leqslant\frac{\lambda R_1V^3}{L},$$

$$(1b, 2b, 2c, 4a)\quad \chi\nabla\cdot\left[\left(\frac{\partial}{\partial t}+\mathbf{V}_\psi\cdot\nabla\right)\mathbf{V}_\chi+\mathbf{V}_\chi\cdot\nabla\mathbf{V}_\psi+\omega\frac{\partial\mathbf{V}_\psi}{\partial p}\right]$$

$$=\nabla\cdot\left\{\chi\left[\left(\frac{\partial}{\partial t}+\mathbf{V}_\psi\cdot\nabla\right)\mathbf{V}_\chi+(\mathbf{V}_\chi\cdot\nabla)\mathbf{V}_\psi+\omega\frac{\partial\mathbf{V}_\psi}{\partial p}\right]\right\}$$

$$-\left[\frac{\partial}{\partial t}\tfrac{1}{2}V_\chi^{\,2}+\mathbf{V}_\chi\cdot(\mathbf{V}_\chi\cdot\nabla\mathbf{V}_\psi)+\mathbf{V}_\chi\cdot(\mathbf{V}_\psi\cdot\nabla\mathbf{V}_\chi)\right.$$

$$\left.+\omega\mathbf{V}_\chi\cdot\frac{\partial\mathbf{V}_\psi}{\partial p}\right] \qquad \leqslant\frac{(\lambda R_1)^2V^3}{L},$$

$$(4b, 2d)\quad \chi\nabla\cdot\left(\omega\frac{\partial\mathbf{V}_\chi}{\partial p}+\mathbf{V}_\chi\cdot\nabla\mathbf{V}_\chi\right)=+\nabla\cdot\left[\chi\left(\omega\frac{\partial\mathbf{V}_\chi}{\partial p}+\mathbf{V}_\chi\cdot\nabla\mathbf{V}_\chi\right)\right]$$

$$-\left(\mathbf{V}_\chi\cdot\nabla+\omega\frac{\partial}{\partial p}\right)\tfrac{1}{2}V_\chi^{\,2} \qquad \leqslant\frac{(\lambda R_1)^3V^3}{L}$$

With respect to typical synoptic waves the first two terms, (5) and (3*a*), are of the same order of magnitude and are sufficient for quasi-geostrophic approximation. On the other hand, the last two sets of terms, (1*b*), (2*b*), (2*c*), (4*a*), (4*b*), and (2*d*), have the smallest magnitudes in synoptic scale motions and normally have been neglected in numerical weather prediction models. If these are omitted and a global average is taken, the result is

$$\int\left[\underset{(2a)\ (3a)}{(\zeta+f)\mathbf{V}_\chi\cdot\nabla\psi}-\underset{(2a)}{\tfrac{1}{2}\mathbf{V}_\chi\cdot\nabla V_\psi^2}-\underset{(5)}{\omega\alpha}-\underset{(5)}{\frac{\partial(\omega\Phi)}{\partial p}}\right]dM=0. \quad (4.38)$$

Note that the time variation of the kinetic energy of the divergent wind does not appear with this approximation. Vertical integration will remove the last term if the boundary condition $\omega = 0$ at $p = p_0$ is assumed along with $\omega = 0$ at $p = 0$. If only terms of similar magnitude ($\lambda R_1 V^3/L$ or greater) in (4.35) had been retained, (4.36) would have had the form

$$-\int\left[\frac{\partial}{\partial t}\tfrac{1}{2}V_\psi^2+(f+\zeta)\mathbf{V}_\chi\cdot\nabla\psi+\omega\frac{\partial}{\partial p}\tfrac{1}{2}V_\psi^2\right]dM=0.$$

Summing the equation above with (4.38) eliminates $(f + \zeta)\mathbf{V}_\chi \cdot \nabla\psi$, giving

$$\int\left[\frac{\partial}{\partial t}\tfrac{1}{2}V_\psi^2+\left(\mathbf{V}_\chi\cdot\nabla+\omega\frac{\partial}{\partial p}\right)\tfrac{1}{2}V_\psi^2+\omega\alpha\right]dM=0.$$

Finally, adding $\frac{1}{2}V_\psi^2(\nabla\cdot\mathbf{V}_\chi + \partial\omega/\partial p) = 0$ to the last equation and applying the divergence theorem removes the advective term, leaving simply

$$\int\frac{\partial K_\psi}{\partial t}\,dM=-\int\omega\alpha\,dM.$$

This result is similar to (4.16), except that only the kinetic energy of the rotational wind appears. The conservation law for adiabatic, frictionless motions, analogous to (4.21), follows from the above equation and (4.19), as before, namely,

$$\frac{\partial}{\partial t}\int(K_\psi+E)\,dM=0.$$

From the above results it is apparent that approximate forms of the vorticity and divergence equations which comprise an energetically consistent pair are constituted of those terms from which the groups of (4.35) and (4.37) of magnitude $\lambda R_1 V^3/L$ or greater, emanated, namely, the vorticity equation,

$$\frac{\partial\zeta}{\partial t}+\mathbf{V}\cdot\nabla(\zeta+f)+\omega\frac{\partial\zeta}{\partial p}+(\zeta+f)\nabla\cdot\mathbf{V}_\chi+\mathbf{k}\cdot\nabla\omega\times\frac{\partial\mathbf{V}_\psi}{\partial p}=0, \quad (4.39a)$$

and the divergence equation,

$$\nabla^2\Phi + \boldsymbol{\nabla}\cdot f\mathbf{k}\times\mathbf{V} + \boldsymbol{\nabla}\cdot[(\mathbf{V}_\psi\cdot\boldsymbol{\nabla})\mathbf{V}_\psi] = 0.$$

Expanding the vector products in the last equation leads to

$$\nabla^2\Phi + \mathbf{k}\times\mathbf{V}\cdot\boldsymbol{\nabla}f - f\zeta - 2J(u_\psi, v_\psi) = 0, \tag{4.39b}$$

which is the so-called *balance equation* [see (3.47)]. Equations 4.39*a*, 4.39*b*, and 4.18 comprise the *balanced system* of equations.

Had all of the terms of (4.35) and (4.37) been retained and the equations integrated and summed, the result would have been the same as obtained from the Newtonian equation (4.16), namely,

$$\frac{\partial\bar{K}}{\partial t} = \frac{-1}{M}\int\omega\alpha\, dM.$$

Continuing the type of reasoning used previously, consider next the terms of (4.35*b*) and (4.37*b*) which are of order $\lambda V^3/L$ or greater, and integrate the corresponding vorticity and divergence equations (after multiplication by ψ and χ respectively) over the entire atmosphere. The results are

$$\frac{\partial}{\partial t}\int\tfrac{1}{2}V_\psi^{\,2}\, dM + \int f\mathbf{V}_\chi\cdot\boldsymbol{\nabla}\psi\, dM = 0$$

and

$$\int\omega\alpha\, dM - \int f\mathbf{V}_\chi\cdot\boldsymbol{\nabla}\psi\, dM = 0, \tag{4.40a}$$

where ω has been taken to be zero at $p = 0$ and $p = p_0$. Summing the two equations to eliminate the second terms gives

$$\frac{\partial}{\partial t}\int\tfrac{1}{2}V_\psi^{\,2}\, dM = -\int\omega\alpha\, dM, \tag{4.40b}$$

a result again consistent with (4.16) provided that only the kinetic energy of the rotational part of the wind is included in the energy total. Combination of (4.40) and (4.19) again gives the conservation law,

$$\bar{K}_\psi + \bar{E} = \text{const.}$$

Hence the approximate forms of the vorticity and divergence equations from which (4.40) result form an energetically consistent pair, namely,

$$\frac{\partial\zeta}{\partial t} + \mathbf{V}_\psi\cdot\boldsymbol{\nabla}(\zeta + f) + \boldsymbol{\nabla}\cdot(f\mathbf{V}_\chi) = 0, \tag{4.41a}$$

$$\nabla^2\Phi + \boldsymbol{\nabla}\cdot(f\mathbf{k}\times\mathbf{V}_\psi) = 0, \tag{4.41b}$$

or, expressed in a slightly different form,

$$\frac{\partial \zeta}{\partial t} + \mathbf{V}_\psi \cdot \boldsymbol{\nabla}(\zeta + f) + \mathbf{V}_\chi \cdot \boldsymbol{\nabla} f = f \frac{\partial \omega}{\partial p}, \tag{4.42a}$$

and

$$\nabla^2 \Phi - \boldsymbol{\nabla} \cdot (f \boldsymbol{\nabla} \psi) = 0, \tag{4.42b}$$

which is the linear balanced system. As seen from (4.41*a*), this system obviously fulfills the integral constraint on vorticity. Note further that both systems (4.39) and (4.42) involve the divergent wind component and also require thermal advection with the divergent wind in (4.18).

A further simplification of the system (4.42*a*) and (4.42*b*), which remains consistent both with respect to energetics and the scale analysis, results from taking a constant value for f in (4.42*b*) and in $\boldsymbol{\nabla} \cdot (f\mathbf{V}_\chi)$ in (4.42*a*), giving the *quasi-geostrophic system*†

$$\frac{\partial \zeta}{\partial t} + \mathbf{V}_\psi \cdot \boldsymbol{\nabla}(\zeta + f) = \bar{f} \frac{\partial \omega}{\partial p},$$

$$\nabla^2 \Phi - \bar{f} \nabla^2 \psi = 0. \tag{4.43}$$

As shown in Section 4.2, this system also fulfills the required integral constraint with regard to vorticity [see (4.9)]. Equation 4.24 is an appropriate thermodynamic equation for this system, provided σ is a constant or at most a function of pressure as discussed in Section 4.4.

The model with maximum possible simplification is the *quasi-geostrophic barotropic* case which results from including only the term labeled (*a*) in (4.36), namely,

$$\frac{\partial}{\partial t} \int \tfrac{1}{2} V_\psi^2 \, dM = 0. \tag{4.44}$$

The corresponding vorticity equation is

$$\frac{\partial \zeta}{\partial t} + \mathbf{V}_\psi \cdot \boldsymbol{\nabla}(\zeta + f) = 0, \tag{4.45}$$

where

$$\mathbf{V}_\psi = \mathbf{k} \times \boldsymbol{\nabla} \psi = \mathbf{k} \times \boldsymbol{\nabla} \Phi / \bar{f},$$

which is equivalent to (4.43).

As do the more sophisticated models represented by (4.41) and (4.43), the barotropic model (4.45) conserves the mean vorticity as well as kinetic energy. Since (4.45) has a single unknown ψ, no additional equations are necessary in this case.

† The basis for this result lies in a further decomposition of the terms involving f in (4.35*a*) and (4.35*b*) and taking into account that the scaled value of $\boldsymbol{\nabla} f$ is f/a or β.

4.6 LORENZ' TREATMENT OF INTEGRAL CONSTRAINTS

The previous discussion of the consistency requirements for large scale meteorological dynamical systems with respect to energy was based in part on the scale analysis treated in Chapter 3. E. N. Lorenz (1960) provided an ingenious systematic treatment of the integral constraints which does not require the scale analysis. This is accomplished in the following manner. First specify those variables which are determinable by the equation of state and the hydrostatic approximation by the subscript 1 as follows:

$$p\alpha_1 = RT_1, \tag{4.46}$$

$$\frac{\partial \Phi_1}{\partial p} + \alpha_1 = 0. \tag{4.47}$$

The potential temperature is also so designated, namely,

$$\theta_1 = T_1\left(\frac{p_0}{p}\right)^{R/c_p}. \tag{4.48}$$

Next the horizontal wind vector is divided into its rotational and divergent parts, which are denoted by the numbers 2 and 3:

$$\mathbf{V}_2 = \mathbf{k} \times \nabla\psi_2, \qquad \mathbf{V}_3 = \nabla\chi_3,$$
$$\zeta_2 = \nabla^2\psi_2, \qquad \delta_3 = \nabla^2\chi_3.$$

By virtue of the continuity equation (in the pressure coordinate system), the vertical velocity is also denoted by the number 3

$$-\delta_3 = \frac{\partial \omega_3}{\partial p}. \tag{4.49}$$

The six dependent variables, θ_1, α_1, Φ_1, ψ_2, χ_3, ω_3 may be considered as a complete set for a dry atmosphere.

The kinetic energy per unit mass, as expressed in terms of ψ_2 and χ_3, is

$$\frac{V^2}{2} = \tfrac{1}{2}\nabla\psi_2 \cdot \nabla\psi_2 + J(\psi_2, \chi_3) + \tfrac{1}{2}\nabla\chi_3 \cdot \nabla\chi_3.$$

Its mean value over the entire atmosphere is

$$\bar{K} = \bar{K}_1 + \bar{K}_2 = \frac{1}{2}\int \nabla\psi_2 \cdot \nabla\psi_2 \, dM + \frac{1}{2}\int \nabla\chi_3 \cdot \nabla\chi_3 \, dM, \tag{4.50}$$

since the Jacobian term vanishes when integrated over the entire atmosphere. The total potential energy E expressed in terms of potential temperature is

$$\int E \, dM = c_p \int \left(\frac{p}{p_0}\right)^{R/c_p} \theta_1 \, dM. \tag{4.51}$$

The vorticity, divergence and thermodynamic equations are now readily expressed as

$$\frac{\partial \zeta_2}{\partial t} = -J(\psi_2, \zeta_2) - J(\psi_2, f) - \nabla \cdot f\mathbf{V}_3 - \mathbf{V}_3 \cdot \nabla\zeta_2 - \zeta_2\delta_3$$

(2, 2) (2) (3) (2, 3) (2, 3)

$$-\omega_3 \frac{\partial \zeta_2}{\partial p} - \nabla\omega_3 \cdot \nabla \frac{\partial \psi_2}{\partial p} - J\left(\omega_3, \frac{\partial \chi_3}{\partial p}\right) \tag{4.52}$$

(2, 3) (2, 3) (3, 3)

$$\frac{\partial \delta_3}{\partial t} = -\nabla^2\Phi_1 + \nabla \cdot (f\nabla\psi_2) - J(f, \chi_3) - \nabla \cdot [(\mathbf{V}_2 \cdot \nabla)\mathbf{V}_2]$$

(1) (2) (3) (2, 2)

$$-\nabla \cdot [(\mathbf{V}_2 \cdot \nabla)\mathbf{V}_3] - \nabla \cdot [(\mathbf{V}_3 \cdot \nabla)\mathbf{V}_2] - \nabla\omega_3 \cdot \frac{\partial \mathbf{V}_2}{\partial p}$$

(2, 3) (2, 3) (2, 3)

$$-\nabla \cdot [(\mathbf{V}_3 \cdot \nabla)\mathbf{V}_3] - \nabla\omega_3 \cdot \frac{\partial \mathbf{V}_3}{\partial p} - \omega_3\nabla \cdot \frac{\partial \mathbf{V}_3}{\partial p}, \tag{4.53}$$

(3, 3) (3, 3) (3, 3)

$$\frac{\partial \theta_1}{\partial t} = -\left[\nabla \cdot (\theta_1\mathbf{V}_2) + \nabla \cdot (\theta_1\mathbf{V}_3) + \frac{\partial}{\partial p}(\theta_1\omega_3)\right]. \tag{4.54}$$

(1, 2) (1, 3) (1, 3)

By use of (4.51) and (4.54) it is easily shown by integrating over the entire atmosphere that

$$\frac{\partial}{\partial t}\int E_1\, dM = -c_p \int \left(\frac{p}{p_0}\right)^{R/c_p} \frac{\partial}{\partial p}(\theta_1\omega_3)\, dM. \tag{4.55}$$

Integrating by parts and using the boundary conditions, $\omega = 0$ at $p = p_0$ and $p = 0$, leads to

$$\frac{\partial \bar{E}_1}{\partial t} = c_p p_0^{-1} \int \left(\frac{p_0}{p}\right)^{c_v/c_p} \theta_1\omega_3\, dM. \tag{4.56}$$

In order to obtain expressions for the mean rate of change of kinetic energy, multiply (4.52) and (4.53) by ψ_2 and χ_3 respectively and integrate over the entire atmosphere. The procedure is similar to that used earlier to obtain (4.36) and (4.38). The results are

$$\frac{\partial \bar{K}_2}{\partial t} = \underset{(2,\,3)}{\int \psi_2 \nabla \cdot f\mathbf{V}_3\, dM} + \underset{(2,\,2,\,3)}{\int \psi_2 \mathbf{V}_3 \cdot \nabla \zeta_2\, dM} + \underset{(2,\,2,\,3)}{\int \psi_2 \zeta_2 \delta_3\, dM}$$

$$+ \underset{(2,\,2,\,3)}{\int \psi_2 \omega_3 \frac{\partial \zeta_2}{\partial p}\, dM} + \underset{(2,\,2,\,3)}{\int \psi_2 \nabla \omega_3 \cdot \nabla \frac{\partial \psi_2}{\partial p}\, dM} + \underset{(2,\,2,\,3)}{\int \psi_2 J\left(\chi_3, \frac{\partial \chi_3}{\partial p}\right) dM}, \quad (4.57)$$

$$\frac{\partial \bar{K}_3}{\partial t} = \underset{(1,\,3)}{\int \chi_3 \nabla^2 \Phi_1\, dM} - \underset{(2,\,3)}{\int \chi_3 \nabla \cdot (f \nabla \psi_2)\, dM}$$

$$+ \underset{(2,\,2,\,3)}{\int \chi_3 \nabla \cdot [(\mathbf{V}_2 \cdot \nabla)\mathbf{V}_2]\, dM} + \underset{(2,\,3,\,3)}{\int \chi_3 \nabla \cdot [(\mathbf{V}_2 \cdot \nabla)\mathbf{V}_3]\, dM} \quad (4.58)$$

$$+ \underset{(2,\,3,\,3)}{\int \chi_3 \nabla \cdot [(\mathbf{V}_3 \cdot \nabla)\mathbf{V}_2\, dM} + \underset{(2,\,3,\,3)}{\int \chi_3 \nabla \omega_3 \cdot \frac{\partial \mathbf{V}_2}{\partial p}\, dM}.$$

The variables denoted by the subscript 1, namely, α, T, Φ, and θ, are directly related to one another by the "diagnostic" equations (4.46) through (4.48). Similarly, ψ_2 and $\mathbf{V}_2$ are directly related, and also χ_3, $\mathbf{V}_3$, and ω_3. However, the three groups are not related diagnostically to one another, but only more generally through the *prognostic* hydrodynamic equations (those equations with time derivatives). Hence it may be expected that, when (4.56), (4.57), and (4.58) are added, the terms would cancel identically group by group (as denoted by the numbers in parentheses). That the total sum of the right-hand sides will vanish may be foreseen from the previous result that the sum of the kinetic and potential energies will vanish under adiabatic, frictionless conditions when integrated over the entire atmosphere.

As an example of group cancellation, consider the terms labeled (2, 3) in (4.57) and (4.58), which may be written in alternative form as follows:

$$\underset{(2,\,3)}{\int \psi_2 \nabla \cdot f\mathbf{V}_3\, dM} = \int [\nabla \cdot (f\mathbf{V}_3 \psi_2) - f\mathbf{V}_3 \cdot \nabla \psi_2]\, dM$$

$$-\underset{(2,\,3)}{\int \chi_3 \nabla \cdot f \nabla \psi_2\, dM} = -\int [\nabla \cdot (f\chi_3 \nabla \psi_2) + f \nabla \psi_2 \cdot \nabla \chi_3]\, dM.$$

When summed, the last terms clearly cancel, while the first integrals vanish by virtue of the divergence theorem.

In a similar fashion the group of terms denoted (2, 3, 3) must cancel one another, etc., with the final result

$$\frac{\partial \bar{E}_1}{\partial t} + \frac{\partial \bar{K}_2}{\partial t} + \frac{\partial \bar{K}_3}{\partial t} = 0.$$

If approximate forms of the thermodynamic, vorticity, and divergence equations are utilized, the terms must be omitted by groups in order to ensure that the mean energy $\overline{K + E}$ remains invariant. First note that, if the divergence equation is made into a diagnostic equation by omitting $\partial\delta_3/\partial t$, then the kinetic energy of the divergent wind component must not be counted in the energy tally, i.e., $\overline{K_2 + E}$ becomes the *new invariant.*

Now consider, in addition, the omission of all terms of the group (2, 3, 3). Examining the vorticity equation (4.52), it may be noted that in this event it must have had the initial approximate form

$$\frac{\partial \zeta_2}{\partial t} = -J(\psi_2, \zeta_2) - J(\psi_2, f) - \nabla \cdot f\mathbf{V}_3 - \mathbf{V}_3 \cdot \nabla\zeta_2 - \zeta_2\delta_3 - \omega_3\frac{\partial \zeta_2}{\partial p} - \nabla\omega_3 \cdot \nabla\frac{\partial \psi_2}{\partial p}. \tag{4.59}$$

Similarly, the corresponding form of the divergence equation would have been

$$\nabla^2\Phi_1 - \nabla \cdot f\nabla\psi_2 + \nabla \cdot [(\mathbf{V}_2 \cdot \nabla)\mathbf{V}_2] = 0, \tag{4.60}$$

which is the *balance equation.* Note that the only term omitted in the vorticity equation is that part of the twisting term involving the divergent wind. Thus *(4-59) and (4-60) comprise an energetically consistent pair.*

If as a further simplification the terms denoted by (2, 2, 3) are omitted from (4.58) and (4.57), then only the first three terms on the right side of (4.52) may be kept, and only the first two terms on the right side of (4.53). This energetically consistent system is identical to (4.42*a*) and (4.42*b*).

Note finally that in both of the systems, (4.59), (4.60), and (4.42), advection of temperature with the divergent wind must be included as was done, for example, in (4.54) or (4.18).

4.7 OBSERVATIONAL EVIDENCE ON ENERGY EXCHANGES

Observational studies of the atmosphere show that the dominant flow of energy in the atmosphere is from zonal potential energy to eddy potential energy to eddy kinetic energy to zonal kinetic energy, though there are exceptions of course. Further partitions can be made to categorize the energy exchanges in greater detail. For example, there are linear processes including the generation of potential energy, conversion between potential and kinetic energies, and dissipation of kinetic energy by friction. Then there are the nonlinear exchanges arising from the nonlinear terms in the thermodynamic and motion equations. These include two-component exchanges such as the interaction between zonal motion and a particular wave component of the eddy motion, between a wave component and its first harmonic, and the flow of zonal available potential energy to the eddies. In addition, there are three-component exchanges between the wave components which redistribute the energy among the waves comprising the eddy motion.

C. H. Yang (1967) has summarized the present state of knowledge regarding energy exchange and made additional computations and analyses. Briefly summarizing, in the troposphere there generally is a transfer of kinetic energy from medium to long waves and somewhat to short waves; whereas the potential-energy transfer is from long to both medium and short waves. Exceptions occur mainly in the summer months when the energy exchange is reduced.

Saltzman and Teweles (1964) studied a nine-year period and found a general transfer of eddy kinetic energy to zonal kinetic energy, and also that the wave numbers 2 and 5 through 10 (the term *wave number* as used here denotes the number of waves around the entire earth for the particular harmonic) act as sources of eddy kinetic energy for the remaining waves. The source at wave number 2 appears to be a forced conversion on the scale of the major continents and oceans.

In the exchange of available potential energy, Yang found that all wave components are fed by the zonal mean, while wave numbers 2–5 and 7 act as sources for the other waves through interaction. The largest exchange of kinetic energy takes place close to the jet stream, and for available potential energy, in the middle troposphere.

In the lower stratosphere the direction of the available potential-energy flux due to interaction appears to be opposite to that of the troposphere. Yang also verified Saltzman's previous findings that the medium waves, numbers 6–10, act as sources of kinetic energy for long waves, 1–5, and short waves, 11–15, in the troposphere; but the gain by the short waves in the lower troposphere was slight.

In the lower stratosphere the long waves appear to be a source of kinetic energy for both the medium and short waves. With respect to available potential energy, the long waves due to nonlinear actions act as a source to the others in the troposphere, but become the recipient of potential energy in the lower stratosphere.

Most of the exchanges between waves are effected by the fast moving waves, while the slow moving and stationary waves become important primarily in exchange with the zonal mean.

chapter five

Numerical Methods

5.1 FINITE DIFFERENCING AND TRUNCATION ERRORS

The previous chapters have presented the fundamental hydrodynamical equations appropriate for meteorological purposes, a discussion of some simple types of atmospheric waves, a systematic method of comparing the various terms comprising the fundamental equations, and finally some basic energy relationships. The primary aim of this text is to describe methods for the utilization of the dynamical equations for predicting the meteorological variables. Since analytical solutions generally are not obtainable, it is necessary to make use of numerical integration techniques. Although no comprehensive treatment of numerical methods is feasible within the framework of this text, it is nevertheless desirable to treat briefly the basic concepts and to highlight some of the important difficulties that may be encountered. This chapter provides such a discussion, after which some specific numerical prediction models will be described.

Consider an arbitrary function $f(x)$ and its expansion in a convergent power series

$$f(x \pm \Delta x) = f(x) \pm f'(x)\Delta x + f''(x)\frac{\Delta x^2}{2!} \pm f'''(x)\frac{\Delta x^3}{3!} \pm \cdots. \quad (5.1)$$

Now solve for the first derivative $f'(x)$, which may be expressed in the form [using the plus sign in (5.1)]

$$f'(x) = \frac{f(x + \Delta x) - f(x)}{\Delta x} + R. \quad (5.2)$$

Here R, referred to as the remainder, represents the remaining terms of the series, which involve Δx to the first and higher powers. In this case the remainder is said to be of order Δx and is denoted by the symbol $0(\Delta x)$. When R is omitted in (5.2), there results a *forward difference approximation* for $f'(x)$ (assuming Δx is positive); and R is referred to as the *truncation error*. Had the negative sign been used in (5.1), the same procedure would have led to a *backward difference* approximation for $f'(x)$ with the same order of truncation error.

If the difference and the sum of the two series (5.1) are taken (one with the plus and the other with the minus sign), there result *central difference* approximations for $f'(x)$ and $f''(x)$, both with truncation errors of order $0(\Delta x^2)$, as follows:

$$f'(x) \doteq \frac{f(x+\Delta x) - f(x-\Delta x)}{2\Delta x}, \qquad 0(\Delta x^2), \tag{5.3}$$

$$f''(x) \doteq \frac{f(x+\Delta x) - 2f(x) + f(x-\Delta x)}{\Delta x^2}, \qquad 0(\Delta x^2). \tag{5.4}$$

In general, the higher the order of the truncation error with respect to Δx, the more accurate is the finite difference approximation.

Since wavelike motions are typical of the atmosphere, it is of interest to apply one of the foregoing finite difference approximations to a simple harmonic wave.

Let

$$f(x) = A \sin \frac{2\pi}{L} x. \tag{5.5}$$

Then

$$f'(x) = A \frac{2\pi}{L} \cos \frac{2\pi}{L} x. \tag{5.6}$$

On the other hand, the finite difference approximation (5.3) gives

$$\begin{aligned} f'(x) &\doteq A \frac{\sin (2\pi/L)(x+\Delta x) - \sin (2\pi/L)(x-\Delta x)}{2\Delta x} \\ &= A \frac{\cos (2\pi/L)x \sin (2\pi/L)\Delta x}{\Delta x} = A \frac{2\pi}{L} \cos \frac{2\pi x}{L} \frac{\sin (2\pi/L)\,\Delta x}{2\pi\,\Delta x/L}. \end{aligned}$$

Thus the simple central difference approximation for the first derivative of the function (5.5) is equal to the true value of the derivative (5.6) multiplied by the factor

$$\frac{\sin (2\pi/L)\,\Delta x}{(2\pi/L)\,\Delta x}.$$

Since the ratio $(\sin \alpha)/\alpha$ approaches 1 as α approaches zero, it is evident that the accuracy of the approximation (5.3) will increase with decreasing Δx. When the ratio $\Delta x/L$ is small, say $L \geq 10\Delta x$, the approximation (5.3) will give a good representation of $f'(x)$. On the other hand, when $\Delta x \geq L/2$, the result is very poor indeed. For example, with $\Delta x = L/2$, the finite difference approximation for $f'(x)$ would always be zero regardless of the value of x.

5.2 LINEAR COMPUTATIONAL INSTABILITY

Truncation error is by no means the only source of error when numerical methods are utilized to solve differential equations. To become acquainted with some of the other difficulties, we shall first examine a specific simple, first-order, linear, partial-differential equation for which the analytic solution is known as well as the exact solutions of some corresponding finite difference equations. In later sections more general principles will be discussed.

Consider now the advective equation

$$\frac{\partial F}{\partial t} + c\,\frac{\partial F}{\partial x} = 0, \tag{5.7}$$

where c is a constant, which may be thought of as the wind velocity and/or the phase velocity of gravity waves.

The most general solution of this equation is $F(\xi)$, where F is an arbitrary function and $\xi = x - ct$. When $t = 0$, F reduces to simply $F(x)$, which is therefore the initial condition. Now since wave motions are of interest here, assume that at the initial time $t = 0$,

$$F(x) = Ae^{i\mu x}, \tag{5.8}$$

where $\mu = 2\pi/L$ and A is a constant. Then the general solution $F(x - ct)$ is

$$F(x - ct) = Ae^{i\mu(x-ct)}, \tag{5.9}$$

which is a single harmonic of the Fourier spectrum with *wavelength* L, *phase speed* c, and a constant *amplitude* A.

The analytic solution (5.9) can also be obtained by the method of separation of variables by writing

$$F(x, t) = G(t)H(x),$$

and substituting it into (5.7), with the result

$$H(x)G'(t) + cG(t)H'(x) = 0.$$

Thus

$$\frac{G'(t)}{G(t)} = -c\,\frac{H'(x)}{H(x)} = -k,$$

where k is a constant. Then

$$G = A_1 e^{-kt}, \qquad H = A_2 e^{kx/c}$$

and

$$F(x, t) = Ae^{k(x-ct)/c}. \tag{5.10}$$

Utilization of the initial condition (5.8) gives $k = i\mu c$, from which (5.9) follows immediately.

Next a numerical solution to the initial value problem will be sought by approximating the derivatives of (5.7) by finite differences.

Let

$$x = m\,\Delta x, \qquad m = 0, \pm 1, \pm 2 \ldots, \qquad t = n\,\Delta t, \qquad n = 0, 1, 2 \ldots.$$

Thus the continuous (x, t) space has been replaced by a mesh or grid of discrete points in this space. Using the central difference approximation (5.3) for both the space and time derivatives in (5.7) leads to

$$\frac{F_{m,n+1} - F_{m,n-1}}{2\Delta t} = -c\,\frac{F_{m+1,n} - F_{m-1,n}}{2\Delta x}, \tag{5.11}$$

where m denotes the space index and n the time index. Apart from the first step, which can be obtained by a forward difference, future values of F can be obtained by solving (5.11) for $F_{m,n+1}$ as follows:

$$F_{m,n+1} = F_{m,n-1} - \frac{c\,\Delta t}{\Delta x}(F_{m+1,n} - F_{m-1,n}). \tag{5.12}$$

Equation (5.12) states that the value of F at any point $m\,\Delta x$ at a future time $(n + 1)\,\Delta t$ can be obtained from present and previous values of F at the given point and adjacent points. Because of the character of the extrapolation formula (5.12) which adds a change to $F_{m,n-1}$ to obtain the value at $F_{m,n+1}$, the central difference scheme is commonly referred to as the *leapfrog* method. Thus an iterative procedure has been established for computing values of F at the gridpoints at any future time, assuming that the initial and boundary values are known. However, there is no guarantee that the values computed by (5.12) will agree with the analytic solution of the differential equation. Some information regarding the numerical solution can be obtained in this case without making actual numerical calculations by finding the analytic solution of the difference equation (5.12).

Assume an exponential solution of (5.12) of the form

$$F_{m,n} = B^{n\,\Delta t} e^{i\mu m\,\Delta x}. \tag{5.13}$$

This form assumes that the variables are separable and presupposes a solution which is harmonic in x to conform with the initial conditions.

Substituting (5.13) into (5.12) gives

$$(B^{(n+1)\Delta t} - B^{(n-1)\Delta t})e^{i\mu m\,\Delta x} = -\lambda B^{n\,\Delta t}(e^{i\mu(m+1)\Delta x} - e^{i\mu(m-1)\Delta x}), \quad (5.14)$$

where

$$\lambda = \frac{c\,\Delta t}{\Delta x}. \quad (5.15)$$

Simplifying (5.14) by multiplying through by $B^{-(n-1)\Delta t}e^{-i\mu m\,\Delta x}$ and application of Euler's formula, $e^{i\theta} = \cos\theta + i\sin\theta$, leads to

$$B^{2\,\Delta t} + 2i\lambda\sin\mu\,\Delta x B^{\Delta t} - 1 = 0. \quad (5.16)$$

Solving this quadratic equation in $B^{\Delta t}$ gives

$$B^{\Delta t} = -i\lambda\sin\mu\,\Delta x \pm \sqrt{1 - \lambda^2\sin^2\mu\,\Delta x}. \quad (5.17)$$

Consider the case where $\lambda \leq 1$. Then the quantity under the radical is real and the two roots of (5.16), which are complex numbers, may be written in polar form as

$$B_1^{\Delta t} = e^{-i\alpha} \qquad \text{and} \qquad B_2^{\Delta t} = e^{i(\pi+\alpha)}$$

where α is a real angle defined by

$$\alpha = \arcsin(\lambda\sin\mu\,\Delta x). \quad (5.18)$$

The complete solution (5.13) then takes the form

$$F_{m,n} = (Me^{-i\alpha n} + Ee^{i(\pi+\alpha)n})e^{i\mu m\,\Delta x}, \quad (5.19)$$

where M and E are arbitrary constants.

At the initial time $n = 0$, the foregoing solution must reduce to the initial condition corresponding to (5.8). Thus

$$F_{m,0} = (M + E)e^{i\mu m\,\Delta x} = Ae^{i\mu m\,\Delta x}$$

and $M = A - E$. Also, $e^{\pi i} = -1$; hence solution (5.19) is expressible in the form

$$F_{m,n} = (A - E)e^{i\mu(m\,\Delta x - \alpha n/\mu)} + (-1)^n Ee^{i\mu(m\,\Delta x + \alpha n/\mu)}. \quad (5.20)$$

The solution (5.20) of the difference equation (5.12) is seen to consist of two waves rather than of a single wave such as (5.9), which is the analytic solution to the original differential equation. The two waves occur because (5.11) is a second-order difference equation which resulted from the use of centered time differences. Comparison of (5.9) and (5.20) shows that the finite difference solution and the analytic solution have the following characteristics:

	Finite Difference Solution		
	First Wave	Second Wave	Analytic Solution
Amplitude	$A - E$	E	A
Phase speed	$\alpha/\mu\,\Delta t$	$-\alpha/\mu\,\Delta t$	c
Phase change	None	Every time step	None

The phase change referred to results from the factor $(-1)^n$. The two wave components of the finite difference solution have constant amplitudes; however, they move in opposite directions at speeds of $\alpha/\mu\,\Delta t$.

Referring back to the definition of α in (5.18), note that, if $\lambda = 1$, then $\alpha = \mu\,\Delta x$; and from (5.15), $c\,\Delta t = \Delta x$. In this event the first finite difference wave travels at the same speed as the analytic wave, that is,

$$\frac{\alpha}{\mu\,\Delta t} \equiv c.$$

Moreover, it is easily shown from (5.18) that $\alpha/\mu\,\Delta t$ approaches c as Δx and Δt approach zero.

For this reason the first wave is referred to as the *physical mode.* On the other hand, the second finite difference wave travels in the opposite direction and, moreover, changes phase at every time step. Since this *spurious* wave has no counterpart in the analytic solution and arises from the finite difference approximation, it is referred to as a *computational mode.* More will be said about the amplitudes of the two modes shortly. It is of interest to point out here that the existence of a computational mode in space may be demonstrated in an analogous manner if F is prescribed for all time at some point, say $x = 0$.

Let us return now to (5.17) and consider the case where $\lambda > 1$. Then there will always exist wavelengths for which

$$\lambda^2 \sin^2 \mu\,\Delta x > 1.$$

In this case the quantity under the radical in (5.17) will be negative; and the two roots will be imaginary numbers with magnitudes

$$|B^{\Delta t}|^2 = 2\lambda^2 \sin^2 \mu\,\Delta x - 1 \pm 2\lambda \sin \mu\,\Delta x(\lambda^2 \sin^2 \mu\,\Delta x - 1)^{1/2}. \quad (5.21)$$

One of these magnitudes will exceed unity.

As a result the factor $(B^{\Delta t})^n$ will increase without bound with increasing time $(n\,\Delta t)$. Thus the finite difference solution will continually amplify, which is at variance with the constant-amplitude analytic solution. Such spurious amplification is referred to as *computational instability* and *must be avoided.* It is apparent from (5.17) through (5.20) that a *sufficient* condition for the solution to remain computationally stable is that

$$\frac{c\,\Delta t}{\Delta x} \leq 1. \tag{5.22}$$

If this condition is not fulfilled, there will always be some wavelengths for which $(c\,\Delta t/\Delta x)\sin \mu\,\Delta x > 1$; and these waves will experience spurious amplification. Even if such waves were not present initially, roundoff errors during calculations could produce them and then the amplification would follow.

Although the condition (5.22) guarantees the computational stability of the finite difference scheme (5.12) for (5.7), the finite difference solution is not free of error. In general the speed and amplitude of the physical mode will differ from the true speed and amplitude, and also the spurious computational mode is present. These errors will be discussed somewhat further now. As noted earlier, the time central differencing scheme is not feasible for the initial step, as may be seen from (5.12); and the values are not known prior to the initial time, $n = 0$, that is, $F_{m,-1}$ is unknown. Hence some other technique must be used for the first time step. The usual procedure in numerical weather prediction is to use a forward difference in time for the first step while retaining the central space difference, namely,

$$F_{m,1} = F_{m,0} - \frac{c\,\Delta t}{2\,\Delta x}(F_{m+1,0} - F_{m-1,0}). \tag{5.23}$$

In the example being considered, $F_{m,0} = Ae^{i\mu m\,\Delta x}$; hence, by (5.23),

$$\begin{aligned} F_{m,1} &= Ae^{i\mu m\,\Delta x} - \frac{\lambda A}{2}(e^{i\mu(m+1)\,\Delta x} - e^{i\mu(m-1)\,\Delta x}) \\ &= Ae^{i\mu m\,\Delta x}\left[1 - \frac{\lambda}{2}(e^{i\mu\,\Delta x} - e^{-i\mu\,\Delta x})\right]. \end{aligned}$$

Use of the Euler formula gives

$$F_{m,1} = A(1 - i\sin\alpha)e^{i\mu m\,\Delta x}. \tag{5.24}$$

This is the value of F at time Δt at an arbitrary gridpoint $x = m\,\Delta x$. Beyond this time the central differencing scheme is used; hence, for continuity, (5.24) should equal the value of F in (5.20) with $n = 1$. Thus

$$A(1 - i\sin\alpha)e^{i\mu m\,\Delta x} = (A - E)e^{i\mu m\,\Delta x - i\alpha} - Ee^{i\mu m\,\Delta x + i\alpha}.$$

Dividing by $Ae^{i\mu m\,\Delta x}$ gives

$$1 - i\sin\alpha = \left(1 - \frac{E}{A}\right)e^{-i\alpha} - \frac{E}{A}\,e^{+i\alpha},$$

which may be reduced to

$$-\frac{E}{A} = \frac{1-\cos\alpha}{2\cos\alpha}. \tag{5.25}$$

Thus E is determined as a function of A and α. Substituting (5.25) into (5.20) gives the complete solution for $F_{m,n}$, namely,

$$F_{m,n} = A\,\frac{1+\cos\alpha}{2\cos\alpha}\,e^{i\mu(m\,\Delta x - \alpha n/\mu)} + (-1)^{n+1}A\,\frac{1-\cos\alpha}{2\cos\alpha}\,e^{i\mu(m\,\Delta x + \alpha n/\mu)}. \tag{5.26}$$

It is clear that as α approaches zero the amplitude of the physical mode approaches the amplitude of the analytic solution A while the amplitude of the computational mode vanishes. Now for a given value of λ, (5.18) shows that α will decrease with decreasing $\mu\,\Delta x = 2\pi\,\Delta x/L$. Thus when the wavelength of the disturbance is large compared to the grid length Δx, the amplitude of the physical mode will be nearly that of the analytic wave, and, moreover, the amplitude of the computational mode will be small. For example, when $\Delta x = L/10$ and $\lambda = 0.75$, the amplitude of the physical mode is about $1.06A$ while that of the computational mode is only about 6% of A. On the other hand, if $\Delta x/L$ is relatively large, say 0.5, $\alpha = \pi$, and the amplitude of the physical mode is zero while the spurious computational mode has amplitude A, which is disastrous. It was noted earlier that for $\lambda = 1$, i.e., when $c\,\Delta t = \Delta x$, the phase speed of the physical mode is identical to that of the analytic wave solution. For $\lambda \neq 1$, the speed of the numerical wave will differ from the true speed; in fact, $\alpha/\mu\,\Delta t < c$, also false dispersion occurs.

5.3 FORWARD TIME DIFFERENCING

The spurious computational mode described in the previous section arose because of the use of central time differences. Moreover, a forward time difference had to be used to initiate the central difference scheme. A question naturally arises as to the desirability of the central difference approximation in spite of the smaller truncation error when compared to the forward difference scheme. In this section the simple forward-in-time scheme will be examined, first with a central space difference and later with forward and backward space differences. Consider then the following difference equation as an approximation to (5.7) and assume the same initial condition (5.8):

$$F_{m,n+1} = F_{m,n} - \lambda'(F_{m+1,n} - F_{m-1,n}), \qquad \lambda' = \frac{c\,\Delta t}{2\,\Delta x}. \tag{5.27}$$

Again assume a solution of the form (5.13) and substitute it into (5.27). After simplifying, the equation reduces to

$$B^{\Delta t} = 1 - 2\lambda' i \sin \mu \,\Delta x.$$

Placing the complex number on the right in polar form gives

$$B^{\Delta t} = \left[1 + \left(\frac{c\,\Delta t}{\Delta x}\right)^2 \sin^2 \mu\,\Delta x\right]^{1/2} e^{-i\theta} \tag{5.28}$$

where

$$\theta = \arctan \frac{c\,\Delta t}{\Delta x} \sin \mu\,\Delta x.$$

Hence the finite difference solution for F may be expressed as

$$F_{m,n} = A\left[1 + \left(\frac{c\,\Delta t}{\Delta x}\right)^2 \sin^2 \mu\,\Delta x\right]^{n/2} e^{i\mu(m\,\Delta x - n\theta/\mu)}. \tag{5.29}$$

It is immediately apparent from (5.29) that the difference scheme which is *forward in time* and *central in space* is *always computationally unstable* for all wavelengths (except when $\mu\,\Delta x$ is a multiple of π) because the amplitude of the finite difference wave grows continuously with time. This occurs because the magnitude of $B^{\Delta t}$ exceeds one, that is,

$$\left[1 + \left(\frac{c\,\Delta t}{\Delta x}\right)^2 \sin^2 \mu\,\Delta x\right]^{1/2} > 1. \tag{5.30}$$

Consider next the scheme which is forward both in time and space,

$$F_{m,n+1} = F_{m,n} - \lambda(F_{m+1,n} - F_{m,n}), \qquad \lambda = \frac{c\,\Delta t}{\Delta x}. \tag{5.31}$$

The equation for $B^{\Delta t}$ is readily found to be

$$B^{\Delta t} = 1 - \lambda(e^{i\mu\,\Delta x} - 1). \tag{5.32}$$

The crucial factor with respect to computational stability is the magnitude of the complex quantity $B^{\Delta t}$. It is clear that the solution will be amplified (unstable), neutral, or damped according to whether $|B^{\Delta t}|$ is greater than, equal to, or less than unity, that is

$$|B^{\Delta t}| \begin{cases} > 1 & \text{amplified,} \\ = 1 & \text{neutral,} \\ < 1 & \text{damped.} \end{cases} \tag{5.33}$$

From (5.32) it may be verified that the square of the magnitude of $B^{\Delta t}$ is expressible in the form

$$|B^{\Delta t}|^2 = 1 + 2\lambda(1 - \cos \mu\,\Delta x)(1 + \lambda). \tag{5.34}$$

Now note that, if λ is positive, which implies a positive c, the right side of (5.34) exceeds one and the solution will *amplify*. Thus the difference scheme (5.31) is computationally *unstable* with a positive c. In this case the space derivative in (5.31) is taken in the direction of c (or "*downstream*" for $c = U$).

On the other hand, if c is negative and if $-1 < \lambda < 0$, then $|B^{\Delta t}| < 1$, and the solution is *damped*. A negative c implies that the space derivative in (5.31) is taken in the direction opposite to c (or "*upstream*" if $c = U$). Although such damping would be undesirable in general, it is sometimes useful to utilize upstream differencing along an outflow boundary where central differences are obviously not feasible.

The case of a backward space difference, rather than the forward space difference on the right side of (5.31), may be examined in a similar manner. The results are again that *upstream differencing gives damping* and *downstream differencing gives amplification.*

5.4 IMPLICIT METHOD

Both the central and forward difference schemes described in the preceding sections have certain problems associated with them. We shall now consider a technique which, in a sense, combines these two schemes and, as a result, some of the earlier limitations are removed. Unfortunately a new difficulty is introduced; nevertheless the method is worthy of attention. Central differencing is used because of its stability and smaller truncation error; however by applying the time difference midway between two time steps, e.g., at $(n + \frac{1}{2})\,\Delta t$, the computational mode does not appear. Since values of the dependent variables are not available at such off-time points, averages of on-time values must be used. Consider then the following difference equation as applied to (5.7):

$$\frac{F_{m,n+1} - F_{m,n}}{\Delta t} = -\frac{c}{2}\left(\frac{F_{m+1,n+1} - F_{m-1,n+1}}{2\Delta x} + \frac{F_{m+1,n} - F_{m-1,n}}{2\Delta x}\right) \tag{5.35}$$

Note that the space derivative on the right, which should be evaluated at the midtime $(n + \frac{1}{2})\,\Delta t$, has been estimated by averaging the central difference approximations for $\partial F/\partial x$ at times $(n + 1)\,\Delta t$ and $n\,\Delta t$. The obvious difficulty associated with solving (5.35) is that more than one value of the function F at time $(n + 1)\,\Delta t$ appears in the equation. Consequently, the simple marching procedure, that characterizes (5.12), (5.27), and (5.31) for stepping forward in time at a given point from known values at previous times, is not possible with (5.35). Associated with each gridpoint, except possibly boundary points, will be an equation of the form (5.35) with three "unknowns," $F_{m-1,n+1}$, $F_{m,n+1}$, and $F_{m+1,n+1}$. Thus for a large grid, such as required to cover the Northern Hemisphere, there will be a very large system of simultaneous equations to solve. Such a system can be solved by inverting the matrix of the coefficients or by successive approximation methods. In any event, the solution of the *implicit* system (5.35) is far more time consuming than the *explicit* system (5.12) where only one value of F at the time level $n + 1$ is present.

To examine the stability of this scheme, assume a solution of the form (5.13) and substitute it into (5.35). After some simplification the result may be written as

$$B^{\Delta t} = \frac{1 - i(c\,\Delta t/2\Delta x)\sin\mu\,\Delta x}{1 + i(c\,\Delta t/2\Delta x)\sin\mu\,\Delta x}. \tag{5.36}$$

Since the numerator and denominator are of the form $a - bi$ and $a + bi$ respectively, $B^{\Delta t}$ may be written simply as

$$B^{\Delta t} = e^{-2i\theta},$$

where θ is the real angle

$$\theta = \arctan\frac{c\,\Delta t}{2\Delta x}\sin\mu\,\Delta x. \tag{5.37}$$

Thus the finite difference solution of (5.35) with the initial condition (5.8) is

$$F_{m,n+1} = Ae^{i\mu(m\,\Delta x - 2n\theta/\mu)} \tag{5.38}$$

It is apparent that the solution is *always computationally stable* regardless of the value of $c\,\Delta t/\Delta x$. Furthermore, the amplitude of the finite difference solution equals that of the analytic solution A and, finally, there is no spurious computational mode. From (5.38) it may be seen that the phase speed of the finite difference wave, say c_f, is given by the relation

$$c_f t = c_f n\,\Delta t = \frac{2n\theta}{\mu}$$

or

$$c_f = \frac{2\theta}{\mu\,\Delta t}. \tag{5.39}$$

In general c_f is less than the true value c; however, if $c\,\Delta t/2\Delta x$ is taken to be exactly 1, then $c_f = c$.

Note that in each of the examples previously discussed the complex number $B^{\Delta t}$ can be expressed as $e^{-i\alpha}$, where α is a complex numaer in general. Consequently, for the purposes of determining the stability of a finite difference scheme, the assumed solution (5.13) could have been put in the form

$$F_{m,n} = Ae^{i(\mu m\,\Delta x - \alpha n)}.$$

Substitution of this function into the difference equation then permits the determination of α. If α is real the numerical solution is stable (neutral). On the other hand, if α is complex, the numerical solution will be damped or amplified according to whether the imaginary part of α is negative or positive, respectively, contrary to the analytic solution of the differential equation. In the amplifying case the solution will grow without bound with increasing n, which may be catastrophic.

If there is a system of equations with more than one dependent variable, there will be more than one function of the above form with constants A_1, A_2, etc. This results in a homogeneous system of equations for the dependent variables from which the parameter α can be calculated and the stability criterion inferred. For example, consider the hyperbolic system of equations†

$$\frac{\partial u}{\partial t} = -U\frac{\partial u}{\partial x} - \frac{\partial \Phi}{\partial x}$$

$$\frac{\partial \Phi}{\partial t} = -U\frac{\partial \Phi}{\partial x} - \overline{\Phi}\frac{\partial u}{\partial x}$$

and the corresponding semi-implicit difference equations

$$\frac{u_{m,n+1} - u_{m,n-1}}{2\Delta t} = -U\frac{u_{m+1,n} - u_{m-1,n}}{2\Delta x} - \frac{\Phi_{m+1,n} - \Phi_{m-1,n} + \Phi_{m+1,n+1} - \Phi_{m-1,n+1}}{4\Delta x}$$

$$\frac{\Phi_{m,n+1} - \Phi_{m,n-1}}{2\Delta t} = -U\frac{\Phi_{m+1,n} - \Phi_{m-1,n}}{2\Delta x} - \overline{\Phi}\frac{u_{m+1,n} - u_{m-1,n} + u_{m+1,n+1} - u_{m-1,n+1}}{4\Delta x}$$

Now substitute the functions, $u = Ae^{i(\mu m\,\Delta x - \alpha n)}$ and $\Phi = Be^{i(\mu m\,\Delta x - \alpha n)}$ into the equation and set to zero the determinant of the coefficients of A and

† To show that this system is hyperbolic, eliminate Φ by cross-differentitaion which will yield a second-order partial differential equation of the general form

$$A\frac{\partial^2 u}{\partial t^2} + B\frac{\partial^2 u}{\partial x\,\partial t} + C\frac{\partial^2 u}{\partial x^2} + D\frac{\partial u}{\partial t} + E\frac{\partial u}{\partial x} + Fu + G = 0$$

This equation is classified as hyperbolic, parabolic, or elliptic according to the sign of $(B^2 - 4AC)$ as follows:

$$(B^2 - 4AC)\begin{cases} > 0 \text{ hyperbolic} \\ = 0 \text{ parabolic} \\ < 0 \text{ elliptic}\end{cases}$$

B in the resulting pair of homogeneous equations. After some manipulation it may be seen that α will be real if

$$\bar{\Phi} \sin^2 \mu \Delta x \geq U^2 \sin^2 \mu \Delta x - (\Delta x/\Delta t)^2$$

This condition is clearly fulfilled if $\bar{\Phi}^{1/2} \geq U$, which provides a criterion for stability.

The procedure just described is essentially the Fourier method of stability analysis introduced by J. von Neumann in which a Fourier series (one term in these examples) is used to represent a solution of the difference equation [see Richtmyer (1967)]. The stability criterion specifies the conditions under which the solution, hence any errors, will remain bounded as t goes to infinity. A somewhat modified procedure for the von Neumann method will be discussed later in Section 5.6.1; and an alternative approach to the concept of stability will be considered in the following section. The remainder of this section will be devoted to some remarks on the application of the stability criteria to the actual practice of numerical weather prediction and a brief discussion of boundary conditions.

The linear advection equation discussed in the preceding sections is far more simple than the nonlinear equations that are used in actual numerical weather prediction. Thus the difficulties cited are compounded in practice. For example, neither wind velocities nor wave speeds are constant; therefore a conservative value of $c\,\Delta t/\Delta x$ or $V\,\Delta t/\Delta x$ must be chosen to insure computational stability of the linear type discussed here. (For two dimensions the stability criterion is easily shown to be $V\,\Delta t/d \leq 2^{-1/2}$.) In this connection it must be remembered that many different wave components are represented in the data fields. Thus a selected ratio of $\Delta t/\Delta x$ may guarantee computational stability for wave speeds up to a particular maximum value, but a faster wave will produce instability. Even in the absence of such fast waves in the initial data, roundoff and truncation errors or actual physical phenomena could produce faster moving waves that could amplify rapidly and completely obscure the meteorological phenomena being forecast.

There is another type of instability which arises from the nonlinearity of the actual advection terms. However it is normally not a serious problem with short-range forecasts using the filtered equations. We will therefore delay the discussion of nonlinear instability to a later chapter in order to discuss some simple prediction models as early as possible. Nevertheless, if the reader prefers, Chapter 12 may be read at this time.

Another source of error in numerical prediction schemes results from the imposition of lateral boundary conditions since most forecasts are not global. A commonly used condition for the filtered models consists of maintaining the initial value of the stream function ψ on the boundaries throughout the forecast period. This implies that there is no net flow of the rotational wind across the lateral boundaries, provided ψ is continuous, because

$$0 = \oint d\psi = \oint \nabla\psi \cdot d\mathbf{r} = -\mathbf{k} \cdot \oint \mathbf{V}_\psi \times d\mathbf{r}.$$

Here $d\mathbf{r}$ is a directed line segment along the boundary. If ψ has the same constant value along the entire boundary, then the normal component of the rotational wind will vanish at the boundary giving a zero flux boundary condition. If the divergent wind component is relevant in the model, the normal component can be made to vanish by requiring the normal derivative of the potential function χ to vanish.

Platzman (1954) investigated the stability of several types of lateral boundary conditions imposed on the simplified vorticity equation expressed in the centered difference form (5.11). When the boundary values of ζ are specified as functions of time and condition (5.22) is fulfilled, the numerical solution is stable. On the other hand, even in the simplest cases when the boundary values of ζ are obtained by extrapolation from one or two rows in from an outflow boundary, the numerical solutions are computationally unstable.

Whatever lateral boundary conditions are imposed, they are unnatural and thus can give rise to various errors, including wave reflection, which can propagate back into the interior of the forecast region and contaminate the forecast. Such errors may be expected to be more severe when the primitive equations are used than with the filtered equations. Usually an attempt is made to place the boundaries in relatively inactive regions in order to minimize the errors caused by these artificial constraints.

Even when flow is permitted to cross the boundaries, such difficulties may occur. In a recent paper, Matsuno (1966) showed that false reflection of a spurious computational mode from an outflow boundary gives rise to errors which propagate back into the interior. The inward propagation is ascribed to upstream dispersion at negative group velocities, which is especially prominent for very short waves. The rate of reflection decreases with increasing wavelength and also decreases as the order of continuity required at the boundary increases. The latter essentially implies an increase in the number of interior points used to extrapolate the boundary value of the advected quantity, for example, vorticity or momentum.

Small-scale irregularities created by gravity waves, boundary effects, truncation errors, etc., can sometimes be controlled; for example (*a*) by occasional explicit smoothing of the forecast fields; (*b*) by the implicit smoothing inherent in certain kinds of finite difference techniques; or (*c*) by the inclusion of diffusion terms in the forecast equations, e.g., a term of the type $\nu\nabla^2 A$ in the prediction equation for A.

Kesel and Winninghoff (see Section 13.7) have controlled undesirable boundary effects in the Navy primitive-equation prediction model by establishing a "sponge" zone adjacent to the lateral boundaries with the following characteristics. South of about 4N the initial, objectively analyzed fields of mass and velocity are held constant, while near 17N the meteorological parameters are permitted their full variability as determined from the governing dynamical equations. Between these latitudes all predicted values are partially restored to their initial values with a restoration coefficient which decreases nonlinearly from one to zero with increasing latitude, slowly at first, then more rapidly.

5.5 DISCRETIZATION, CONVERGENCE, STABILITY, TRUNCATION ERROR, COMPATIBILITY

When a differential equation is to be solved by numerical methods, it may be first replaced by a difference equation and then numerical calculations are performed. During the calculations, errors result from approximating coefficients in one fashion or another, roundoff during calculations, etc.; consequently the numerical solution, say F_N, differs from the exact solution of the difference equation. Moreover, the solution of the difference equation, say F_D, will differ from that of the differential equation. If F represents the solution of the differential equation, the difference $F - F_D$ is referred to as the *discretization error*, while the difference $F_D - F_N$ is sometimes called the *stability error*. Thus the *total error* is the sum of errors due to discretization and roundoff during the numerical calculation.

In general, the discretization error can be decreased by appropriately decreasing the finite difference increments in space and time and by using higher order approximations for the derivatives, both of which increase computation time. The finite difference solution is said to be *convergent* if F_D approaches F as the finite difference increments in time and space approach zero in a given domain of integration. The problem of convergence is usually difficult to investigate, because the expression for the discretization error generally involves unknown derivatives for which upper and lower bounds are unavailable.

If it were possible to carry out all of the calculations to an infinite number of decimal places, an exact solution of the difference equation could be obtained since there would be no roundoff errors. Some recent texts on numerical methods state that a set of finite difference equations is *stable* when the cumulative effect of all roundoff errors is negligible and, hence, very importantly, that these errors do not increase exponentially with the number of time steps taken (or space steps depending on the independent variables). According to another definition of stability (Richtmyer, 1967), the solutions of the difference equation must be uniformly bounded functions of the initial state for all sufficiently small Δt and for $n\,\Delta t$ less than some finite value T, i.e., for $0 < \Delta t < \tau$ and $0 \le n\,\Delta t \le T$. In either case the conditions for stability are essentially the same. Continuing with some basic concepts, it is sometimes possible to approximate an initial-value type of partial-differential equation with a difference scheme which is stable but does not reduce to the given differential equation when the mesh lengths are reduced to zero. Such a difference scheme is said to be *incompatible* (or *inconsistent*) with the original differential equation. In general it appears that compatibility (or consistency) and stability imply convergence, but this has been rigorously proved for only certain classes of differential equations.

The *truncation error* T is defined to be the difference between the difference equation and the differential equation. For example, consider the differential equation

$$\frac{\partial F}{\partial t} - \frac{\partial^2 F}{\partial x^2} = 0$$

and the corresponding difference equation

$$\frac{F_{m,n+1} - F_{m,n}}{\Delta t} - \frac{F_{m+1,n} - 2F_{m,n} + F_{m-1,n}}{\Delta x^2} = 0.$$

Now if G is any function with continuous partial derivatives, the truncation error for G is defined by

$$\frac{G_{m,n+1} - G_{m,n}}{\Delta t} - \frac{G_{m+1,n} - 2G_{m,n} + G_{m-1,n}}{\Delta x^2} - \left(\frac{\partial G}{\partial t} - \frac{\partial^2 G}{\partial x^2}\right)_{m,n} = T.$$

Next express $G_{m,n+1}$, $G_{m+1,n}$, and $G_{m-1,n}$ in terms of power series about the point (m, n) and substitute their equivalents into the previous equation, giving

$$T = \frac{\Delta t}{2}\frac{\partial^2 G}{\partial t^2} - \frac{\Delta x^2}{12}\frac{\partial^4 G}{\partial x^4} + \frac{\Delta t^2}{6}\frac{\partial^3 G}{\partial t^3} - \frac{\Delta x^4}{360}\frac{\partial^6 G}{\partial x^6}\cdots.$$

Now when Δt and Δx approach zero, T approaches zero, which shows that the solution of the difference equation is compatible with the differential equation. When G represents the actual solution of the differential equation, the truncation error is of order Δt and Δx^2. However, if we choose $\Delta t/2 = \Delta x^2/12$, the first two terms cancel with a resulting improvement in accuracy.

5.6 MATRIX METHOD OF STABILITY ANALYSIS

There are two well-known methods of stability analysis: the first is the finite Fourier series method, and the second is the matrix method. In the former, developed by J. von Neumann, the error is represented by a finite Fourier series, which need be only one term for linear equations. This method will be illustrated in the next section. A third method which may be termed the energy method has been developed by A. Arakawa (1966).

The matrix method has the advantage that the boundary conditions can be included in the analysis. To illustrate this method, suppose the solution at the $(n + 1)$th time step can be expressed in terms of the nth step as follows:

$$\mathbf{U}_{n+1} = A\mathbf{U}_n \tag{5.40}$$

where the $\mathbf{U}_n$ is a column vector comprised of the values of U at all the grid points in a given coordinate direction, and A is the matrix representing the differencing system. For example, if the corresponding differential and difference equations are

$$\frac{\partial F}{\partial t} = \frac{\partial^2 F}{\partial x^2} \tag{5.41}$$

and

$$F_{m,n+1} = F_{m,n} + r(F_{m+1,n} - 2F_{m,n} + rF_{m-1,n}), \qquad r = \frac{\Delta t}{\Delta x^2}$$

or

$$F_{m,n+1} = rF_{m-1,n} + (1 - 2r)F_{m,n} + F_{m+1,n}, \tag{5.42}$$

and for simplicity, the boundary values are $F_{0,n} = F_{J,n} = 0$; then (5.42) may be written in the matrix form (5.40), namely,

$$\begin{bmatrix} F_{1,n+1} \\ F_{2,n+1} \\ \cdot \\ \cdot \\ \cdot \\ F_{J-1,n+1} \end{bmatrix} = \begin{bmatrix} 1-2r & r & 0 & & \cdots & 0 \\ r & 1-2r & r & 0 & \cdots & 0 \\ 0 & r & 1-2r & r & \cdots & 0 \\ \cdot & \cdot & \cdot & \cdot & \cdots & \cdot \\ \cdot & \cdot & \cdot & \cdot & \cdots & \cdot \\ 0 & 0 & \cdots & & & 1-2r \end{bmatrix} \begin{bmatrix} F_{1,n} \\ F_{2,n} \\ \cdot \\ \cdot \\ \cdot \\ F_{J-1,n} \end{bmatrix}. \tag{5.43}$$

It follows from (5.40) that

$$\mathbf{F}_n = A\mathbf{F}_{n-1} = A(A\mathbf{F}_{n-2}) = \ldots = A^n\mathbf{F}_0. \tag{5.44}$$

It is apparent that the solution after n time steps will depend on the character of the matrix A. Now if, for example, there are errors at the gridpoints at the initial time $t = 0$, so that $\mathbf{F}_0$ is given by $\mathbf{F}_0'$, then the value of $\mathbf{F}$ after n time steps is given by

$$\mathbf{F}_n' = A^n\mathbf{F}_0', \tag{5.45}$$

assuming no additional errors are introduced in the calculations. The error, $\boldsymbol{\epsilon}_n = \mathbf{F}_n - \mathbf{F}_n'$, after n time steps is found by subtracting (5.45) from (5.44), namely,

$$\boldsymbol{\epsilon}_n = A^n\boldsymbol{\epsilon}_0. \tag{5.46}$$

The difference scheme is termed stable when $\boldsymbol{\epsilon}_n$ remains bounded as n increases indefinitely. To derive the stability criterion it must be possible to express the vector $\boldsymbol{\epsilon}_0$ as a linear combination of a complete set of linearly independent vectors (orthogonal-normal set). The latter are obtainable as the so-called *eigenvectors* of the matrix A, which are defined as follows. First the eigenvalues of A are defined to be the roots λ_k, $k = 1, \ldots, J - 1$, of the equation

$$|A - \lambda I| = 0, \tag{5.47}$$

where I is the identity matrix and the vertical bars represent a determinant. Then the eigenvectors $\mathbf{v}_k$ are the solutions of the homogeneous system of equations

$$(A - \lambda_k I)\mathbf{v}_k = 0$$

or

$$A\mathbf{v}_k = \lambda_k \mathbf{v}_k. \tag{5.48}$$

Under fairly general conditions the eigenvectors of A form an orthonormal set and $\boldsymbol{\epsilon}_0$ is expressible as

$$\boldsymbol{\epsilon}_0 = \sum_{k=1}^{J-1} C_k \mathbf{v}_k, \tag{5.49}$$

where the C_k are constant. Substituting (5.49) into (5.46) gives

$$\boldsymbol{\epsilon}_n = \sum_{k=1}^{J-1} C_k A^n \mathbf{v}_k = \sum_{k=1}^{J-1} C_k A^{n-1} A \mathbf{v}_k = \sum_{k=1}^{J-1} C_k A^{n-1} \lambda_k \mathbf{v}_k.$$

Repeating this procedure gives

$$\boldsymbol{\epsilon}_n = \sum_{k=1}^{J-1} C_k \lambda_k^n \mathbf{v}_k.$$

It can be inferred from this result that the errors will not increase exponentially with n, provided that the largest magnitude of the eigenvalues is less or equal to unity, i.e.,

$$|\lambda_k| \leq 1, \qquad k = 1, 2, \ldots.$$

Thus the crux of determining stability conditions is finding the eigenvalues. In general, the problem of determining the eigenvalues may be quite difficult. However, there are known specific formulas for certain types of matrices and also general theorems describing bounds of the eigenvalues. Finally, there are numerical methods for calculating the eigenvalues on a computer.

In (5.43), the matrix A can be expressed as the sum of the unit matrix I

and a *tridiagonal matrix* T as follows:

$$A = I + rT,$$

$$T = \begin{bmatrix} -2 & 1 & 0 & & \dots & 0 \\ 1 & -2 & 1 & & \dots & 0 \\ 0 & 1 & -2 & 1 & \dots & 0 \\ \dots & & & & & \dots \\ 0 & \dots & \dots & \dots & & -2 \end{bmatrix}.$$

It is easily shown that if a *matrix* A is a *rational function* of a second *matrix* B, then the eigenvalues of A are the same function of the eigenvalues of B, i.e., if

$$A = f(B),$$

then

$$\alpha_k = f(\beta_k),$$

where α_k and β_k are the eigenvalues of A and B respectively. Also the eigenvalues of a $(J - 1) \times (J - 1)$ tridiagonal matrix, with elements a, b, c in that order, are

$$\lambda_k = b + 2(\sqrt{ac}) \cos \frac{k\pi}{J}, \qquad k = 1, \dots, J - 1.$$

The two previous results can be used to obtain the eigenvalues of A which are

$$\lambda_k = 1 - 4r \sin^2 \frac{k\pi}{2J}, \qquad k = 1, \dots, J - 1.$$

It is readily seen that the magnitudes of all of the λ_k will be no greater than unity, provided $r \le \frac{1}{2}$; hence the difference scheme will be stable when $2\Delta t/\Delta x^2 \le 1$.

When the boundary values are not zero or there are boundary conditions on the derivatives, the procedure is the same; however, the matrix and column vectors are augmented to include the boundary conditions. Similarly, when there are more than two time levels involved or when the problem to be solved involves a system of differential equations, the matrix must again be augmented accordingly.

Digressing for the moment, note that this stable numerical integration scheme (5.42) for the diffusion equation (5.41) is forward in time and central in space, while a central time differencing scheme would be unstable here. Earlier it was noted that with the advective equation (5.7), the reverse was true, namely, the leapfrog scheme was stable, while the forward time scheme was unstable. When the meteorological equations contain both advection and diffusion terms, the leapfrog scheme may be used successfully, provided the diffusion terms are evaluated at the previous time step, say $n - 1$, while the advective term and perhaps others are evaluated at n. As a specific example consider the equation

$$\frac{\partial F}{\partial t} = A + D,$$

where A represents advective, pressure force and coriolis terms, and D is a diffusion term or a damping term, $-C_D F$. Then a stable difference scheme is

$$F_{n+1} = F_{n-1} + 2\Delta t A_n + 2\Delta t D_{n-1},$$

provided that A and D are evaluated with central space differences and Δt is sufficiently small compared to Δx. A central difference scheme is stable for the system 2.34 if $f\,\Delta t \leq 1$; and for the system, (2.25, 2.26), if

$$|U \pm (gH)^{1/2}|\ \Delta t/\Delta x \leq 1,$$

(student exercises).

5.6.1 Von Neumann Method

The von Neumann method, though less general than the matrix method just described, is simpler to apply. As an example, consider the advective equation (5.7) and form the difference equation with centered differences in both space and time as in (5.11). Since three time levels are involved, (5.11) is often referred to as a *three-level formula*. It is convenient to introduce a new variable $G_{m,n} = F_{m,n-1}$, which will permit a formal vector representation in terms of only two levels as follows:

$$F_{m,n+1} = G_{m,n} - \frac{c\,\Delta t}{\Delta x}(F_{m+1,n} - F_{m-1,n}), \tag{5.50}$$

$$G_{m,n+1} = F_{m,n}.$$

Now since the equations are linear, separate solutions are additive; thus only a single Fourier term need be utilized to represent F and G, instead of a Fourier series. Hence let

$$F_{m,n} = {}_1B_n e^{i\mu m\,\Delta x}, \qquad G_{m,n} = {}_2B_n e^{i\mu m\,\Delta x}. \tag{5.51}$$

Substituting these functions into the system (5.50) gives

$${}_1B_{n+1}e^{i\mu m\,\Delta x} = {}_2B_n e^{i\mu m\,\Delta x} - {}_1B_n \frac{c\,\Delta t}{\Delta x}(e^{i\mu(m+1)\,\Delta x} - e^{i\mu(m-1)\,\Delta x}),$$

$${}_2B_{n+1}e^{i\mu m\,\Delta x} = {}_1B_n e^{i\mu m\,\Delta x}.$$

Simplifying leads to

$${}_1B_{n+1} = {}_2B_n - {}_1B_n \frac{2ic\,\Delta t}{\Delta x}\sin \mu\,\Delta x,$$

$${}_2B_{n+1} = {}_1B_n,$$

which may be placed in the following matrix form

$$\begin{pmatrix} {}_1B_{n+1} \\ {}_2B_{n+1} \end{pmatrix} = \begin{pmatrix} \dfrac{-2ic\,\Delta t}{\Delta x}\sin\mu\,\Delta x & 1 \\ 1 & 0 \end{pmatrix} \begin{pmatrix} {}_1B_n \\ {}_2B_n \end{pmatrix} \tag{5.52}$$

or

$$\mathbf{B}_{n+1} = A\mathbf{B}_n = A^2\mathbf{B}_{n-1} = A^3\mathbf{B}_{n-2} = \ldots .$$

Note that the propagation of initial errors for this system obeys (5.52), the same matrix equation, as shown earlier for the system (5.44), (5.46). In any event, the eigenvalues of A must be examined, which are the roots of the equation

$$\begin{vmatrix} -\dfrac{2ic\,\Delta t}{\Delta x}\sin\mu\,\Delta x - \lambda & 1 \\ 1 & -\lambda \end{vmatrix} = 0 \tag{5.53}$$

or

$$\lambda^2 + \lambda\left(\frac{2ic\,\Delta t}{\Delta x}\sin\mu\,\Delta x\right) - 1 = 0,$$

giving

$$\lambda = -i\frac{c\,\Delta t}{\Delta x}\sin\mu\,\Delta x \pm \left[1 - \left(\frac{c\,\Delta t}{\Delta x}\right)^2 \sin^2\mu\,\Delta x\right]^{1/2}. \tag{5.54}$$

Clearly both eigenvalues have a magnitude of unity provided that

$$\frac{c\,\Delta t}{\Delta x} \leq 1,$$

which is the stability condition derived earlier for the leapfrog scheme, namely (5.22).

The *von Neumann necessary condition* for *stability* is somewhat less restrictive than implied by the preceding examples, and may be expressed as follows. All of the eigenvalues λ_k of the amplification matrix A must be such that for all wave numbers

$$|\lambda_k| \leq 1 + 0(\Delta t), \qquad 0 < \Delta t < \tau, \tag{5.55}$$

where it is assumed that Δx is also a function of Δt. The term of order Δt, namely $0(\Delta t)$, can permit exponential growth of the solution of the difference equation when legitimate in accordance with the original differential equation. Thus if $\lambda = 1 + C\,\Delta t$, where C is a constant and $n\,\Delta t = T$, then

$$\lambda^n = (1 + C\,\Delta t)^n = \left(1 + \frac{CT}{n}\right)^n \underset{n\to\infty}{\longrightarrow} e^{CT}.$$

Richtmyer gives several different sufficient conditions for stability [Richtmyer (1967)]:

1. If the condition (5.55) is fulfilled for the eigenvalues of the product of the amplification matrix and its conjugate transpose, then this condition is also sufficient for stability.
2. If A is a normal matrix, that is one which commutes with its conjugate transpose, it will have a complete set of orthonormal vectors and (5.55) will be sufficient as well as necessary.
3. If the determinant Δ of the normalized eigenvectors of A fulfills the condition $|\Delta| \geq a > 0$, where a is a positive constant, then the von Neumann condition is sufficient.
4. If the elements of A, which is a function of Δt and μ, are bounded for all μ and $0 < \Delta t < \tau$, and all the eigenvalues have a magnitude less than some constant which is less than unity, then (5.55) is sufficient for stability.

It may be noted that both the "matrix" method and the von Neumann method involve the determination of the eigenvalues of a matrix. Nevertheless, in the von Neumann method, the matrix is smaller and usually the process is less difficult. It is also less complete since the boundary conditions are not included, as mentioned earlier.

A final comment concerns the relationship between convergence, consistency, and stability. According to the Lax equivalence theorem [see Richtmyer (1967)], given a properly posed initial, boundary-value problem and a finite difference approximation to it that satisfies the consistency condition, then stability is a necessary and sufficient condition for convergence.

5.7 RELAXATION METHODS

Many prediction equations that have been developed for barotropic and baroclinic models may be written in schematic form as

$$\nabla^2 T - MT = F'(x, y), \tag{5.56}$$

where T represents a tendency e.g., $\partial z/\partial t$, and F and M are known functions, that is, determinable from observed meteorological data. The simplest finite difference approximation for the Laplacian is

$$\nabla^2 T_{i,j} \doteq \frac{T_{i-1,j} + T_{i,j+1} + T_{i+1,j} + T_{i,j-1} - 4T_{i,j}}{d^2/m_{ij}^2} \equiv \frac{m_{ij}^2 \nabla^2 T_{ij}}{d^2}, \tag{5.57}$$

where m_{ij} is the ratio of map distance to true distance. The identity in (5.57) defines the symbol ∇^2.

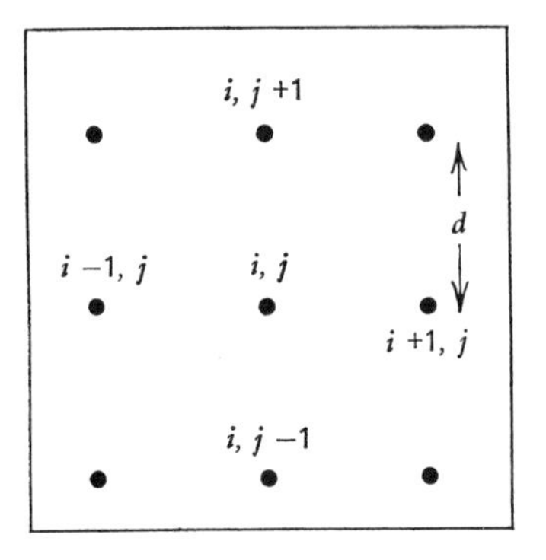

Figure 5.1. Stencil of grid-points for simple centered difference approximations.

The function $F(x, y)$ generally includes advection terms of the form $\mathbf{V}_\psi \cdot \nabla B$, which are expressible as Jacobians. The latter, in turn, are frequently approximated by simple centered differences in the form

$$J(A, B) = \frac{\partial A}{\partial x}\frac{\partial B}{\partial y} - \frac{\partial A}{\partial y}\frac{\partial B}{\partial x} \doteq \frac{m_{ij}^2}{4d^2}\,\mathbb{J}(A_{ij}, B_{ij})$$

$$\equiv \frac{(A_{i+1,j} - A_{i-1,j})(B_{i,j+1} - B_{i,j-1}) - (A_{i,j+1} - A_{i,j-1})(B_{i+1,j} - B_{i-1,j})}{4d^2/m_{ij}^2}. \tag{5.58}$$

Equation 5.56 is an "elliptic" (also Helmholtz) type of equation, which therefore can be written in the difference form as

$$\nabla^2 T_{ij} - \lambda_{ij}^2 T_{ij} = F_{ij}. \tag{5.59}$$

Now consider a grid similar to the above but much larger, for example, covering the map projection of a large part (or the whole) of the Northern Hemisphere, and suppose there are $n + 1$ points on each side.

Then if the pressure height is known at some initial time over the entire grid, it is possible to compute the geostrophic vorticity $(g/\bar{f})\nabla^2 z$ (or more generally $\nabla^2\psi$) over the interior points of the grid by a formula of the type (5.57), that is, $(gm^2/\bar{f}d^2)\nabla^2 z$ (subscripts have been omitted for simplicity). Values of the vorticity can be specified or computed at the boundary points by partially using one-sided differences. Similarly, the function F_{ij} (including the advection of vorticity) may be computed at all interior points of the region by utilizing (5.58). Hence $(n - 1)^2$ equations of the type (5.59) may be written for the $(n - 1)^2$ interior points of the grid. These simultaneous equations involve $(n + 1)^2$ "unknowns" T_{ij}. If the $4n$ boundary values of T_{ij} are specified (usually the tendency is assumed to be zero at the boundary), then the system of equations for the T_{ij} is complete and can be solved. Now for large n, typical of grids used in numerical forecasting, exact methods of solution such as elimination, matrix inversion, etc., are too time consuming, even on an electronic computer. However, when the elements along the main diagonal of the matrix of the coefficients of the system are larger than the other elements, it is possible to solve the equation by a technique of successive approximations called the *relaxation method.* In this method an initial guess is made and then the error of the guess is reduced by an improved guess. The cycle is repeated over and over until the error at every point is reduced below some preassigned value.

5.7.1 Simultaneous Relaxation

To illustrate the general procedure, assume that the mth guess of the tendency at point (i, j) is T_{ij}^m. Then (5.59) may be written in the form

$$\nabla^2 T_{ij}^m - \lambda_{ij}^2 T_{ij}^m - F_{ij} = R_{ij}^m, \tag{5.60}$$

where R_{ij}^m, called the *residual*, is a measure of the error of the guess. Clearly, if the residuals were zero at every point of the grid, the solution of the system would have been found; however, a guess field is unlikely to be the correct solution. Hence the left side will not be zero at every point of the grid, perhaps at none at all. The object of the relaxation method is to provide a systematic way of reducing the residuals to successively smaller values. Equation 5.60 may be expanded to

$$T_{i-1,j}^m + T_{i+1,j}^m + T_{i,j-1}^m + T_{i,j+1}^m - 4T_{ij}^m - \lambda_{ij}^2 T_{ij}^m - F_{ij} = R_{ij}^m. \tag{5.61}$$

Now let us determine the change in the value of T_{ij}^m which will reduce the residual R_{ij}^m to zero. The adjacent values of T will be left unmodified for the present.

$$T_{i-1,j}^m + T_{i+1,j}^m + T_{i,j-1}^m + T_{i,j+1}^m - (4 + \lambda_{ij}^2)T_{ij}^{m+1} - F_{ij} = 0. \tag{5.62}$$

Subtracting (5.62) from (5.61) leads to

$$T_{ij}^{m+1} = T_{ij}^m + \frac{R_{ij}^m}{4 + \lambda_{ij}^2}. \tag{5.63}$$

This result shows that a $(m + 1)$th guess of T, which is increased over the previous value by the quantity $R_{ij}^m/(4 + \lambda_{ij}^2)$, will reduce the residual to zero at that particular point. For example, if $R_{ij}^m = 25$ and $\lambda_{ij}^2 = 1$, the correction to T_{ij}^m would be 5.

Unfortunately the process does not stop there, since such a correction at a particular point also changes the residuals at the adjacent points. For example consider the point $(i, j + 1)$, where the mth residual is

$$T_{i+1,j+1}^m + T_{i-1,j+1}^m + T_{i,j+2}^m + T_{i,j}^m - (4 + \lambda_{i,j+1}^2)T_{i,j+1}^m - F_{i,j+1} = R_{i,j+1}^m. \tag{5.64}$$

Now substitute the value of T_{ij}^{m+1} from (5.63) for T_{ij}^m into (5.61) written for the point $(i, j + 1)$ without modifying any other values of T as follows:

$$T_{i+1,j+1}^m + T_{i-1,j+1}^m + T_{i,j+2}^m + T_{i,j}^m + \frac{R_{ij}^m}{4 + \lambda_{ij}^2} - (4 + \lambda_{i,j+1}^2)T_{i,j+1}^m - F_{i,j+1} = R_{i,j+1}^{m'}. \quad (5.65)$$

Subtracting (5.64) from (5.65) gives

$$R_{i,j+1}^{m'} = R_{i,j+1}^m + \frac{R_{ij}^m}{4 + \lambda_{ij}^2}. \quad (5.66)$$

Thus while the $(m + 1)$th guess for the (i, j) point given by (5.63) reduces the residual to zero at that point, the residuals at the four neighboring points are increased by precisely the correction which was applied to T_{ij}^m. Nevertheless, each step represents overall progress and the method converges (with appropriate mathematical conditions) toward the true solution for the tendencies. The procedure is halted when none of the residual magnitudes exceeds some preassigned small value, say ϵ, which insures any desired accuracy for the tendency.

A little experimentation and reflection soon make it clear that, where the residuals in a given area are generally of the same sign over a region including many gridpoints, it is better to *overrelax*, that is, to apply a larger correction and "overshoot." Thus (5.63) is replaced by

$$T_{ij}^{m+1} = T_{ij}^m + \frac{\alpha R_{ij}^m}{4 + \lambda_{ij}^2}, \quad (5.67)$$

where α is an overrelaxation factor, generally somewhere between 1 and 2. The optimum value for a particular equation is usually found by numerical experimentation.

The relaxation procedure may be applied systematically over the entire grid, then the cycle is repeated until all residuals are less than the preassigned ϵ. The procedure as described in (5.61) and (5.67) implies a guess of the entire field, then a modification of the entire field based on that guess. This technique may be referred to as *simultaneous relaxation*.

5.7.2 Sequential Relaxation

In practice the method of *sequential or Liebmann relaxation* is usually found to converge more rapidly and is more economical with respect to computer storage. In this scheme each new estimate or guess of the function T_{ij}^{m+1}, as determined by (5.67), is immediately used in computing the new guess of the adjacent point, say $T_{i+1,j}^{m+1}$, by the same formula. Thus if the index j denotes the rows and i the columns, and if the order of iteration is row by row in order of increasing i and j, then the residual R_{ij}^m for use in (5.67) is computed by the formula

$$T_{i+1,j}^m + T_{i-1,j}^{m+1} + T_{i,j+1}^m + T_{i,j-1}^{m+1} - (4 + \lambda_{ij}^2)T_{ij}^m - F_{ij} = R_{ij}^m \quad (5.68)$$

The order in which the relaxation proceeds need not follow rows or columns; for example, it was recently noted that convergence was more rapid over a Northern Hemisphere grid when the sequence of points followed a pattern which begins at the outer boundary and spirals in toward the center (North Pole).

It can be shown that specifying either the value of the function (Dirichelet condition) or the value of its normal derivative (Neumann condition) along the closed boundary of a region is sufficient to determine a solution to an elliptic equation on the bounded region.

chapter six

Barotropic Model

6.1 EQUIVALENT BAROTROPIC MODEL

In this chapter the so-called barotropic numerical prediction model will be discussed. This model is of historical interest because it was used in the first successful attempt at numerical weather prediction. The name stems from the fact that the forecast equation involves but one dependent variable and is applied at a single pressure level in the vertical. In a *barotropic* atmosphere the density is a function of pressure alone; therefore the density, temperature, and pressure surfaces all coincide. Moreover, with barotropy, all of the pressure surfaces are parallel; hence only one level need be forecast. In practice the barotropic vorticity equation has been applied at 500 mb and with considerable success.

To derive the prediction equation, consider the approximate form of the vorticity equation (3.44*b*):

$$\frac{\partial \zeta}{\partial t} + \mathbf{V} \cdot \nabla(\zeta + f) = \bar{f}\frac{\partial \omega}{\partial p}, \tag{6.1}$$

where $\mathbf{V}$ and ζ are to be evaluated geostrophically or with a stream function. Next it will be assumed that the wind direction is constant in the vertical; however, the speed will be allowed to vary. Accordingly the wind is expressible in the form

$$\mathbf{V} = A(p)\bar{\mathbf{V}}, \tag{6.2}$$

where A is an empirical function of pressure and the bar over the velocity represents the integral mean with respect to pressure, namely,

$$\overline{(\quad)} = \frac{1}{p_0}\int_0^{p_0} (\quad)\, dp. \tag{6.3}$$

If the bar operator is applied to (6.2), the result $\bar{\mathbf{V}} = \bar{A}\bar{\mathbf{V}}$. It follows at once that $\bar{A} = 1$. Climatological data indicate that in the mean in middle latitudes A is less than one from the surface to about a level of 600 mb; A is approximately 1 at 600 mb, and then continues to increase with height within the troposphere.

From (6.2) it is evident that the vorticity is

$$\zeta = A(p)\bar{\zeta}. \tag{6.4}$$

Substitution of (6.2) and (6.4) into (6.1) and applying the integral operator (6.3) leads to the following result:

$$\frac{\partial\bar{\zeta}}{\partial t} + \overline{A^2}\bar{\mathbf{V}}\cdot\nabla\bar{\zeta} + \bar{\mathbf{V}}\cdot\nabla f = \frac{f\omega_0}{p_0}, \tag{6.5}$$

where ω_0 represents the "vertical velocity" at the lower boundary p_0. At the upper boundary ($p = 0$) ω equals zero. Next expand $w = g^{-1}\, d\Phi/dt$ in terms of partial derivatives, giving

$$gw = \frac{\partial\Phi}{\partial t} + \mathbf{V}\cdot\nabla\Phi + \omega\frac{\partial\Phi}{\partial p}.$$

Now substitute from the hydrostatic relation and solve for ω:

$$\omega = \rho\left(\frac{\partial\Phi}{\partial t} + \mathbf{V}\cdot\nabla\Phi - gw\right). \tag{6.6}$$

If the lower boundary is level, the kinematic boundary condition requires that $w_0 = 0$. Moreover, if the geostrophic approximation is used, it follows that $\mathbf{V}_g\cdot\nabla\Phi = 0$. With these two conditions the value of ω at the lower boundary reduces to simply

$$\omega_0 = \rho_0\left(\frac{\partial\Phi}{\partial t}\right)_{p=p_0} = \rho_0 A(p_0)\frac{\partial\bar{\Phi}}{\partial t}. \tag{6.7}$$

The last relation follows directly from (6.2) and the geostrophic wind approximation.

The form of (6.5) can be further simplified by defining

$$\mathbf{V}^* = \overline{A^2}\bar{\mathbf{V}} \qquad \text{and} \qquad \zeta^* = \overline{A^2}\bar{\zeta}. \tag{6.8}$$

Now multiplication of (6.5) by $\overline{A^2}$ and utilization of (6.7) and (6.8) leads to

$$\frac{\partial \zeta^*}{\partial t} + \mathbf{V}^* \cdot \nabla(\zeta^* + f) = M' \frac{\partial \Phi^*}{\partial t}, \tag{6.9}$$

where

$$M' = \frac{\bar{f} p_0 A_0}{p_0} = \frac{\bar{f} A_0}{RT}.$$

Note that (6.9) has essentially one dependent variable or unknown, namely Φ^*, since both $\mathbf{V}^*$ and ζ^* are geostrophically related to Φ^*. Equation 6.9 should be applied at the level p^* for which $A(p^*) = \overline{A^2}$, as estimated from climatological data. Since $\overline{A^2}$ is about 1.25, p^* is somewhat higher in altitude than 600 mb and turns out to be near 500 mb. If the atmosphere were barotropic, $A(p)$ would be identically 1, and (6.5) would have the same form as (6.9) and apply to all levels. Consequently, p^* is known as the *equivalent barotropic level* and (6.9) as the *equivalent barotropic vorticity* equation. A slightly different form was originally derived by J. G. Charney (1948, 1949) and numerically integrated by Charney, Fjortoft, and von Neumann (1950). In spite of the many simplifying assumptions, barotropic forecasts at 500 mb have been remarkably successful, and the increase in accuracy achieved by many more sophisticated models at 500 mb has been surprisingly little. The term $M' \, \partial\Phi^*/\partial t$ in (6.9) represents the effect of velocity divergence on vorticity generation only to a limited degree through the resultant surface tendency; however, velocity divergence is usually rather small at 500 mb in any case. A term of this type can appear through other considerations as well, for example, by simulating the presence of the stratosphere. Because of the various simplifications made in deriving the prediction equation(s) for the barotropic and other models, optimum values of coefficients, such as M', are normally determined through the statistical verification of actual forecasts with observed data.

In the initial application of the equivalent barotropic model to numerical forecasting, the so-called Helmholtz term $M' \, \partial\Phi^*/\partial t$ in (6.9) was omitted. A Fourier analysis by P. M. Wolff (1958) showed that the planetary waves of numbers 1, 2, and 3 rapidly retrogressed in the numerical forecasts, whereas verifying maps of actual data showed them to be nearly stationary. Merely holding these ultralong waves stationary resulted in a significant improvement in forecasting. The rapid retrogression is not surprising, since a linear analysis of the simpler model with the M' term omitted shows that the phase velocity of harmonic waves is the Rossby speed C_R, [see Chapter 2, (2.58)].

$$C_R = U - \frac{\beta L^2}{4\pi^2}. \tag{6.10}$$

When L is large, (6.10) gives large negative values for C_R.

Now consider the linearized version of (6.9) based on a constant zonal current U and perturbations which are independent of latitude, which gives (dropping the * for convenience):

$$\frac{\partial}{\partial t}\frac{\partial v}{\partial x} + U\frac{\partial^2 v}{\partial x^2} + \beta v - M'\frac{\partial \Phi}{\partial t} = 0. \tag{6.11}$$

Assuming v to be geostrophic and harmonic in x and t implies that

$$\Phi = \Phi_0 e^{i\mu(x-ct)}, \qquad v = \frac{1}{\bar{f}}\frac{\partial \Phi}{\partial x} = \frac{i\mu}{\bar{f}}\Phi_0.$$

Substitution into (6.11) yields

$$i\mu^3 c\Phi_0 - Ui\mu^3\Phi_0 + i\mu\beta\Phi_0 + \bar{f}M'i\mu c\Phi_0 = 0$$

or

$$c = \frac{U - \beta L^2/4\pi^2}{1 + M'\bar{f}L^2/4\pi^2}. \tag{6.12}$$

It is apparent by a comparison of (6.12) and (6.10) that the addition of the term $M'\,\partial\Phi/\partial t$ reduces the speed of sinusoidal waves, especially long waves. Note that (6.12) reduces to (6.10) if $M' = 0$. Thus the inclusion of velocity divergence of the particular form $M'\partial\Phi/\partial t$ in (6.9) leads to better control of the ultralong or planetary waves, but has not increased the number of unknowns.

The latter point may be made more evident by utilizing the simplified geostrophic approximation in (6.9) as follows:

$$\mathbf{V}_g = \frac{g}{\bar{f}}\mathbf{k}\times\nabla z \qquad \text{and} \qquad \zeta_g = \frac{g}{\bar{f}}\nabla^2 z. \tag{6.13}$$

Thus

$$(\nabla^2 - M)\frac{\partial z}{\partial t} + J\left[z, \left(\frac{g}{\bar{f}}\nabla^2 z + f\right)\right] = 0, \tag{6.14}$$

where

$$M = f^2\frac{A_0}{RT_0}, \qquad \Phi = gz, \tag{6.15}$$

or in finite difference form,

$$\left(\frac{m_{ij}^2}{d^2}\nabla^2 - M\right)\left(\frac{\partial z}{\partial t}\right)_{ij} = -\frac{m_{ij}^2}{4d^2}\,\mathbb{J}(z_{ij}, \eta_{ij}).$$

The coefficient M is normally treated as a constant; moreover in order to satisfy integral constraints regarding vorticity and kinetic energy as described in Chapter 4, the wind velocity should be nondivergent in (6.1). This has been achieved by using a constant value of the coriolis parameter in (6.13).

As discussed in Section 4.1, the use of the geostrophic wind in (6.9) led to spurious anticyclogenesis which can be avoided by using a strictly non-divergent wind. More general stream functions ψ, derivable from the *balance equation* [see (3.47) or (3.48)] have also been used in (6.9) for operational forecasts. In this case,

$$\mathbf{V} = \mathbf{k} \times \nabla\psi \qquad \text{and} \qquad \zeta = \nabla^2\psi. \tag{6.16}$$

The resulting vorticity equation is similar in form to (6.14) except that ψ essentially replaces z.

The preparation of numerical forecasts using the equivalent barotropic model requires the solution of (6.14) over the forecast region. Equation 6.14 is of the *Helmholtz* type for $M \geq 0$, namely,

$$\nabla^2 T - MT = F(x, y), \tag{6.17}$$

where T is the height tendency and F is a known function over the region. When $M = 0$, (6.17) is a *Poisson*-type equation. Both types are generally solvable by relaxation methods discussed in Chapter 5 if T is known on the boundary. If the region is sufficiently large, the tendency may be taken as zero on the boundary without seriously affecting the forecast in the interior of the region. When the tendency $T = \partial z/\partial t$ has been obtained by solution of (6.17), the height at a future time may be obtained, for example, by a centered difference extrapolation as follows:

$$z_{t+\Delta t} = z_{t-\Delta t} + \left(\frac{\partial z}{\partial t}\right)_t 2\Delta t. \tag{6.18}$$

Having obtained a new height field, the process may be repeated as often as needed to yield a forecast of any desired length. The central difference extrapolation represented by (6.18) is obviously not feasible for the first step from the initial conditions, hence a forward extrapolation is usually used, that is,

$$z_{\Delta t} = z_0 + \left(\frac{\partial z}{\partial t}\right)_0 \Delta t. \tag{6.19}$$

6.2 VERTICAL VELOCITY IN THE EQUIVALENT BAROTROPIC MODEL

Although the vertical velocity has been removed by vertical integration in the equivalent barotropic model, it is possible to deduce the distribution of the ω implied by the assumed wind field. For this purpose substitute the expressions for $\mathbf{V}$ and ζ, (6.2) and (6.4), back into (6.1), giving

$$A\frac{\partial \bar{\zeta}}{\partial t} + A\bar{\mathbf{V}} \cdot \nabla f + A^2\bar{\mathbf{V}} \cdot \nabla\bar{\zeta} = \bar{f}\frac{\partial \omega}{\partial p}.$$

Next multiply (6.5) by A and subtract it from the previous equation, giving

$$\bar{f}\frac{\partial \omega}{\partial p} = (A^2 - A\overline{A^2})\bar{\mathbf{V}} \cdot \nabla\bar{\zeta} + \frac{A\bar{f}\omega_0}{p_0},$$

where ω_0 is the vertical velocity at p_0. Integrating this equation from an arbitrary pressure p to the surface pressure p_0 gives ω as a function of pressure

$$\omega = B(p)\omega_0 + p_0\bar{f}^{-1}C(p)\bar{\mathbf{V}} \cdot \nabla\bar{\zeta}, \tag{6.20}$$

where

$$B(p) = 1 - p_0^{-1}\int_p^{p_0} A(p)\,\delta p, \tag{6.21}$$

$$C(p) = \overline{A^2}[1 - B(p)] - p_0^{-1}\int_p^{p_0} A^2\,\delta p. \tag{6.22}$$

It is evident that $B(p)$ is maximum at the surface and decreases with altitude, thus the contribution to ω due to the lower boundary value ω_0 decreases with height, as would be expected. Over level terrain the value of ω at the lower boundary is given by (6.7), i.e., $\omega_0 = \partial p_0/\partial t = g\rho_0\,\partial z_0/\partial t$, where z_0 is the height of the pressure surface p_0. The lower boundary condition over mountainous terrain will be discussed later, as well as the low-level vertical motions induced by surface friction.

Besides the effects of the lower boundary, vertical motions are associated with the horizontal advection of vorticity in accordance with the second term on the right of equation (6.20). A plot of the function $C(p)$ based on climatological data indicates that such dynamically induced vertical velocities tend to be maximum near mid-troposphere with upward (downward) motions associated with positive (negative) vorticity advection, as for example, east (west) of the trough. The computed vertical velocities are useful for making cloud and precipitation forecasts.

6.3 ENERGETICS OF THE BAROTROPIC GEOSTROPHIC MODEL

The simple barotropic-geostrophic model is representable by the single equation

$$\nabla^2\frac{\partial \Phi}{\partial t} = -J(\Phi, \eta). \tag{6.23}$$

Multiplying each member of this equation by Φ gives

$$-\Phi\nabla^2\left(\frac{\partial \Phi}{\partial t}\right) = -\nabla \cdot \left(\Phi\nabla\frac{\partial \Phi}{\partial t}\right) + \nabla\Phi \cdot \nabla\frac{\partial \Phi}{\partial t},$$

$$-\Phi J(\Phi, \eta) = \tfrac{1}{2}\nabla \cdot (\eta\mathbf{k} \times \nabla\Phi^2).$$

Then integrating over a global pressure surface and applying the divergence theorem eliminates the Jacobian term, leaving

$$\frac{\partial}{\partial t}\int \frac{1}{2}(\nabla\Phi \cdot \nabla\Phi)\, dS = 0, \tag{6.24}$$

where dS is an element of area.

Thus the mean kinetic energy (geostrophic) is constant with respect to time, as discussed earlier in Chapter 4 [see (4.44) and (4.45)].

If the term $M\, \partial\Phi/\partial t$ is included in (6.23) [see (6.14)], the integrand of the conservation equation (6.24) contains an additional term of the form $M\Phi^2/2$, which is a kinetic-energy sink.

6.4 BAROTROPIC INSTABILITY

It is of interest to ascertain whether disturbances can amplify or dissipate when a barotropic model is used for numerical weather prediction. Since the pressure and temperature surfaces coincide under barotropy, all pressure surfaces will be parallel and the wind velocity will not vary in the vertical. Thus vertical integration of (6.1) yields

$$\frac{\partial\zeta}{\partial t} + \mathbf{V} \cdot \nabla(\zeta + f) = 0. \tag{6.25}$$

For simplicity, the vertical boundary conditions have been taken here as $\omega = 0$ at $p = p_0$ and $p = 0$. Integration of the continuity equation with these boundary conditions gives

$$\nabla \cdot \mathbf{V} = 0,$$

hence

$$\mathbf{V} = \mathbf{k} \times \nabla\psi.$$

To investigate the question of dynamic instability, consider a zonal current which varies only with latitude, that is, $U = U(y)$. Then with the β-plane approximation the linearized vorticity equation (6.25) is

$$\frac{\partial\zeta'}{\partial t} + U\frac{\partial\zeta'}{\partial x} + v'\left(\frac{\partial\bar{\zeta}}{\partial y} + \beta\right) = 0, \tag{6.26}$$

where $\bar{\zeta} = -\partial U/\partial y$ and the primes denote perturbation quantities. Replacing ζ' by $\nabla^2\psi$ leads to

$$\frac{\partial\, \nabla^2\psi}{\partial t} + U\frac{\partial\, \nabla^2\psi}{\partial x} + \left(\beta - \frac{\partial^2 U}{\partial y^2}\right)\frac{\partial\psi}{\partial x} = 0. \tag{6.27}$$

Now consider perturbations of the form $\psi = \Psi(y)e^{i\mu(x-ct)}$ (referred to as normal mode solutions) and substitute into (6.27) giving

$$(U - c)\left(\frac{d^2\Psi}{dy^2} - \mu^2\Psi\right) - \left(\frac{d^2U}{dy^2} - \beta\right)\Psi = 0. \tag{6.28}$$

The current will be taken to be of finite width, centered at $y = 0$ and with rigid boundaries at $y = \pm d$. Since the normal velocities must vanish at these boundaries,

$$v(\pm d) = \frac{\partial\psi}{\partial x}(\pm d) = -i\mu\Psi(\pm d)e^{i\mu(x-ct)} = 0.$$

Hence the boundary conditions are fulfilled if

$$\Psi(d) = \Psi(-d) = 0. \tag{6.29}$$

For perturbations of the type considered here to be unstable, the phase velocity must be complex; that is, $c = c_r + ic_i$, and the amplitude function Ψ will also be complex in general. Next multiply (6.28) by the complex conjugate of Ψ, say Ψ^*,

$$(U - c)\left(\Psi^*\frac{d^2\Psi}{dy^2} - \mu^2\Psi^*\Psi\right) - \left(\frac{d^2U}{dy^2} - \beta\right)\Psi^*\Psi = 0. \tag{6.30}$$

The first term may be written in the form

$$\Psi^*\frac{d^2\Psi}{dy^2} = \frac{d}{dy}\left(\Psi^*\frac{d\Psi}{dy}\right) - \frac{d\Psi^*}{dy}\frac{d\Psi}{dy}.$$

Also the product of a complex quantity and its conjugate is the square of the absolute value of the quantity; hence,

$$\Psi\Psi^* = |\Psi|^2 \qquad \text{and} \qquad \frac{d\Psi^*}{dy}\frac{d\Psi}{dy} = \left|\frac{d\Psi}{dy}\right|^2.$$

Utilizing these results, dividing (6.30) by $(U - c)$, and integrating between $\pm d$ yields

$$\int_{-d}^{d}\left[\frac{d}{dy}\left(\Psi^*\frac{d\Psi}{dy}\right) - \mu^2|\Psi|^2 - \left|\frac{d\Psi}{dy}\right|^2\right]dy = \int_{-d}^{d}\frac{(d^2U/dy^2 - \beta)\,|\Psi|^2}{(U - c)}\,dy. \tag{6.31}$$

The boundary conditions require that Ψ vanish at $\pm d$; hence its real and imaginary parts must vanish separately. Thus Ψ^* as well as Ψ is zero at the boundaries, and the first term of the left side of (6.31) integrates to zero, leaving

$$\int_{-d}^{d}\left(\mu^2|\Psi|^2 + \left|\frac{d\Psi}{dy}\right|^2\right) dy = -\int_{-d}^{d}\frac{(d^2U/dy^2 - \beta)(U-c)^*|\Psi|^2}{|U-c|^2}\, dy.$$

Note that the right-hand integrand has been multiplied above and below by $(U - c)^*$. Next replace c with $c_r + ic_i$ and take the conjugate as indicated, giving

$$\int_{-d}^{d}\left(\mu^2|\Psi|^2 + \left|\frac{d\Psi}{dy}\right|^2\right) dy = -\int_{-d}^{d}\frac{(d^2U/dy^2 - \beta)(U-c_r)|\Psi|^2}{|U-c|^2}\, dy$$
$$- ic_i\int_{-d}^{d}\frac{(d^2U/dy^2 - \beta)|\Psi|^2}{|U-c|^2}\, dy.$$

Equating real and imaginary parts in this equation requires the coefficient of i to vanish, since the other two integrals are real. Consequently,

$$c_i\int_{-d}^{d}\frac{(d^2U/dy^2 - \beta)|\Psi|^2}{|U-c|^2}\, dy = 0. \tag{6.32}$$

If amplified waves exist, $c_i \neq 0$. Therefore (6.32) requires that the integral vanish. For this to occur, it is apparent that the *quantity* $d^2U/dy^2 - \beta$ *must change sign at least once in the region*, $-d < y < d$. Thus a *necessary condition* for *barotropic instability* is that at some value(s) of y, say y_k,

$$\left(\frac{d^2U}{dy^2} - \beta\right)_{y_k} = 0, \qquad -d < y_k < d. \tag{6.33}$$

This theorem, originally derived by Lord Rayleigh for a nonrotating system, was extended by H. L. Kuo (1951) for meteorological applications to a rotating earth by the addition of the β term. Condition (6.33) may be written in the form

$$\frac{d}{dy}\left(-\frac{dU}{dy} + f\right) = 0, \qquad \text{or} \qquad \frac{d\zeta_a}{dy} = 0 \quad \text{at } y_k,$$

which states that the absolute vorticity must be a maximum or minimum at some point(s) in the basic current.

Upper-air observations normally show a jetlike structure to the zonal wind belts with respect to the horizontal and the vertical as well. To simulate these conditions, Kuo considered a symmetrical jet of the type shown in Figure 6.1. His investigation showed that for this profile (6.33) is both necessary and sufficient for the existence of amplifying waves, and there exists a spectrum of neutral, amplified, and damped waves. The principal results are as follows.

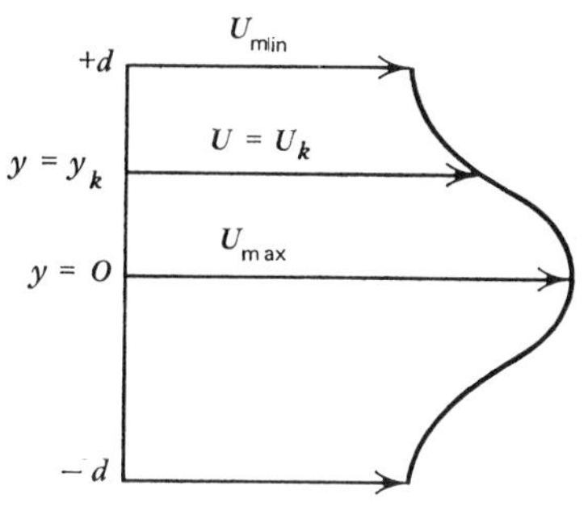

Figure 6.1. Horizontal jet-type wind profile.

The phase velocity of neutral waves can never exceed the maximum wind speed but may be less than the minimum wind speed; these are Rossby or Haurwitz-type waves. To have a neutral wave with a phase velocity between the maximum and minimum wind speeds, the absolute vorticity must have a maximum or minimum at some point in the wind belt. If there is just one point y_k (in each half of the current), then there is just one neutral wave and its phase velocity is U_k (which makes y_k a singular point of the differential equation).

For amplified waves to exist, the absolute vorticity must have a maximum or minimum within the belt and, if no such point exists, all waves with a phase velocity greater than the minimum wind speed will be damped. Waves with phase velocities between $U_{\min}$ and U_k, that is, $U_{\min} < c_r < U_k$, are *amplified* and $L > L_k$. On the other hand, the faster moving waves, $c_r > U_k$, are *damped* and $L < L_k$. These characteristics are illustrated graphically in Figure 6.2.

When the waves are amplified, kinetic energy is fed from the basic current into the perturbations; moreover, the troughs and ridges are oriented from SE to NW south of the point where the absolute vorticity is minimum and SW to NE north of the point of maximum absolute vorticity. Neutral waves are oriented NS. As a final remark, it should be mentioned that the *normal*

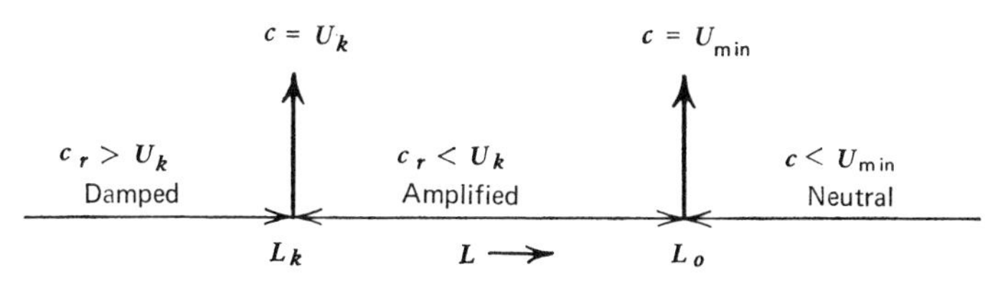

Figure 6.2. Distribution of damped, amplified, and neutral waves as related to zonal velocity and wavelength for a barotropic jet.

mode solutions described in this section usually constitute only a *discrete* set, and to accommodate an arbitrary initial distribution the continuous spectrum similar to those waves discussed in the next section must be included.

6.4.1 Linear Shear

It is natural to ask whether instability is possible in a current with linear horizontal shear. Obviously the approach would have to be different from that taken in the preceding treatment, because in the case of linear shear there are no amplifying *normal modes* according to the Rayleigh criterion (6.33).

An alternative approach is to treat the issue as an initial-value problem. To simplify the discussion, disregard the rotational effect of the earth (β term) in (6.27) and assume $U = U_0 + Sy$, in which case (6.27) reduces to

$$\frac{\partial \nabla^2 \psi}{\partial t} + U \frac{\partial \nabla^2 \psi}{\partial x} = 0. \tag{6.34}$$

As discussed in Chapter 5, the solution of this equation is of the form

$$\nabla^2 \psi = F(x - Ut),$$

where F is an arbitrary function and $F(x)$ is the perturbation vorticity at $t = 0$.

Next consider an initial-perturbation velocity field consisting simply of

$$v = -v_0 \sin \mu x.$$

Then

$$\nabla^2 \psi = \zeta = -\mu v_0 \cos \mu x \qquad \text{at} \qquad t = 0,$$

and at time t,

$$\nabla^2 \psi = -\mu v_0 \cos \mu(x - Ut). \tag{6.35}$$

To find the stream function, assume it has the same functional form as the vorticity, that is,

$$\psi = A \cos \mu(x - Ut). \tag{6.36}$$

Then

$$\nabla^2 \psi = -A\mu^2 \cos \mu(x - Ut) - AS^2 t^2 \mu^2 \cos \mu(x - Ut).$$

Equating the right sides of this equation and (6.35) determines the value of A; hence,

$$\psi = \frac{v_0 \cos \mu(x - Ut)}{\mu(1 + S^2 t^2)}. \tag{6.37}$$

The total velocity components are:

$$u = U - \frac{\partial \psi}{\partial y} = U - \frac{v_0 \mu S t \sin \mu(x - Ut)}{\mu(1 + S^2 t^2)},$$

$$v = \frac{\partial \psi}{\partial x} = -\frac{v_0 \sin \mu(x - Ut)}{1 + S^2 t^2}.$$

It is apparent that ψ and v are inversely proportional to the square of t, and u varies inversely as the first power of t. Since the perturbation moves with velocity, $U = U_0 + Sy$, the waves become tilted in the direction of the shear as well as being damped with time, which generally go hand in hand. Note that there is a solution for each value of U and, hence, an infinite set, which is referred to as the *continuous spectrum.*

The analytical treatment here also applies to the vertical plane for an incompressible, homogeneous fluid or an adiabatic atmosphere. The damping characteristics may partially account for the destruction of cumulus clouds under conditions of strong vertical shear of the horizontal wind.

6.4.2 Concluding Remarks

The purpose of Section 6.4 is to provide some insight as to whether development and dissipation of synoptic disturbances are possible with a barotropic prediction model. It was found that with an appropriate horizontal structure of the zonal westerlies, the normal-mode barotropic disturbances will amplify, in which case their growth takes place at the expense of the kinetic energy of mean current. Damped disturbances, on the other hand, feed energy into the mean zonal flow. The continuous modes described in Section 6.4.1 were damped and constitute a source of energy to maintain the zonal flow. Although no general analytical treatment exists which describes the dynamic stability of zonal flow with an arbitrary horizontal-vertical variation in velocity, several theoretical studies (see also Section 7.4.3) as well as some observational evidence seem to suggest that the atmosphere is usually barotropically stable. The damped disturbances help to maintain the zonal flow, which must receive energy in the mean to compensate for the loss by frictional dissipation. Early 500-mb barotropic models, which contained no mechanism for the vertical redistribution of momentum nor for frictional dissipation, overpredicted the strength of the westerly jet and the trades. Also, since conditions for barotropic instability are apparently relatively uncommon, the barotropic models rarely predicted occurrences of strong intensification of 500-mb troughs and ridges. To account adequately for such development and dissipation, it is necessary to have prediction models which have the capability of converting potential to kinetic energy, that is, baroclinic models.

chapter seven

Baroclinic Models

7.1 INTRODUCTION

In the equivalent barotropic model, advection of temperature is impossible because the wind is always parallel to the isotherms. Moreover the absolute vorticity is conserved (aside from the rather minor effect of the Helmholtz term); hence absolute vorticity centers will be essentially maintained at the same intensity. The relative vorticity patterns may vary somewhat; however, by and large, only limited changes in the intensity of the pressure systems occur. As was shown in Chapter 4 and discussed in Chapter 6, the total kinetic energy is conserved in a barotropic model. Thus the barotropic model does not allow adequately for the very important processes of development and decay of pressure systems. The former, in particular, is generally associated with strong temperature advection; hence a natural extension of the equivalent barotropic model would be a vertical wind structure in which advection is possible. According to the thermal wind relation, a horizontal temperature gradient implies vertical shear and a turning of the wind with height, which must, therefore, be an inherent feature of any baroclinic model.

7.1.1 Integrated Two-Parameter Model

The simplest generalization of (6.2) is a wind structure of the form

$$\mathbf{V} = \bar{\mathbf{V}} + B(p)\mathbf{V}_T, \tag{7.1}$$

where $\mathbf{V}_T$ is the thermal wind between two arbitrarily selected levels, here taken as 1,000 mb, and the level of the mean wind $\bar{\mathbf{V}}$, say $\bar{p} = 600$ mb. The bar again denotes the integral operator (6.3). While (7.1) is obviously less restrictive than (6.2), it is by no means completely representative of the atmosphere; for example, a consequence of (7.1) is that the vertical hodograph of the wind vector is a straight line. The function $B(p)$ may be determined from empirical data; however, some obvious features of $B(p)$ are $\bar{B}(p) = 0$, and since $\mathbf{V}_T = \mathbf{V}_{600} - \mathbf{V}_{1,000}$, $B(1,000) = -1$. From (7.1) it follows immediately that

$$\zeta = \bar{\zeta} + B(p)\zeta_T, \qquad \zeta_T = \mathbf{k} \cdot \nabla \times \mathbf{V}_T. \tag{7.2}$$

Here ζ_T is called the thermal vorticity and geostrophically equals $g\bar{f}^{-1}\nabla^2 h$, where h is the thickness of the layer defining $\mathbf{V}_T$. Substitution of (7.1) and (7.2) into (6.1) and integration in the vertical by means of the operator (6.3) gives the result

$$\frac{\partial \bar{\zeta}}{\partial t} + \bar{\mathbf{V}} \cdot \nabla(\bar{\zeta} + f) + \overline{B^2}\mathbf{V}_T \cdot \nabla\zeta_T = 0. \tag{7.3}$$

For simplicity, the boundary conditions on ω have been taken to be $\omega(p_0) = \omega(0) = 0$.

With the geostrophic approximation, (7.3) contains two unknowns, $\bar{z}(x, y, t)$, the height of the pressure surface $\bar{p}$, and $h(x, y, t)$, the thickness between the two pressure levels defining $\mathbf{V}_T$. Evidently one or more equations are needed to form a complete system. Since thickness changes imply temperature changes, it is necessary to utilize the thermodynamic equation

$$c_p \frac{dT}{dt} - \alpha\omega = Q.$$

Use of the hydrostatic equation $\partial z/\partial p = -RT/gp$, the equation of state and the definition of potential temperature, $\theta = T(1,000/p)^{R/c_p}$, leads to

$$\frac{\partial}{\partial t}\frac{\partial z}{\partial p} + \mathbf{V} \cdot \nabla \frac{\partial z}{\partial p} + \sigma'\omega = -\frac{RQ}{pgc_p}, \tag{7.4}$$

where σ' is a measure of the static stability,

$$\sigma' = \frac{1}{\theta}\frac{\partial z}{\partial p}\frac{\partial \theta}{\partial p}. \tag{7.5}$$

Assuming adiabatic flow ($Q = 0$), treating σ' as a constant (note the integral constraints of Chapter 4), using the geostrophic approximation, and

† The word parameter as used in this context is synonymous with dependent variable.

integrating (7.4) through the layer defining $\vec{\mathbf{V}}_T$ gives the following result:

$$\frac{\partial h}{\partial t} + \bar{\mathbf{V}} \cdot \nabla h - \sigma\tilde{\omega} = 0. \tag{7.6}$$

Here h is the 1,000 to 600-mb thickness, $\tilde{\omega}$ is the integral mean of the vertical velocity in the 1,000 to 600-mb layer, and $\sigma = (p_0 - \bar{p})\sigma'$.

Equations 7.3 and 7.6 contain three "unknowns" $\frac{\partial \bar{z}}{\partial t}, \frac{\partial h}{\partial t}, \tilde{\omega}$, and consequently do not constitute a complete system as yet. This will be remedied shortly; however, it is of interest to mention here an approximation which has been used in operational forecasting with a fair degree of success. Dropping the $\tilde{\omega}$ term in (7.6) and partially compensating for its omission by decreasing the advection term with a factor $k < 1$ reduces the thermodynamic equation to

$$\frac{\partial h}{\partial t} + k\bar{\mathbf{V}} \cdot \nabla h = 0. \tag{7.7}$$

The factor k, which is determined empirically to maximize forecasting skill, is usually between 0.5 and 0.75; moreover it may be varied with the prevailing synoptic conditions, such as the type of advection (warm or cold), the nature of the underlying surface (sea or land), etc. The obvious advantage of the above approximation is that Eqs (7.3) and (7.7) then form a complete system of two equations in two unknowns, $\frac{\partial \bar{z}}{\partial t}$ and $\frac{\partial h}{\partial t}$. This system is easily solved by numerical methods on digital computers.

The more general system including vertical motions requires an additional equation in $\bar{z}$, h, and $\tilde{\omega}$. Actually two equations are used; the first is obtained by integrating (7.1) through the layer p_0 to $\bar{p}$, and the second by multiplying by p and then integrating through the same layer. After considerable manipulation, the vertical velocity may be eliminated from the two derived equations, together with (7.3) and (7.6) to yield

$$\frac{\partial \zeta_T}{\partial t} - M_1\frac{\partial h}{\partial t} + M_2\bar{\mathbf{V}} \cdot \nabla h + (M_3\mathbf{V}_T + \bar{\mathbf{V}}) \cdot \nabla\zeta_T + \mathbf{V}_T \cdot \nabla(\bar{\zeta} + f) = 0. \tag{7.8}$$

The coefficients M_1, M_2, and M_3 involve the function $B(p)$ which may be estimated from observational wind data.

The system (7.3) and (7.8) comprises two equations in two unknowns, $\bar{z}$ and h, which may be solved numerically to give forecasts of the 600- and 1,000-mb levels. Alternatively, the 500-mb level can be used throughout instead of the 600-mb level; of course, the constants would differ somewhat.

There is a slight inconsistency in the assumptions of the constant static stability parameter σ and the wind distribution (7.1). Berkofsky removed the inconsistency in a three-parameter model, which is considerably more complicated, however, and apparently has never been used in operational numerical prediction.

7.2 TWO-LEVEL MODEL

Another two-parameter baroclinic model may be obtained in a somewhat simpler fashion by replacing the divergence $\partial\omega/\partial p$ with a finite difference approximation. The atmosphere will be divided into four layers of thickness Δp, as shown in Figure 7.1; for example, the levels could be 0, 250, 500, 750 and 1,000 mb, or 200, 400, 600, 800 and 1,000 mb.

The vorticity equation (6.1) is now written for levels 1 and 3:

$$\frac{\partial \zeta_1}{\partial t} + \mathbf{V}_1 \cdot \nabla(\zeta_1 + f) = \bar{f}\left(\frac{\partial \omega}{\partial p}\right)_1 = \bar{f}\,\frac{\omega_2 - \omega_0}{2\Delta p}, \tag{7.9}$$

$$\frac{\partial \zeta_3}{\partial t} + \mathbf{V}_3 \cdot \nabla(\zeta_3 + f) = \bar{f}\left(\frac{\partial \omega}{\partial p}\right)_3 = \bar{f}\,\frac{\omega_4 - \omega_2}{2\Delta p}. \tag{7.10}$$

At the upper boundary, $\omega_0 = 0$, which, for simplicity again, will also be taken as the lower boundary condition, i.e., $\omega_4 = 0$. Next define

$$\mathbf{V} = \tfrac{1}{2}(\mathbf{V}_1 + \mathbf{V}_3) \quad \text{and} \quad \mathbf{V}_T = \frac{\mathbf{V}_1 - \mathbf{V}_3}{2},$$
$$\zeta = \tfrac{1}{2}(\zeta_1 + \zeta_3) \quad \text{and} \quad \zeta_T = \frac{\zeta_1 - \zeta_3}{2}. \tag{7.11}$$

Substituting these expressions into (7.9) and (7.10) and then summing and differencing these equations leads to

$$\frac{\partial \zeta}{\partial t} + \mathbf{V} \cdot \nabla(\zeta + f) + \mathbf{V}_T \cdot \nabla\zeta_T = 0, \tag{7.12}$$

$$\frac{\partial \zeta_T}{\partial t} + \mathbf{V} \cdot \nabla\zeta_T + \mathbf{V}_T \cdot \nabla(\zeta + f) = \frac{\bar{f}\omega_2}{2\Delta p}. \tag{7.13}$$

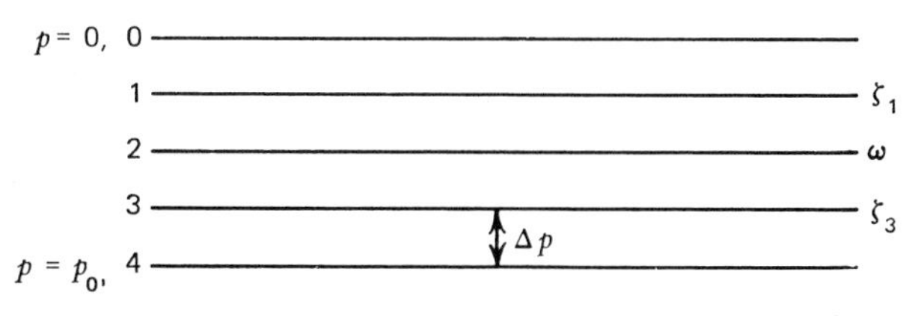

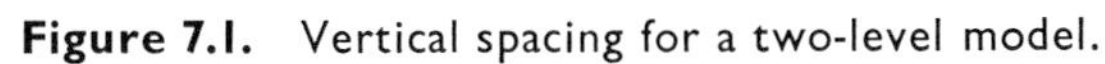

Figure 7.1. Vertical spacing for a two-level model.

With the geostrophic approximation, (7.12) and (7.13) contain three unknowns z_1, z_3, and ω_2, hence an additional equation is needed to complete the system. We again utilize the thermodynamic equation (7.4), assuming adiabatic conditions and a constant static stability parameter σ. Applying this equation at level 2 and approximating the derivative $\partial z/\partial p$ by

$$\frac{z_3 - z_1}{2\Delta p} = -\frac{h}{\Delta p}, \qquad h = \frac{z_1 - z_3}{2},$$

and taking $\mathbf{V}_2 \doteq \mathbf{V}$ leads to a result similar to (7.6);

$$\frac{\partial h}{\partial t} + \mathbf{V} \cdot \nabla h - \sigma\omega_2 = 0. \tag{7.14}$$

Elimination of ω_2 between (7.13) and (7.14) gives

$$\frac{\partial \zeta_T}{\partial t} + \mathbf{V} \cdot \nabla\zeta_T + \mathbf{V}_T \cdot \nabla(\zeta + f) - \frac{\bar{f}}{2\Delta p\sigma}\left(\frac{\partial h}{\partial t} + \mathbf{V} \cdot \nabla h\right) = 0. \tag{7.15}$$

With the geostrophic approximation, (7.12) and (7.14) comprise a system of two equations and two unknowns, $\partial z/\partial t$ and $\partial h/\partial t$, which can be solved by relaxation methods from initial fields of z and h. The previous equation is similar in form to that of the integrated two-parameter model (7.8).

An important feature of this model is the direct availability of vertical motions from (7.14), once $\partial h/\partial t$ has been calculated by solution of (7.15). The vertical motions are useful for making cloud and precipitation forecasts. A diagnostic equation for ω_2 can be easily derived by taking the Laplacian of (7.14) and subtracting this equation from (7.13). The result is

$$\sigma\nabla^2\omega_2 - \frac{\bar{f}^2\omega_2}{2g\,\Delta p} = \nabla^2(\mathbf{V} \cdot \nabla h) - \frac{\bar{f}}{g}[\mathbf{V} \cdot \nabla\zeta_T + \mathbf{V}_T \cdot \nabla(\zeta + f)].$$

This is a Helmholtz-type equation which can be directly solved for ω_2 by the relaxation methods described in Chapter 5.

Note from (7.3) and (7.12) that the absolute vorticity is not conserved as in the simple barotropic models, (6.25). On the other hand, the advection of thermal vorticity by the thermal wind $\mathbf{V}_T \cdot \nabla\zeta_T$ in (7.22) and (7.12) can give rise to intensification or weakening of pressure systems. Specifically, the *absolute vorticity will increase in regions where the thermal wind is advecting relatively greater values of thermal vorticity.* These values are positive in thermal troughs and cold centers, and negative in thermal ridges and warm centers. Hence with a wavelike temperature pattern, the absolute vorticity will tend to increase between the thermal trough and the downwind thermal ridge. It follows that, when the *thermal trough lags behind the pressure trough*, the thermal advection of thermal vorticity will promote the *intensification of the pressure wave, deepening the trough and intensifying the associated ridge.* On the other hand, if the *thermal wave precedes the pressure wave*, the resulting decrease in absolute vorticity between the trough and downwind ridge tends to decrease the amplitude of the pressure wave.

It is of interest to express the system (7.9) and (7.10) in a different form by substituting for ω_2 from (7.14) with the following result:

$$\frac{\partial(\zeta_1 - \lambda' h)}{\partial t} + \mathbf{V}_1 \cdot \nabla(\zeta_1 + f - \lambda' h) = 0, \tag{7.16}$$

$$\frac{\partial(\zeta_3 + \lambda' h)}{\partial t} + \mathbf{V}_3 \cdot \nabla(\zeta_3 + f + \lambda' h) = 0, \tag{7.17}$$

where $\lambda' = \bar{f}/2\sigma\,\Delta p$.

These equations show that in the two-level baroclinic model, a quantity of the form $\zeta + f \pm \lambda' h$ is conserved at levels 3 and 1. Such a quantity is commonly referred to as *potential vorticity*, as mentioned earlier. The pair (7.16) and (7.17) can be solved to obtain $\partial z_1/\partial t$ and $\partial z_3/\partial t$, similarly to the system (7.12) and (7.15).

7.2.1 Dynamics of the Two-Level Model

Some of the characteristics of the two-level model described earlier may be established by a dynamical analysis of a linearized version of (7.12) and (7.15). For this purpose it is convenient to introduce stream functions

$$\psi = \frac{gz}{\bar{f}} \qquad \text{and} \qquad \psi_T = \frac{gh}{\bar{f}}.$$

Then

$$\begin{aligned} \mathbf{V} &= \mathbf{k} \times \nabla\psi, & \mathbf{V}_T &= \mathbf{k} \times \nabla\psi_T, \\ \zeta &= \nabla^2\psi, \quad \text{and} & \zeta_T &= \nabla^2\psi_T. \end{aligned} \tag{7.18}$$

Next the undisturbed wind field will be represented by a constant mean wind U and thermal wind U_T. The thermal wind is related to the mean thickness $H = (\bar{Z}_1 - \bar{Z}_3)/2$, which is a linear function of y, by the relation $U_T = -(g/\bar{f})\,\partial H/\partial y$. Assuming the perturbations are independent of the lateral direction y, the linearized forms of (7.12) and (7.15) are found to be

$$\frac{\partial}{\partial t}\frac{\partial^2\psi}{\partial x^2} + U\frac{\partial^3\psi}{\partial x^3} + \beta\frac{\partial\psi}{\partial x} + U_T\frac{\partial^3\psi_T}{\partial x^3} = 0, \tag{7.19}$$

$$\frac{\partial}{\partial t}\frac{\partial^2\psi_T}{\partial x^2} - \lambda\frac{\partial\psi_T}{\partial t} + U\frac{\partial^3\psi_T}{\partial x^3} - \lambda U\frac{\partial\psi_T}{\partial x} + \lambda U_T\frac{\partial\psi}{\partial x} + U_T\frac{\partial^3\psi}{\partial x^3} + \beta\frac{\partial\psi_T}{\partial x} = 0, \tag{7.20}$$

where the ψ and ψ_T are perturbation quantities and

$$v = \frac{\partial \psi}{\partial x}, \qquad v_T = \frac{\partial \psi_T}{\partial x}, \qquad \lambda = \frac{f^2}{2g\,\Delta p\sigma}. \tag{7.21}$$

Next assume a harmonic solution of the form

$$\psi = \psi_0 e^{i\mu(x-ct)}, \tag{7.22}$$

$$\psi_T = \psi_{T0} e^{i\mu(x-ct)}, \tag{7.23}$$

where $\mu = 2\pi/L$, and c is the phase speed as before. The amplitudes ψ_0 and ψ_{T0} are constants, but they may be complex, implying a possible phase difference between the stream wave and the thermal wave. The phase velocity may also be complex which will permit amplified and/or damped waves. When (7.22) and (7.23) are substituted into (7.19) and (7.20), the following system results:

$$\begin{aligned} \left(c - U + \frac{\beta}{\mu^2}\right)\psi_0 - U_T\psi_{T0} &= 0, \\ U_T\left(\frac{\lambda}{\mu^2} - 1\right)\psi_0 + \left[(c - U)\left(1 + \frac{\lambda}{\mu^2}\right) + \frac{\beta}{\mu^2}\right]\psi_{T0} &= 0. \end{aligned} \tag{7.24}$$

This pair of homogeneous equations in ψ_0 and ψ_{T0} has a nonzero solution if and only if the determinant of the coefficients is equal to zero, which gives the frequency equation

$$\left(1 + \frac{\lambda}{\mu^2}\right)(c - C_R)^2 - \frac{\beta\lambda}{\mu^4}(c - C_R) - U_T^2\left(1 - \frac{\lambda}{\mu^2}\right) = 0, \tag{7.25}$$

where C_R is the Rossby wave speed,

$$C_R = U - \beta/\mu^2. \tag{7.26}$$

Solving for $c - C_R$ gives

$$c - C_R = \frac{\beta\lambda/\mu^4 \pm \sqrt{\beta^2\lambda^2/\mu^8 + 4U_T^2(1 - \lambda^2/\mu^4)}}{2(1 + \lambda/\mu^2)}. \tag{7.27}$$

Now consider the phase velocity (7.27) for some special cases. In a statically stable atmosphere, $\gamma < \gamma_d$, hence σ and λ are positive. As the lapse rate approaches the dry adiabatic value, λ becomes infinite, as seen from (7.21). In this case the vertical motions do not affect the temperature structure nor the thickness h. This is similar to the advective model (7.7), where thickness changes can result only from horizontal advection. As λ approaches infinity, the limiting phase velocity becomes

$$c = C_R + \frac{1}{2}\left[\frac{\beta}{\mu^2} \pm \left(\frac{\beta^2}{\mu^4} - 4U_T^2\right)^{1/2}\right]. \tag{7.28}$$

For a given $U_T \neq 0$, the quantity inside the root will eventually become negative as the wavelength decreases (μ increases), and the phase velocity then becomes complex. Recall that, when $c = c_r + ic_i$, the stream function wave may be written as

$$\psi = \psi_0 e^{\mu c_i t} e^{i\mu(x - c_r t)}.$$

It follows that the stream wave will be amplified with time if $c_i > 0$ and damped if $c_i < 0$. Returning now to (7.28), it may be inferred that waves shorter than some critical length will have an amplified and a damped component. This is a shortcoming of the advective model; for in the general case, when $\gamma \neq \gamma_d$, the shortest wavelengths are neutral, as will be shown below.

Continuing with the case $\gamma = \gamma_d$, assume that the atmosphere is barotropic and thus has zero thermal wind, $U_T = 0$. Then the phase velocity (7.28) reduces to the Rossby value $c = C_R$ or the trivial solution, $c = U$.

Returning to the general formula, (7.27), it may be noted that, if the quantity under the radical is positive or zero, the waves will be stable (neutral); and if this quantity is negative, unstable (amplified and damped) waves occur. Thus with

$$R = \frac{\beta^2 \lambda^2}{\mu^8} + 4U_T^2\left(1 - \frac{\lambda^2}{\mu^4}\right). \tag{7.29}$$

$R \geq 0$ gives stable (neutral) waves; $R < 0$ gives unstable (amplified and damped) waves. As the wavelength decreases toward zero, μ approaches infinity; hence U_T^2 eventually dominates in (7.29) and *sufficiently short waves are stable.* On the other hand, for sufficiently long waves (μ small), the term $\beta^2\lambda^2/\mu^8$ dominates and again stable waves occur.

For intermediate wavelengths, $\mu^4 < \lambda^2$, and when

$$4U_T^2 > \frac{\beta^2\lambda^2/\mu^4}{\lambda^2 - \mu^4}, \tag{7.30}$$

R becomes negative, c becomes complex, $c = c_r + ic_i$, and unstable waves occur. It is evident from the inequality (7.30) that a large thermal wind (or vertical wind shear), high latitude, and small static stability are favorable for baroclinic instability.

Moreover, for a given set of parameters, β, U_T and λ, the rate of amplification $e^{\mu c_i t}$ will be maximum for some particular wavelength. This wavelength of maximum growth rate may be determined by setting the derivative $\partial(\mu c_i)/\partial L$ to zero and solving for L.

Figure 7.2 shows a typical stability diagram with wavelength in thousands of km plotted as the abscissa and the thermal wind as the ordinate. The lapse rate is approximately the standard atmosphere value and the latitude is 45°. The solid curve separates the regions of stable and unstable waves, and the

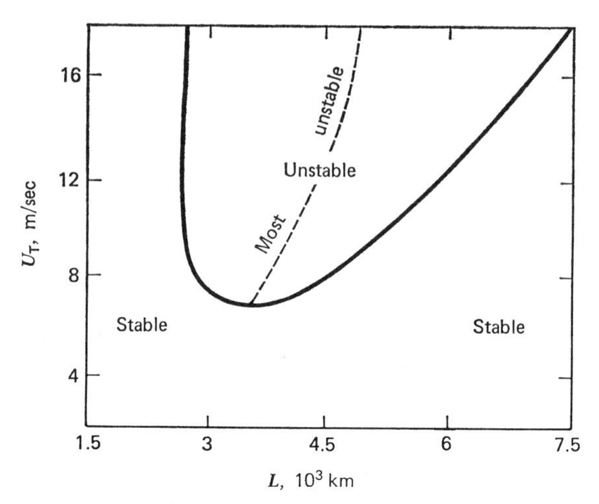

Figure 7.2. Dynamic stability as a function of wavelength and thermal wind for the two-level model.

dashed line shows the most unstable wavelength as a function of thermal wind. Note that the most unstable waves are in the range 3,000 to 6,000 km, which is in agreement with the observed wavelengths of typical synoptic waves.

The system (7.24) not only gives the dynamic stability characteristics of these baroclinic waves through the condition of a vanishing determinant but also relates the thermal wave to the pressure wave. There are two roots to the frequency equation, say c_+ and c_-, corresponding to the plus and minus signs in (7.27). The two roots give two pairs of values, (ψ_+, ψ_{T+}) and (ψ_-, ψ_{T-}), which may be complex. The members of each pair are related by one of the equations of the set (7.24) (the two equations are dependent for either c_+ or c_-). Since the initial differential equations are linear, a linear combination of the two solutions represents the general solution of the system as follows:

$$\begin{aligned} \psi(x, t) &= \psi_{0+} e^{i\mu(x - c_+ t)} + \psi_{0-} e^{i\mu(x - c_- t)}, \\ \psi_T(x, t) &= \psi_{T0+} e^{i\mu(x - c_+ t)} + \psi_{T0-} e^{i\mu(x - c_- t)}. \end{aligned} \tag{7.31}$$

The coefficients ψ_{0+} and ψ_{0-} may be considered arbitrary, while ψ_{T0+} and ψ_{T0-} may be determined from them by means of (7.24). The amplitudes of the stream and thermal waves are fixed by the initial conditions

$$\psi(x, 0) = Ae^{i\mu x}, \tag{7.32}$$

$$\psi_T(x, 0) = A_T e^{i(\mu x + \alpha_0)}, \tag{7.33}$$

where α_0 represents the phase lag of the thermal wave behind the stream wave. From (7.31) through (7.33) it follows immediately that

$$\psi_{0+} + \psi_{0-} = A, \tag{7.34}$$

$$\psi_{T0+} + \psi_{T0-} = A_T e^{i\alpha_0}. \tag{7.35}$$

The relationship between ψ_0 and ψ_{T0} is obtained by the use of one of the pair (7.24); for example, the first equation gives

$$(c_+ - C_R)\psi_{0+} - U_T\psi_{T0+} = 0, \tag{7.36}$$

$$(c_- - C_R)\psi_{0-} - U_T\psi_{T0-} = 0. \tag{7.37}$$

The set of four equations (7.34) through (7.37) comprises a complete system from which ψ_{0+}, ψ_{0-}, ψ_{T0+}, ψ_{T0-} can be determined in terms of the initial amplitudes and phase lag, namely, A, A_T, α_0, and the other parameters such as the zonal wind, thermal wind, latitude, wavelength, and static stability. The solution of the system is easily found to be

$$\begin{aligned} \psi_{0+} &= \frac{U_T A_T e^{i\alpha_0} - A(c_- - C_R)}{c_+ - c_-}, & \psi_{T0+} &= \frac{c_+ - C_R}{U_T}\psi_{0+} \\ \psi_{0-} &= \frac{-U_T A_T e^{i\alpha_0} + A(c_+ - C_R)}{c_+ - c_-}, & \psi_{T0-} &= \frac{c_- - C_R}{U_T}\psi_{0-}. \end{aligned} \tag{7.38}$$

The phase velocities may be written as the complex conjugates

$$c_+ = c_r + ic_i, \qquad c_- = c_r - ic_i, \tag{7.39}$$

where c_r is the speed of propagation and μc_i gives the amplification or damping rate. When (7.38) and (7.39) are substituted into (7.31), the solutions are completely determined.

7.2.2 Unstable Waves

The solution will first be examined for unstable waves, i.e., when c is complex. As seen from (7.31), the complete solution for the stream wave is composed of two components—the first being amplified and the second damped—with both components propagating individually at the same speed [see (7.27)]. A similar situation holds for the thermal wave. It is apparent that eventually the amplified components will dominate as the damped components disappear; hence the ultimate or limiting condition will be characterized by the amplifying stream and thermal-wave components propagating at the same speed and with a constant phase difference that is favorable for growth. Since the only mechanism here for growth of the disturbance is the transformation of potential energy into kinetic energy, the conversion process must take place in accordance with relations (4.31), (4.16), and (4.19); that is, the vertical velocity ω and the temperature must be negatively correlated. At this point an equation for the vertical velocity could be obtained for this model in order to establish the vertical velocity pattern; however, this will be deferred until the next chapter when a more general vertical velocity equation will be derived. It suffices to state at this time that upward velocities (negative ω) are normally found between the trough and the downwind ridge and downward velocities (positive ω) between the trough and upwind ridge. It can be inferred, therefore, that ultimately the thermal trough must lag the pressure trough in the case of the unstable wave, since this will give the negative correlation between the vertical velocity ω and the temperature (or thickness), which is necessary for a kinetic-energy increase and concomitant amplification of the pressure wave. These inferences are easily verified by numerical computations with (7.31), utilizing (7.27), (7.38), and (7.39).

The limiting conditions described previously do not generally hold in the early stages during which the stream function perturbation (or pressure wave) may amplify or dampen, depending on the relative location of the thermal wave. It should be mentioned here that, although the perturbation stream function has two components, it can be expressed as a single sinusoidal wave with an amplitude and propagation speed that are functions of time, but which ultimately approach those of the amplified component of (7.31). The same is true of the thermal wave. Now, as might be expected from the foregoing discussion, amplification takes place immediately if the thermal wave initially lags behind the pressure wave. The maximum growth occurs with a phase lag of about one-quarter wavelength, which is roughly the limiting phase difference for an adiabatic, frictionless model. On the other hand, if the thermal wave initially precedes the pressure wave, the potential energy will grow at the expense of the kinetic energy, and the pressure perturbation will be damped. However, in this case, the thermal wave tends to move rapidly into a lagging position with respect to the stream wave, especially when the amplitude of the thermal wave is the lesser of the two. In any case, when the positive lag position is reached, the baroclinic wave will begin to amplify.

There are, of course, differences in behavior which depend on the various governing basic parameters of thermal wind, latitude, static stability, etc. and also the initial amplitudes and phase angles. For example, when the thermal wave lags, say 90°, and its amplitude is less than that of the stream wave, there is warm (cold) advection mainly between the trough and the downwind (upwind) ridge. A similar lagging thermal wave with an amplitude greater than that of the pressure wave gives rise to cold advection over most of the pressure trough and warm advection over the ridge. A relatively large thermal wave preceding the pressure wave is characterized by warm advection in the trough and cold advection in the ridge. These variations tend to complicate the vertical motion pattern; but nevertheless, by and large, they result in upward motions ahead of the trough and downward motions between the trough and upwind ridge. Finally, the reader should note that the foregoing discussion regarding amplification and damping of unstable waves is in agreement with the earlier discussion on page 132 on the effects of $-\mathbf{V}_T \cdot \nabla \zeta_T$ on development.

7.2.3 Stable Waves

When the baroclinic wave is stable, that is, when R of (7.29) is positive, (7.27) gives two different real roots. Hence the two wave components of the perturbation stream function ψ have constant amplitudes but propagate at different speeds. Their sum can be expressed as a single sinusoidal wave of length L, whose amplitude varies periodically with time. The amplitude variation may be several-fold, with a period of a few days to a week or more for very long waves. The thermal wave behaves in a similar fashion and its phase relative to the pressure wave varies periodically.

7.2.4 Energy in the Two-Level Quasi-Geostrophic Baroclinic Model

The equations for the two-level model may be expressed in the form [see (7.9), (7.10), and (7.14)]:

$$\nabla^2 \frac{\partial \psi_1}{\partial t} + \mathbf{V}_1 \cdot \nabla(\zeta_1 + f) = k\omega_2, \tag{7.40}$$

$$\nabla^2 \frac{\partial \psi_3}{\partial t} + \mathbf{V}_3 \cdot \nabla(\zeta_3 + f) = -k\omega_2, \tag{7.41}$$

$$\frac{\partial}{\partial t}(\psi_1 - \psi_3) + \mathbf{V} \cdot \nabla(\psi_1 - \psi_3) = \sigma\omega_2, \tag{7.42}$$

where k is a constant, σ is a static stability parameter, and $\mathbf{V} = (\mathbf{V}_1 + \mathbf{V}_3)/2$. Next multiply (7.40) by ψ_1, and (7.41) by ψ_3, add, and then integrate over the entire atmosphere. As shown for the barotropic case, terms involving the Jacobian vanish when integrated globally. Now

$$\psi\mathbf{V} \cdot \nabla\eta = \psi J(\psi, \eta) = \nabla \cdot (\psi\eta\mathbf{V}) - \psi\eta\nabla \cdot \mathbf{V} + \eta\mathbf{V} \cdot \nabla\psi,$$

and the local change term can be written as

$$-\psi\nabla^2 \frac{\partial \psi}{\partial t} = -\nabla \cdot \left(\psi\nabla \frac{\partial \psi}{\partial t}\right) + \nabla\psi \cdot \nabla \frac{\partial \psi}{\partial t}.$$

The first term on the right side above vanishes when integrated, while the other may be written as $\partial/\partial t(\frac{1}{2}\nabla\psi \cdot \nabla\psi)$. The final result is

$$\frac{\partial}{\partial t}\int \frac{V_1^2 + V_3^2}{2}\, dS = -k\int \omega_2(\psi_1 - \psi_3)\, dS. \tag{7.43}$$

Next multiply (7.42) by $k(\psi_1 - \psi_3)/\sigma$ and integrate, leading to

$$\frac{k}{\sigma}\frac{\partial}{\partial t}\int \frac{(\psi_1 - \psi_3)^2}{2}\, dS = k\int \omega_2(\psi_1 - \psi_3)\, dS. \tag{7.44}$$

Adding (7.43) and (7.44) gives

$$\frac{\partial}{\partial t}\int \left[\frac{V_1^2 + V_3^2}{2} + \frac{k}{\sigma}\frac{(\psi_1 - \psi_3)^2}{2}\right] dS = 0. \tag{7.45}$$

The first term in the integrand is obviously the sum of kinetic energies per unit mass at levels 1 and 3, while the second term involves the thickness (which is proportional to the mean temperature in the layer) and thus represents the available potential energy. Equation 7.45 shows that the sum of these two forms of energy is conserved in the model, whereas (7.43) and (7.44) show, respectively, how kinetic and potential energies are generated through vertical motions, as previously discussed in Chapter 4. From (7.43) it follows that kinetic energy will increase if, in the mean, ascending air is relatively warm and descending air is cold.

7.3 FRICTIONAL EFFECTS ON BAROCLINIC INSTABILITY

In Section 7.2.1 the dynamics of a frictionless, adiabatic model showed that decreasing static stability, increasing thermal wind (or vertical shear), and increasing latitude all contribute to dynamic instability. The influence of surface friction may be determined by adjusting the vertical velocity for the lower boundary value of ω_4 in (7.10) [see (8.11)].

As shown by Haltiner and Caverly (1965), the friction not only reduces the growth of the unstable waves but also affects the limiting phase difference between pressure and thermal waves.

Figure 7.3 shows the stream function amplitude after 36 hr as a function of surface drag coefficient and wavelength for a particular example of dynamic instability. The initial amplitude is 20 units, the latitude is 45°, and the thermal wind is 20 m/sec. The diagram clearly shows the effect of friction in reducing amplification. There is also a slight tendency for the wavelength of maximum amplification to shift toward shorter values with increasing friction.

Figure 7.4 gives the limiting phase difference (abscissa) between the thermal and pressure waves as function of drag coefficient (ordinate) and wavelength (units 10^3 km); the latter is denoted at the upper end of each curve. Differences of 40° may be noted between the frictionless case ($C_D = 0$) and the maximum drag coefficient shown. Even greater differences are found between the various wavelengths.

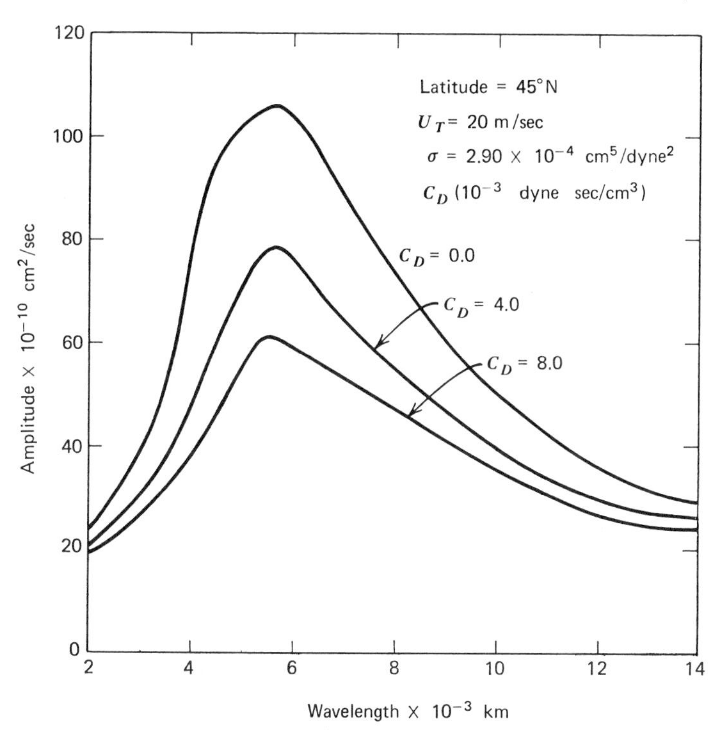

Figure 7.3. Stream function amplitude as a function of surface drag coefficient and wavelength for a particular case of dynamic instability [*after Haltiner and Caverly* (*1965*).]

7.3.1 Sensible Heat-Exchange Effects on Instability

The adiabatic models provide the major characteristics of middle latitude disturbances, at least for periods of a few days; however, the effects of diabatic heating are by no means insignificant. Differences of amplitude and phase velocity of 10 to 20% are possible during a day or two as a result of sensible heat exchange or latent heat release.

To explore the influence of sensible heat exchange, Haltiner (1967) assumed the heat transfer to be proportional to the temperature difference between the air and the underlying surface as follows:

$$Q = A(T_s - T_a)\left(\frac{p}{p_s}\right)^r. \tag{7.46}$$

Here T_s and T_a are the temperatures of the sea (or ground) and the air respectively, and A and r are constants, the latter specifying the distribution of heat with height. This heating function was used in a two-level model with variable static stability developed by Lorenz. The variable σ was considered

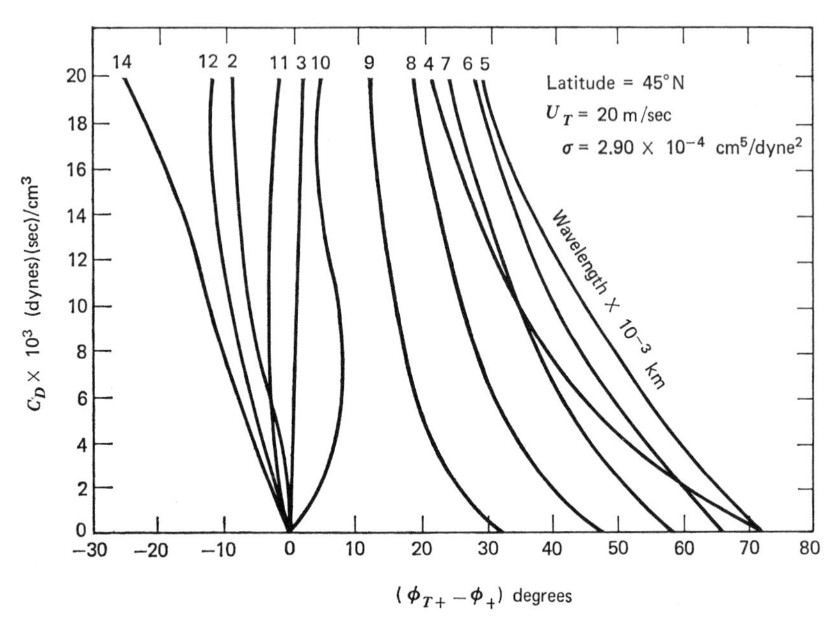

Figure 7.4. Limiting phase difference (abscissa) between the thermal and pressure waves as a function of drag coefficient (ordinate) and wavelength (labeled at the upper end of each curve). [*After Haltiner and Caverly (1965).*]

desirable since heating tends to change the static stability, which, in turn, affects the dynamic stability characteristics of harmonic waves. The linear balance equation (4.41*b*), the vorticity equation, (4.41*a*), and thermodynamic equation, $d\theta/dt = Q$, are applied at the 250- and 750-mb levels. Summing and differencing these equations in a manner similar to the treatment of the quasi-geostrophic two-level model of Section 7.2 leads to the following equations:

$$\frac{\partial \nabla^2 \psi}{\partial t} + J(\psi, \nabla^2 \psi + f) + J(\tau, \nabla^2 \tau) = 0,$$

$$\frac{\partial \nabla^2 \tau}{\partial t} + J(\tau, \nabla^2 \psi + f) + J(\psi, \nabla^2 \tau) - \boldsymbol{\nabla} \cdot (f \boldsymbol{\nabla} \chi) = 0,$$

$$\frac{\partial \theta}{\partial t} + J(\psi, \theta) + J(\tau, \sigma) - \boldsymbol{\nabla} \cdot (\sigma \boldsymbol{\nabla} \chi) = \frac{Q_1 + Q_3}{2}, \quad (7.47)$$

$$\frac{\partial \sigma}{\partial t} + J(\tau, \theta) + J(\psi, \sigma) - \boldsymbol{\nabla} \theta \cdot \boldsymbol{\nabla} \chi = \frac{Q_1 - Q_3}{2},$$

$$2^{-(1+\kappa)} R \nabla^2 \theta - \boldsymbol{\nabla} \cdot (f \boldsymbol{\nabla} \tau) = 0,$$

where

$$\psi = \frac{\psi_1 + \psi_3}{2}, \qquad \tau = \frac{\psi_1 - \psi_3}{2},$$

$$\theta = \frac{\theta_1 + \theta_3}{2}, \qquad \sigma = \frac{\theta_1 - \theta_3}{2},$$

$$\chi = \frac{\chi_3 - \chi_1}{2}, \qquad \kappa = \frac{R}{c_p}.$$

An investigation of a linearized version of the previous system of equations shows that the diabatic heating, represented by (7.46), reduces the instability of short and medium waves and shifts the maximum instability toward shorter wavelengths as compared to the corresponding adiabatic model; however, the instability of long waves is increased. The limiting lag of the thermal wave behind a 4,000-km stream wave is 85°, while the stream wave lags the vertical velocity (ω) and static stability waves by about 90 and 110° respectively. Thus upward motions as well as the minimum static stability are found ahead of the trough. The inclusion of diabatic heating does not alter these results. However, for an 8,000-km wave, the thermal wave ultimately lags the stream wave by only about 30°, and the stream wave, in turn, lags the static stability and vertical velocity waves by 140° under adiabatic conditions. With sensible heat exchange the corresponding figures are 25° and 105°.

Figure 7.5 shows the real and imaginary parts of the complex phase velocity for the adiabatic model with a thermal wind of 10 m/sec at a latitude of 45°. In the unstable region the propagation speeds c_{r+} and c_{r-} of the amplified and damped wave components propagate at the same speed, i.e., $c_{r+} = c_{r-}$. However, the propagation speeds of the neutral (stable) wave components are different. The superimposed dashed curves are the corresponding values of c_{i+} for the amplified wave in two cases with diabatic heating present, one with double the heating rate of the other. The heating causes a reduction in the amplification rate at intermediate wavelengths but produces instability at longer wavelengths.

The effect of sensible heat exchange in reducing the dynamic instability of intermediate synoptic waves (up to roughly 7,000 km in length) can be explained as follows. The warm-air advection ahead of the trough results in a loss of heat from the air to the cooler surface; while behind the trough, in the area of cold advection, the air gains sensible heat from the warm underlying surface. The overall effect is a tendency to shift the temperature wave to a position ahead of the pressure wave, or at least reduce its amplitude when it lags behind the pressure wave, both of which, as noted in Section 7.2.2 reduce the growth of baroclinic waves.

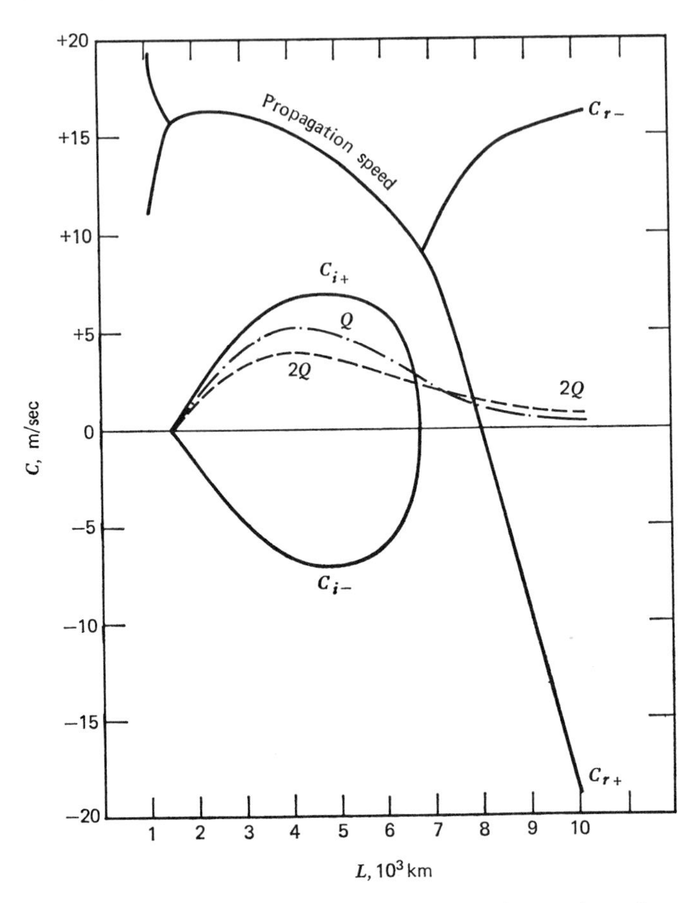

Figure 7.5. Real and imaginary parts of complex phase velocity for the adiabatic model with a thermal wind of 10 m/sec at latitude 45° (solid lines). The dashed lines represent the corresponding value of c_{i3} when sensible heat exchange is included. [*After Haltiner* (*1967*).]

7.3.2 Latent Heat

In general, the inclusion of latent heat tends to increase the dynamic instability of baroclinic waves and also shifts the maximum instability to somewhat shorter wavelengths. The physical basis for the increased instability is that the latent heat release occurs where upward motion exists, usually between the trough and downwind ridge, which therefore tends to place the temperature wave in a favorable position for baroclinic growth of the pressure wave. The latent heat effect can increase the vertical velocities several fold over the adiabatic values.

The scale analysis of Chapter 2 can be generalized to include diabatic heating. It suffices here to say that, with the usual approximations made in baroclinic models for large-scale numerical prediction, the previous results show that the flux of sensible heat should be limited to about 10^3 cal/cm^2 per day, or an equivalent release of latent heat corresponding to about 1.5 cm of precipitation per day.

7.3.3 Effects of Horizontal Shear on Baroclinic Instability

In earlier sections barotropic and baroclinic dynamic instability have been treated separately. The problem of determining the stability characteristics of zonal flows having both vertical and horizontal shear is very difficult indeed and will be discussed only very briefly.

Pedlosky (1964) considered a two-layer model with a stationary lower layer and an upper layer with a parabolic jet, that is,

$$u_1 = u_0(1 - ay^2), \qquad u_2 = 0, \qquad -1 \leq y \leq 1, \quad 0 \leq a \leq 1. \tag{7.48}$$

Since $\partial^2 u/\partial y^2$ is constant, the flow in upper layers would be barotropically stable according to Kuo's criterion; however, the vertical shear permits baroclinic instability. Pedlosky's analysis showed that such a flow releases potential energy to the disturbances which, in turn, generate horizontal Reynold stresses and transfer energy to the zonal flow in the central region of the current, leading to a sharpening of the jet.

When compared to the case of only vertical shear (i.e., $a = 0$), the presence of the horizontal jet reduces the growth rate of the unstable perturbations and also decreases the wavelength of maximum instability. However the value of the minimum vertical shear required for baroclinic instability remains unchanged. Figure 7.6 from Pedlosky's article shows the neutral stability curves for vertical shear combined with zero horizontal shear, $a = 0$, and with a parabolic horizontal jet ($a = 0.5$). With the latter note the smaller unstable zone and the shift of this zone toward shorter wavelengths.

Pedlosky also studied a sinusoidal (inflected), horizontal, zonal wind profile for the upper layer which would permit barotropic instability according to Kuo's results. Thus both baroclinic and barotropic instabilities were possible. It was inferred that slowly moving waves for both inflected and non-inflected flows produce Reynolds stresses that convert eddy kinetic energy to the mean flow while drawing on the potential energy of the mean state. On the other hand, those fast-moving waves, whose existence is provided for only by the vanishing of the potential vorticity gradient within a single layer, release the kinetic energy of the mean flow to the perturbations.

A numerical initial-value study of flows [John A. Brown (1969)] with both horizontal and vertical shears representative of those observed in the earth's atmosphere yielded results which were similar to Pedlosky's but differed in

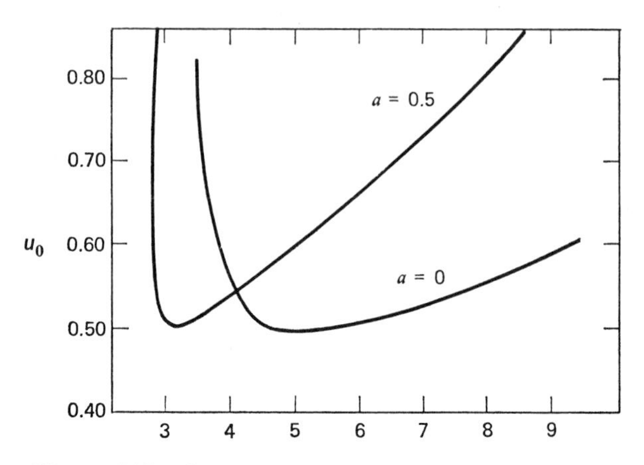

Figure 7.6. Curve of neutral stability for the parabolic jet, $u_1 = u_0(1 - ay^2)$, for the cases $a = 0.5$ and $a = 0$. The units of u_0 (ordinate) are 10 m/sec, the abscissa is wavelength in units of 10^3 km. [*From Pedlosky (1964)*.]

some important respects. Westerly currents characterized by a latitudinally symmetric jet containing absolute vorticity maxima amplify perturbations of some wavelengths through a predominantly baroclinic mechanism and amplify perturbations of other wavelengths through a dominating barotropic mechanism. The unstable perturbations of relatively *short zonal wavelengths convert zonal available potential energy into perturbation energy and simultaneously increase the kinetic energy of the basic zonal flow*. The unstable perturbations of relatively *long wavelengths reduce both the kinetic energy and available potential energy of the basic zonal current with the former dominating*. For certain zonal currents there may be two distinct wavelengths of maximum instability. Brown also considered flows which were similar in structure but contained no vanishing meridional vorticity gradient, and found baroclinically amplifying perturbations with a tendency toward barotropic damping. Unstable waves of this category converted relatively large amounts of available zonal potential energy into perturbation energy and simultaneously returned small amounts of perturbation kinetic energy to the basic flow—a characteristic similar to that of the earth's general circulation.

The question might be raised as to the relevance of distinguishing the effects of horizontal and vertical shears on the development of disturbances since these influences would be automatically accounted for in a numerical solution of the hydrodynamical equations by using observed initial data over a hemispheric or global grid at several levels in the vertical. The principal purpose here is to gain some insight into the physical nature of the atmospheric motions and thus be able to design better numerical prediction models and diagnose any shortcomings. Such knowledge will be helpful in selecting the horizontal and vertical grid lengths over which finite difference approximations to derivatives are formed. In this connection it must be borne in mind that the time required to prepare a prediction is always a critical factor in operational forecasting. Halving the horizontal grid length, for example, quadruples the number of gridpoints and also implies halving the time step to maintain computational stability, thus increasing the time required to prepare a given forecast by a factor of eight, more or less.

Phenomena which have scales near and below the gridlength obviously cannot be resolved on that grid, and are generally referred to as *subgrid* processes. Nevertheless, it is well known that small-scale phenomena, such as turbulence, convection, radiative transfer, etc., may exert significant and even critically important influences on synoptic weather disturbances and the general circulation of the atmosphere. This was, of course, implied in the preceding discussions on friction, sensible heat exchange, evaporation, etc. It is therefore desirable and sometimes necessary to include the effects of these subgrid processes in the large-scale prediction models, especially for extended integration periods. This may be accomplished by approximating or "parameterizing" the statistical effects of these subgrid-scale processes in terms of the large-scale parameters and then including these parametric functions as sources or sinks in the prediction equations for the large-scale dependent variables. While such a procedure is not feasible for all fluid actions, it is practicable in meteorological forecasting because strong nonlinear interactions between certain adjacent scales appear to be suppressed in the atmosphere most of the time. Specific formulations for friction terms, boundary-layer fluxes of heat and water vapor, solar and terrestrial radiation, and convective transfers will be presented in later chapters.

chapter eight

Multilevel Models

8.1 INTRODUCTION

The two-level geostrophic model discussed in Chapter 7 is easily extended to more levels for greater vertical resolution. This may be accomplished by solving the vorticity equation for the height tendency at the odd levels 1, 3, ..., $n - 1$, as shown in Figure 8.1, and a thermal vorticity equation at the even levels, 2, 4, ..., $n - 2$. Alternatively, a diagnostic ω equation may be derived and solved over the three-dimensional grid giving ω at the even numbered levels. The values at the top and bottom of the atmosphere are determined from the vertical boundary conditions which are discussed in more detail shortly. Similarly, lateral boundary conditions must be assigned to ω (except for global integrations); usually ω is taken to be zero on lateral boundaries. After the ω fields have been obtained, the term $\partial\omega/\partial p$ in the vorticity equation [e.g., (8.1)] can be calculated by centered differences, and the equation can be solved for the height tendency by the relaxation techniques described in Chapter 5. Finally, forward extrapolation in time may be accomplished according to (6.18) or some other appropriate technique at each level.

8.2 THE QUASI-GEOSTROPHIC ω EQUATION

The diagnostic equation for ω for the general quasi-geostrophic system will now be derived. The appropriate vorticity equation is (6.1), namely,

$$\frac{g}{\bar{f}}\nabla^2\frac{\partial z}{\partial t} + \mathbf{V}\cdot\nabla\eta = \bar{f}\frac{\partial\omega}{\partial p}. \tag{8.1}$$

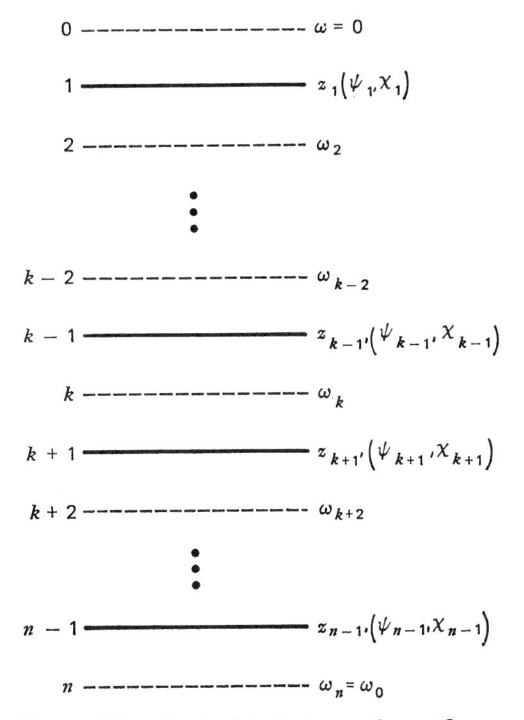

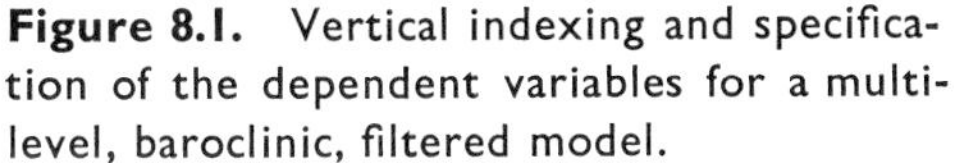

Figure 8.1. Vertical indexing and specification of the dependent variables for a multi-level, baroclinic, filtered model.

The associated thermodynamic equation is (7.4), which will be written in slightly simpler form as

$$\frac{\partial}{\partial t}\frac{\partial z}{\partial p} + \mathbf{V} \cdot \boldsymbol{\nabla} \frac{\partial z}{\partial p} + \sigma_s \omega = -Q, \tag{8.2}$$

where σ_s is a function of p only and Q represents diabatic heating. The time derivatives in (8.1) and (8.2) may be eliminated by differentiating (8.1) with respect to p, taking the Laplacian of (8.2) and then taking the difference of the two equations. The result is

$$\begin{aligned}\sigma_s \nabla^2 \omega + \frac{\bar{f}^2}{g}\frac{\partial^2 \omega}{\partial p^2} &= \frac{\bar{f}}{g}\frac{\partial}{\partial p}(\mathbf{V} \cdot \boldsymbol{\nabla} \eta) - \nabla^2\left(\mathbf{V} \cdot \boldsymbol{\nabla} \frac{\partial z}{\partial p}\right) - \nabla^2 Q \\ &= \frac{\bar{f}}{g}\frac{\partial}{\partial p}(\mathbf{V} \cdot \boldsymbol{\nabla} \eta) + \frac{R}{gp}\nabla^2(\mathbf{V} \cdot \boldsymbol{\nabla} T) - \nabla^2 Q. \end{aligned} \tag{8.3}$$

The right side of (8.3) contains the vertical variation of vorticity advection, the Laplacian of horizontal temperature advection, and the Laplacian of the heating function.

Some qualitative conclusions regarding ω may be inferred from (8.3). It is apparent, for example, that a relative maximum of warm advection or diabatic heating tends to give a relative minimum of ω (that is, upward motions) and, similarly, a region of cooling would contribute to downward motions. The term involving the vertical variation of vorticity is more complex but some simple situations may be considered. If, for example, the horizontal advection of vorticity increases upward in the atmosphere, this term will contribute to upward motions. Thus in a typical trough in the westerlies which tilts westward with increasing height, upward velocities are found between the trough and the downwind ridge. In this situation the thermal trough will lag the pressure trough by a quarter wavelength or so, and there will be conversion of potential into kinetic energy in accordance with (4.31) and (4.32), tending to intensify the baroclinic wave.

The ω equation (8.3) may be expressed in finite difference form utilizing the notation of Section 5.7 and Figure 8.1 as follows:

$$\begin{aligned}
\frac{m_{ij}^2 \sigma_k \nabla^2 \omega_{ijk}}{d^2} &+ \frac{\bar{f}^2}{4g(\Delta p)^2}(\omega_{i,j,k+2} - 2\omega_{ijk} + \omega_{i,j,k-2}) \\
&= \frac{m_{ij}^2}{8d^2\,\Delta p}\left[\mathbb{J}(z_{i,j,k+1}, \zeta_{i,j,k+1} + f_{ij}) - \mathbb{J}(z_{i,j,k-1}, \zeta_{i,j,k-1} + f_{ij})\right] \\
&\quad - \frac{m_{ij}^2 g}{16 d^4 \bar{f}\,\Delta p} \nabla^2 m_{ij}^2\, \mathbb{J}(z_{i,j,k+1} + z_{i,j,k-1}, z_{i,j,k+1} - z_{i,j,k-1}) \\
&\quad - \frac{m_{ij}^2}{d^2} \nabla^2 Q_{ijk}.
\end{aligned} \tag{8.4}$$

Further expansion of $\nabla^2\omega$ and rearrangement permits (8.4) to be placed in a form similar to (5.61), except that the forcing function involves ω_{k+2} and ω_{k-2} which change during the relaxation procedure. This equation can usually be solved without difficulty by using the sequential overrelaxation technique described in Chapter 5 to sweep point by point through the three-dimensional grid. The lateral and vertical boundary conditions for ω must be specified; these are normally taken to be zero except for ω_0 at the earth's surface, which will be discussed in more detail in the next several sections.

8.2.1 Surface Boundary Condition for ω, Mountain Influence

If the expression for $d\Phi/dt$ is expanded in isobaric coordinates, it becomes

$$\frac{d\Phi}{dt} = \frac{\partial \Phi}{\partial t} + \mathbf{V} \cdot \nabla \Phi + \omega \frac{\partial \Phi}{\partial p}.$$

Solving for ω, using $gw = d\Phi/dt$ and the hydrostatic equation, leads to

$$\omega = \rho \frac{\partial \Phi}{\partial t} + \rho \mathbf{V} \cdot \nabla \Phi - g\rho w. \tag{8.5}$$

When applied at a level lower boundary, w is zero, and with geostrophic flow, $\mathbf{V} \cdot \nabla \Phi$ also vanishes; hence ω_0 reduces to the so-called Helmholtz term discussed earlier. However, when the terrain is mountainous, w_0 is no longer zero because of upslope and downslope winds. In this case it is easily seen that w_0 may be expressed in the form $V_n \tan \alpha_t$, or

$$w_0 = \mathbf{V}_0 \cdot \nabla H_t,$$

where $H_t(x, y)$ is the height of the terrain, $\tan \alpha_t$ is the slope of the terrain, $\mathbf{V}_0$ is the surface horizontal wind, and V_n is the wind component normal to the terrain contours. The last term in (8.5) then becomes

$$-g\rho_0 \mathbf{V}_0 \cdot \nabla H_t. \tag{8.6}$$

Alternatively, since $\omega = dp/dt$, the terrain effect may be expressed simply as

$$\omega_0 = \mathbf{V}_0 \cdot \nabla p_t, \tag{8.7}$$

where $p_t(x, y)$ is the pressure at terrain height. The standard atmosphere value will normally suffice for this purpose.

According to the scale analysis of Chapter 3, the large-scale vertical velocity is about 1 cm/sec for $V \sim 10$ m/sec; hence there is a limit to the slope of the mountains permitted in (8.6) and (8.7) in the quasi-geostrophic models. The maximum slope permitted is roughly 10^{-3}; consequently, the actual mountain heights must be considerably smoothed for utilization in large-scale numerical weather prediction.

8.2.2 Surface Friction

The influence of friction may be incorporated into numerical weather prediction models in several ways. The most obvious fashion is simply to include friction terms in the equations of motions from which the model is developed, but this leads to considerable complexity in the prediction equations. The effect of surface drag, however, may be approximated by a simple, but reasonable, technique as follows. The equations of motion utilized for developing the Ekman spiral for the friction layer are [see Haltiner and Martin (14.17)]:

$$0 = -\frac{\partial \Phi}{\partial x} + fv - g\frac{\partial \tau_{zx}}{\partial p}, \tag{8.8}$$

$$0 = -\frac{\partial \Phi}{\partial y} - fu - g\frac{\partial \tau_{zy}}{\partial p}. \tag{8.9}$$

Treating f as a constant and differentiating (8.8) with respect to y, and (8.9) with respect to x, and then forming the difference leads to

$$-f\nabla \cdot \mathbf{V} = g\frac{\partial}{\partial p}\left(\frac{\partial \tau_{zy}}{\partial x} - \frac{\partial \tau_{zx}}{\partial y}\right).$$

Next replace $\nabla \cdot \mathbf{V}$ by $-\partial\omega/\partial p$ from the continuity equation and integrate from the surface to the top of the friction layer, where the stress is assumed to vanish, giving

$$f(\omega_F - \omega_0) = -g\left(\frac{\partial \tau_{zy}}{\partial x} - \frac{\partial \tau_{zx}}{\partial y}\right)_0 = -g\mathbf{k} \cdot \nabla \times \boldsymbol{\tau}_0. \tag{8.10}$$

Here ω_F is the vertical velocity at the top of the friction layer and $\boldsymbol{\tau}_0$ is the surface stress. The latter has been found to be proportional to the square of the wind at a height of a few meters; however, as a rough approximation, a linear relation will be used here for simplicity,

$$\boldsymbol{\tau}_0 = \rho_0 C_D \mathbf{V}_0.$$

Here C_D is a drag coefficient. If it is assumed constant, the vertical velocity at the top of the friction layer in (8.10) becomes

$$\omega_F = -\frac{g\rho_0 C_D}{f}\zeta_0 + \omega_0 \tag{8.11}$$

This result shows that the vertical velocity induced by surface friction is directly proportional to the surface vorticity and drag coefficient, but inversely proportional to $\sin\varphi$. As expected, (8.11) indicates ascending motion around lows due to frictional convergence and descending motion in the vicinity of high pressure areas.

The vertical velocity induced by friction may be combined with upslope motion due to terrain and, as such, would represent the vertical velocity near the earth's surface. This can then be utilized as a lower boundary condition in the solution of the system of hydrodynamical equations, that is,

$$\omega_F = \rho_0\left[\frac{\partial \Phi_0}{\partial t} + \mathbf{V}_0 \cdot \nabla(\Phi_0 - \Phi_t) - \frac{gC_D\zeta_0}{f}\right] \tag{8.12}$$

Some obvious refinements to the foregoing analysis are the use of a variable drag coefficient, a cross-isobar angle for the surface wind, the square relationship between surface stress and wind velocity, and application of ω_F at the top of the friction layer rather than, say, 1,000 mb, which is what is usually done. With the square relationship between stress and velocity

$$\boldsymbol{\tau}_0 = \rho_0 C_D V_0 \mathbf{V}_0,$$

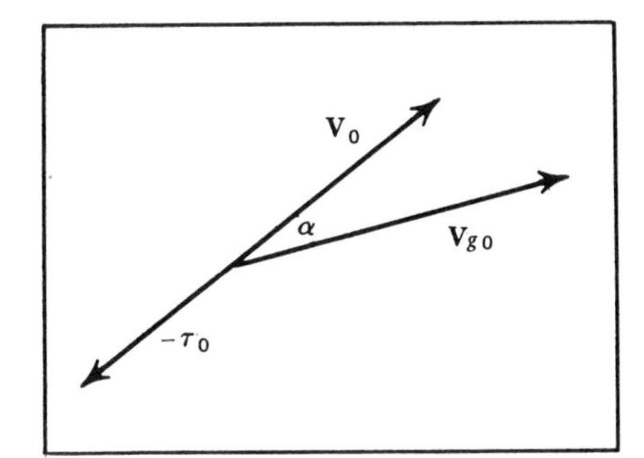

Figure 8.2. Relationships between geostrophic wind $\mathbf{V}_{g0}$, surface wind $\mathbf{V}_0$, and surface stress τ_0.

the vertical velocity becomes

$$\omega_F = \frac{g}{f}\left[\frac{\partial}{\partial y}(\rho_0 C_D u_0 V_0) - \frac{\partial}{\partial x}(\rho_0 C_D v_0 V_0)\right]. \tag{8.13}$$

In addition, the surface wind speed may be taken to be a fraction of the geostrophic value, say,

$$V_0 = rV_{g0},$$

and the surface wind direction α degrees to the left of the geostrophic wind vector, as shown in Figure 8.2.

Smagorinsky (1963) derived a representation for nonlinear lateral diffusion of momentum (also applicable to heat and water vapor) based on the Heisenberg similarity theory. As utilized by Kasahara and Washington in the NCAR general circulation model, the expressions for the horizontal friction forces for the momentum equations in spherical coordinates become

$$F_x = \frac{\partial \tau_\lambda}{\partial z} + \frac{1}{a \cos \varphi}\frac{\partial}{\partial \lambda}(\rho K_{MH} D_T) + \frac{1}{a}\frac{\partial}{\partial \varphi}(\rho K_{MH} D_S),$$

$$F_\varphi = \frac{\partial \tau_\varphi}{\partial z} + \frac{1}{a \cos \varphi}\frac{\partial}{\partial \lambda}(\rho K_{M\dot{H}} D_S) - \frac{1}{a}\frac{\partial}{\partial \varphi}(\rho K_{MH} D_T),$$

where D_T and D_S are the tension and shearing strains:

$$D_T = \frac{1}{a \cos \varphi}\left[\frac{\partial u}{\partial \lambda} - \frac{\partial}{\partial \varphi}(v \cos \varphi)\right],$$

$$D_S = \frac{}{a \cos \varphi}\left[\frac{\partial v}{\partial \lambda} + \frac{\partial}{\varphi \partial}(u \cos \varphi)\right],$$

and

$$\tau_\lambda = \rho K_{MV}\frac{\partial u}{\partial z}, \qquad \tau_\varphi = \rho K_{MV}\frac{\partial v}{\partial z}.$$

Here K_{MV} and K_{MH} are the vertical and horizontal kinematic coefficients of eddy viscosity and

$$K_{MH} = 2k_0{}^2 l^2 D,$$

where k_0 is 0.4,

$$D = (D_T^2 + D_S^2)^{1/2} \qquad \text{and} \qquad l = a\,\Delta\lambda \cos\varphi.$$

8.3 ENERGY IN A GENERAL QUASI-GEOSTROPHIC SYSTEM

The equation for the quasi-geostrophic system may be written

$$\frac{\partial \zeta}{\partial t} + \mathbf{V}\cdot\boldsymbol{\nabla}(\zeta + f) = \bar{f}\frac{\partial \omega}{\partial p} + \left(\frac{\partial F_y}{\partial x} - \frac{\partial F_x}{\partial y}\right), \tag{8.14}$$

where the wind velocity and vorticity are evaluated geostrophically, that is, $\mathbf{V} = \bar{f}^{-1}\mathbf{k} \times \boldsymbol{\nabla}\Phi$, $\zeta = \bar{f}^{-1}\nabla^2\Phi$, and F is a friction term. The terms omitted in the previous vorticity equation are all at least one order of magnitude less than the terms included, as shown in Chapter 3. The thermodynamic equation is

$$\frac{\partial}{\partial t}\frac{\partial \Phi}{\partial p} + \mathbf{V}\cdot\boldsymbol{\nabla}\left(\frac{\partial \Phi}{\partial p}\right) + \sigma\omega + \frac{RQ}{pc_p} = 0. \tag{8.15}$$

Here Q is a diabatic heating function, and the static stability parameter $\sigma = (\partial\Phi/\partial p)(\partial \ln \theta/\partial p)$ is restricted to be a function of p only.

Equations 8.14 and 8.15 form a complete system of two equations in two unknowns, Φ and ω. If (8.15) is differentiated with respect to p after multiplying by σ^{-1}, and then $\partial\omega/\partial p$ is eliminated between the two equations, a prognostic equation for $\partial\Phi/\partial t$ is obtained as follows:

$$\frac{1}{\bar{f}}\nabla^2\frac{\partial \Phi}{\partial t} + \frac{\partial}{\partial p}\left(\frac{\bar{f}}{\sigma}\frac{\partial}{\partial p}\frac{\partial \Phi}{\partial t}\right) + \mathbf{V}\cdot\boldsymbol{\nabla}(\zeta + f) + \frac{\partial}{\partial p}\left[\frac{\bar{f}}{\sigma}\mathbf{V}\cdot\boldsymbol{\nabla}\frac{\partial \Phi}{\partial p}\right]$$
$$= \left(\frac{\partial F_y}{\partial x} - \frac{\partial F_x}{\partial y}\right) - \frac{\partial}{\partial p}\left(\frac{\bar{f}}{\sigma}\frac{RQ}{pc_p}\right). \tag{8.16}$$

Except for a factor of $\bar{f}$, the left side may also be written as $\partial P/\partial t + \mathbf{V}\cdot\boldsymbol{\nabla}P$ where P is the *potential vorticity* for this model:

$$P = \zeta + f + \frac{\partial}{\partial p}\left(\frac{\bar{f}}{\sigma}\frac{\partial \Phi}{\partial p}\right). \tag{8.17}$$

Hence the potential vorticity is conserved except for friction and diabatic heating. The foregoing expression is a special case of a more general form of potential vorticity $\rho^{-1}(\nabla_3 \times \mathbf{V}_3 + 2\mathbf{\Omega}) \cdot \nabla\theta$, which is conserved for three-dimensional, adiabatic, frictionless flow.

The boundary conditions appropriate to (8.16) are that $\omega = 0$ at $p = 0$, and at the lower boundary, $\omega = \omega_0 = (\partial p/\partial t + \mathbf{V} \cdot \nabla p + w\,\partial p/\partial z)_0$. Substitution of $\partial p/\partial z = -g\rho$, $\partial p/\partial t = \rho\,\partial\Phi/\partial t$, and $\mathbf{V}_g \cdot \nabla p = 0$ gives [see (8.5)]

$$\frac{\partial \Phi_0}{\partial t} = \frac{\omega_0}{\rho_0} + g w_0. \tag{8.18}$$

The ω's at the upper and lower boundaries may be replaced by means of the thermodynamic equation (8.15).

Next, the energy equation of the general geostrophic system will be derived by multiplying (8.16) by $-\Phi/\bar{f}^2$ and integrating over the entire atmosphere. The first term on the left can be written as in previous cases

$$-\frac{\Phi}{\bar{f}^2}\nabla^2\frac{\partial\Phi}{\partial t} = -\frac{1}{\bar{f}^2}\left(\nabla \cdot \Phi\nabla\frac{\partial\Phi}{\partial t} - \nabla\Phi \cdot \nabla\frac{\partial\Phi}{\partial t}\right)$$

and

$$\nabla\Phi \cdot \nabla\frac{\partial\Phi}{\partial t} = \frac{1}{2}\frac{\partial}{\partial t}(\nabla\Phi \cdot \nabla\Phi) = \tfrac{1}{2}\bar{f}^2\frac{\partial V^2}{\partial t}.$$

Thus

$$-\iint \frac{\Phi}{\bar{f}^2}\nabla^2\frac{\partial\Phi}{\partial t}\,dS\,dp = \frac{\partial}{\partial t}\iint \tfrac{1}{2}V^2\,dS\,dp,$$

where $dx\,dy\,\rho\,dz = -dS\,dp/g = dM$ is the element of mass.

The second term may be integrated by parts to give

$$-\iint \Phi\frac{\partial}{\partial p}\left(\frac{1}{\sigma}\frac{\partial}{\partial p}\frac{\partial\Phi}{\partial t}\right)dS\,dp = \int\left(\frac{\Phi}{\sigma}\frac{\partial}{\partial p}\frac{\partial\Phi}{\partial t}\right)_{p=p_0}^{p=0}dS + \iint \frac{1}{\sigma}\frac{\partial\Phi}{\partial p}\frac{\partial}{\partial p}\frac{\partial\Phi}{\partial t}\,dS\,dp.$$

The last integral is easily seen to be

$$\iint \frac{1}{2\sigma}\frac{\partial}{\partial t}\left(\frac{\partial\Phi}{\partial p}\right)^2 dS\,dp.$$

The middle term may be evaluated at p and $p = 0$ by use of the boundary conditions $\omega(0) = 0$ and (8.18) together with (8.15). The friction term (see 8.16) may be transformed as follows:

$$-\frac{1}{\bar{f}}\iint \Phi\left(\frac{\partial F_y}{\partial x} - \frac{\partial F_x}{\partial y}\right)dS\,dp = -\frac{1}{\bar{f}}\iiint\left(\frac{\partial F_y\Phi}{\partial x} - \frac{\partial F_x\Phi}{\partial y}\right)dx\,dy\,dp$$
$$+\frac{1}{\bar{f}}\iiint\left(F_y\frac{\partial\Phi}{\partial x} - F_x\frac{\partial\Phi}{\partial y}\right)dx\,dy\,dp. \tag{8.19}$$

The first integral on the right vanishes since the integrand $\mathbf{k} \cdot \nabla \times \Phi\mathbf{F} = \nabla \cdot (\Phi\mathbf{F} \times \mathbf{k})$. Also, the geostrophic wind may be substituted into the last integral, finally giving a simple form to the friction term

$$\iint \mathbf{V} \cdot \mathbf{F} \, dS \, dp,$$

where

$$\frac{\partial \Phi}{\partial x} = v\bar{f}, \qquad \frac{\partial \Phi}{\partial y} = -u\bar{f}, \qquad dS = dx \, dy.$$

Next

$$\iint \Phi \frac{\partial}{\partial p} \frac{RQ}{\sigma p c_p} \, dp \, dS = \int \left(\frac{RQ\Phi}{\sigma p c_p} \right)_{p=0}^{p=p_0} dS - \iint \frac{RQ}{\sigma p c_p} \frac{\partial \Phi}{\partial p} \, dp \, dS. \tag{8.20}$$

Combining all of the terms and taking a horizontal surface boundary for simplicity leads to the energy equation corresponding to system (8.14) and (8.15):

$$\begin{aligned} &\frac{\partial}{\partial t} \iint \left[\frac{1}{2} V^2 + \frac{1}{2\sigma} \left(\frac{\partial \Phi}{\partial p} \right)^2 \right] dS \, dp, \\ &= \iint \mathbf{V} \cdot \mathbf{F} \, dS \, dp - \iint \frac{RQ}{\sigma p c_p} \frac{\partial \Phi}{\partial p} \, dS \, dp. \end{aligned} \tag{8.21}$$

The terms on the left of (8.21) represent kinetic and potential energy changes in the geostrophic system; and the terms on the right are the energy sources (or sinks) of friction and diabatic heating.

8.4 NONGEOSTROPHIC BAROCLINIC MODELS

In Chapters 3 and 4 a hierarchy of filtered models was developed which were consistent with respect to scale analysis and energy conservation. The simplest baroclinic model is the quasi-geostrophic system represented by (8.1) and (8.3).

The next simplest consistent system is the linear balance system [see Chapter 4, (4.41) and (4.42)] is:

$$\frac{\partial \zeta}{\partial t} + \mathbf{V}_\psi \cdot \mathbf{V}\zeta + (\mathbf{V}_\psi + \mathbf{V}_\chi) \cdot \nabla f = f \frac{\partial \omega}{\partial p}, \tag{8.22}$$

$$\nabla \cdot (f \nabla \psi) = \nabla^2 \Phi, \tag{8.23}$$

$$\mathbf{V} = \mathbf{V}_\psi + \mathbf{V}_\chi, \qquad \mathbf{V}_\psi = \mathbf{k} \times \nabla \psi, \qquad \mathbf{V}_\chi = \nabla \chi,$$

together with the continuity (1.16), thermodynamic (3.52), and hydrostatic (1.18) equations. The ω equation for this model has been derived previously, namely, (3.53):

$$\nabla^2(\sigma\omega) + f^2 \frac{\partial^2 \omega}{\partial p^2} = f \frac{\partial}{\partial p} [\mathbf{V}_\psi \cdot \nabla(\zeta + f)]$$
$$- \nabla^2\left(\mathbf{V} \cdot \nabla \frac{\partial \Phi}{\partial p}\right) + f\nabla f \cdot \frac{\partial \mathbf{V}_\chi}{\partial p} - \nabla f \cdot \nabla \frac{\partial^2 \psi}{\partial p\, \partial t}. \quad (3.53)$$

The solution of (3.53) is more complicated than in the geostrophic case because of the presence of the time derivative $\partial^2\psi/\partial p\, \partial t$ and other terms involving the potential function χ. An iterative procedure may be used to solve the system (8.22), (8.23), (1.16), and (3.53) as follows. The first step consists of solving (8.23) for the stream function ψ from the Φ field which is assumed known. Then the ω equation (3.53) is solved in approximate form by omitting the term involving $\partial^2\psi/\partial p\, \partial t$ and the terms involving the χ function as well. With this first approximation to ω, the continuity equation

$$\nabla \cdot \mathbf{V} = \nabla^2\chi = -\frac{\partial \omega}{\partial p}$$

may be expressed in finite difference form, for example, at the $(k + 1)$ level:

$$\frac{m_{ij}^2}{d^2} \nabla^2 \chi_{i,j,k+1} = -\frac{\omega_{i,j,k+2} - \omega_{i,j,k}}{2\Delta p} \quad (8.24)$$

and solved by relaxation for $\chi_{i,j,k+1}$. Next, bearing in mind that $\partial\zeta/\partial t = \nabla^2\, \partial\psi/\partial t$, the vorticity equation may be numerically solved at level $(k + 1)$ for $\partial\psi/\partial t$ by using the first estimates of ω and χ. The usual lateral boundary conditions for the solution of both (8.22) and (8.24) are that $\partial\psi/\partial t$ and χ are assumed to be zero. Now the cycle is repeated by solving (3.53) with the first estimates of $\partial\psi/\partial t$ and χ, leading to a second estimate of ω. Solution of (8.24) and (8.23), in that order, leads to new estimates of χ and $\partial\psi/\partial t$, and so forth. This iterative procedure normally converges and may be repeated until the desired accuracy is attained. However, it is obviously rather time consuming and has been shortened for operational forecasting by simply omitting the terms involving χ and the time derivative in (3.53). In spite of these inconsistencies, the abbreviated model appears to behave satisfactorily for short-range forecasts.

The next energetically consistent system in the hierarchy is [see Chapter 4 (4.39) for further detail]

$$\frac{\partial \zeta}{\partial t} + \nabla \cdot \left(\mathbf{V}\eta + \omega \frac{\partial \nabla \psi}{\partial p}\right) = 0,$$

$$\nabla \cdot [\eta \nabla \psi - \nabla(\tfrac{1}{2} V_\psi^2)] = \nabla^2 \Phi.$$

The vorticity equation is virtually complete, the only approximation being the elimination of $\mathbf{V}_\chi$ in the twisting term. When expanded, the foregoing system appears in the more familiar form

$$\frac{\partial \zeta}{\partial t} + \mathbf{V} \cdot \nabla \eta + \omega \frac{\partial \zeta}{\partial p} - \eta \frac{\partial \omega}{\partial p} + \mathbf{k} \cdot \nabla \omega \times \frac{\partial \mathbf{V}_\psi}{\partial p} = 0, \tag{8.25}$$

$$\nabla \cdot (f \nabla \psi) + 2\left[\frac{\partial^2 \psi}{\partial x^2}\frac{\partial^2 \psi}{\partial y^2} - \left(\frac{\partial^2 \psi}{\partial x\, \partial y}\right)^2\right] = \nabla^2 \Phi. \tag{8.26}$$

The last equation, called the balance equation as noted earlier, has also been used in forecasting with the barotropic vorticity equation (6.14), with ψ in place of geopotential. The balanced pair (8.25), (8.26), is obviously not complete and requires the hydrostatic, continuity, and thermodynamic equations. In both pairs, (8.22), (8.23) and (8.25), (8.26), no essential simplifications of (3.52) are possible without violating the energy constraints; hence we have

$$\frac{\partial T}{\partial t} + \mathbf{V} \cdot \nabla T + \omega\left(\frac{\partial T}{\partial p} - \frac{\alpha}{c_p}\right) = 0. \tag{8.27}$$

Here the advection of temperature includes both the nondivergent and divergent wind components. It may be recalled that in the quasigeostrophic model, temperature advection with only the nondivergent wind is included, and the static stability parameter is generally taken either as a constant or at most a function of p [see Section 4.3 and Chapter 3, (3.50)]. The quasigeostrophic thermodynamic equation is appropriate for the system (8.1) and (8.3).

Equation 8.26 may be elliptic, hyperbolic, or parabolic in various parts of the domain and no general method of solution is available. However, usually a rather slight modification of the initial geopotential field, together with restrictions on subsequent approximations of ψ, permits the solution of (8.26) as an elliptic equation by relaxation techniques such as discussed in Chapter 5, provided ψ is known on the boundary of the region. For quite general second-order partial-differential equations of the type

$$F(p, q, r, s, t, x, y, \psi) = 0,$$

where

$$r = \psi_{xx}, \qquad s = \psi_{xy}, \qquad t = \psi_{yy}, \qquad p = \psi_x, \qquad q = \psi_y,$$

the condition for ellipticity is

$$4F_r F_t - F_s^2 > 0.$$

For (8.26), this criterion can be written

$$(\tfrac{1}{2}f + \psi_{xx})(\tfrac{1}{2}f + \psi_{yy}) - \psi_{xy}^2 > 0. \tag{8.28}$$

This requires that $\frac{1}{2}f + \psi_{xx}$ and $\frac{1}{2}f + \psi_{yy}$ at least have the same sign (plus or minus) throughout the region. The plus and minus signs correspond to two different solutions for given boundary conditions. Since the absolute vorticity is normally positive in large-scale motions, only the plus signs would be of interest here. Combining (8.28) with (8.26) also requires that

$$\nabla^2\Phi + \tfrac{1}{2}f^2 - \nabla f \cdot \nabla\psi > 0. \tag{8.29}$$

This condition is more readily imposed on the initial guess field, except for the term $\nabla f \cdot \nabla\psi$, which is relatively small and can be approximated with the geopotential field. At gridpoints where (8.29) is not fulfilled, the ϕ field may be successfully altered by spreading the negative residual of (8.29) over the four surrounding points until the ellipticity condition is satisfied over the entire grid.

Subsequent approximations of the stream function could require, for example,

$$\tfrac{1}{2}f + \psi_{xx}^{(n)} > \epsilon \qquad \text{and} \qquad \tfrac{1}{2}f + \psi_{yy}^{(n)} > \epsilon, \tag{8.30}$$

as necessary conditions (here n denotes the nth guess) as well as

$$(\tfrac{1}{2}f + \psi_{xx}^{(n)})(\tfrac{1}{2}f + \psi_{yy}^{(n)}) - (\psi_{xy}^{(n)})^2 > \tfrac{1}{2}\epsilon^2,$$

where ϵ is a small positive quantity. These criteria are rather difficult to impose and, after some experimentation (see Arnason, 1957), it was found that the following condition was satisfactory in practice:

$$\psi_{xx}^{(n)} + \psi_{yy}^{(n)} + f > \epsilon, \tag{8.31}$$

which simply requires that the absolute vorticity be positive and exceed a prescribed minimum value.

In the systems (8.22), (8.23) and (8.25), (8.26) described previously, no gravity waves are possible. The essential filtering mechanism for gravity waves, as shown by P. D. Thompson (1956), is the omission of the total derivative of the divergence $d\delta/dt$ in the divergence equation (1.21).

The ω equations associated with the systems, (8.22), (8.23) and (8.25), (8.26) are considerably more complex than the quasi-geostrophic ω equation (8.3). For (8.22), (8.23) the ω equation is (3.53); while for (8.25), (8.26) the ω equation is

$$\begin{aligned}\nabla^2(\sigma\omega) + f^2\frac{\partial^2\omega}{\partial p^2} &= f\frac{\partial}{\partial p}J(\psi,\eta) + \frac{\alpha}{\theta}\nabla^2 J(\psi,\theta) - 2\frac{\partial^2}{\partial t\,\partial p}J\left(\frac{\partial\psi}{\partial x},\frac{\partial\psi}{\partial y}\right)\\ &\quad - f\frac{\partial}{\partial p}(\zeta\nabla^2\chi) + f\frac{\partial}{\partial p}\left(\omega\frac{\partial\zeta}{\partial p}\right) + f\frac{\partial}{\partial p}\left(\nabla\omega\cdot\nabla\frac{\partial\psi}{\partial p}\right)\\ &\quad - f\frac{\partial}{\partial p}(\nabla\chi\cdot\nabla\eta) - \frac{\alpha}{\theta}\nabla^2(\nabla\chi\cdot\nabla\theta) - \nabla f\cdot\nabla\frac{\partial^2\psi}{\partial p\,\partial t}.\end{aligned} \tag{8.32}$$

The solution of this equation is so time consuming and complicated that some of the advantages of such a filtered system over the primitive equations, such as a longer time step, have vanished. Moreover, the balance equation is not always solvable without altering the initial pressure height data.

A point to consider, however, is that strict consistency with respect to energetics is required mainly for long-term integrations, and may not be an absolute necessity in short-range forecasting for periods of a few days. For this purpose some simplifications of the balance and ω equations may be possible which will reduce the time required for solution and yet provide an improvement over the geostrophic system in terms of short-range forecasting skill. Such simplifications, however, can only be established by very careful numerical testing over a large statistical sample.

As a final remark it is desirable to mention that filtered sytems do have an important advantage of not being as sensitive to small errors in the initial state as are the primitive equations. In fact, the filtered equations have been frequently used to determine the initial conditions for integration of the primitive equations.

chapter nine

The Inclusion of Moisture and Radiation in Numerical Forecasting

9.1 MOISTURE CONSERVATION EQUATION

Probably the most important effect of the inclusion of moisture in the dynamics of atmospheric motions is the transfer of latent heat, although information on clouds and precipitation naturally are of vital interest to the forecaster. For dry adiabatic motions a complete system of equations consisted of the vector equation of motion, equation of state, first law of thermodynamics, and the equation of continuity; or, alternatively, the equations of motion in the x- and y-directions can be replaced by the vorticity and divergence equations.

When moisture is included, an equation for the conservation of water must be added, as well as an appropriate modification of the first law of thermodynamics. To obtain the former, consider a fixed infinitesimal element of volume $\delta x\, \delta y\, \delta z$. The net gain of water vapor per unit time per unit volume as a result of flow through the $\delta y\, \delta z$ faces is $-\partial/\partial x(\rho_v u)$, where ρ_v is the density of water vapor (see Figure 9.1).

Consideration of all faces of the volume leads to

$$\frac{\partial \rho_v}{\partial t} = -\nabla_3 \cdot \rho_v \mathbf{V}_3 + S, \tag{9.1}$$

where S represents any sources or sinks of water vapor in mass per unit volume per unit time, and $\mathbf{V}_3$ is the three-dimensional wind velocity. If ρ

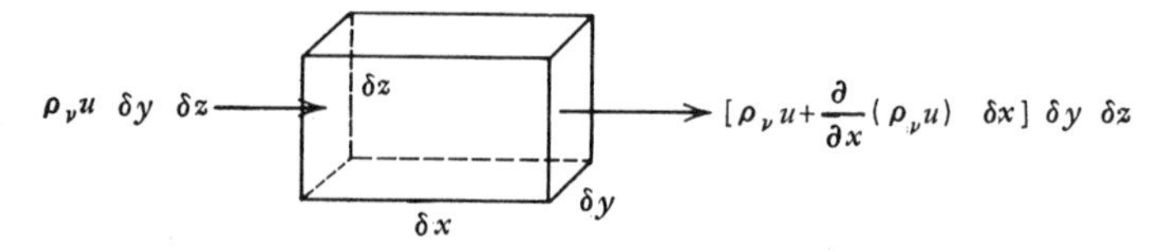

Figure 9.1. Illustrating conservation of water vapor.

is the air density and q is the specific humidity, then $\rho_v = \rho q$; and (9.1) may be written in flux form as

$$\frac{\partial \rho q}{\partial t} = -\nabla_3 \cdot (\rho q \mathbf{V}_3) + S. \tag{9.2}$$

Multiplying the equation of continuity

$$\frac{\partial \rho}{\partial t} = -\nabla_3 \cdot (\rho \mathbf{V}_3) \tag{9.3}$$

by q and subtracting it from (9.2) gives

$$\frac{dq}{dt} = \frac{\partial q}{\partial t} + \mathbf{V}_3 \cdot \nabla_3 q = \frac{S}{\rho}. \tag{9.4}$$

If the only loss of moisture is through condensation, then

$$\frac{S}{\rho} = \frac{dq_s}{dt}, \tag{9.5}$$

where q_s is the saturation specific humidity, given by the approximate formula

$$q_s \doteq \frac{0.622 e_s}{p}, \tag{9.6}$$

e_s being the saturation vapor pressure. On the other hand, if turbulent diffusion of water vapor is considered, then (9.4) must contain terms of the form $A_v \nabla^2 q$ and $\partial/\partial z(K_v\ \partial q/\partial z)$, representing the differential horizontal and vertical fluxes of water vapor. There may also be other sources or sinks of water vapor as evaporation from the surface, although such evaporation or condensation would be the vertical flux at the surface. Differentiation of (9.6) gives

$$\frac{1}{q_s}\frac{dq_s}{dt} \doteq \frac{1}{e_s}\frac{de_s}{dt} - \frac{\omega}{p}, \tag{9.7}$$

where $\omega = dp/dt$. Clapeyron's equation relating the saturation vapor pressure and temperature is

$$\frac{de_s}{e_s} = \frac{L\,dT}{R_v T^2}, \tag{9.8}$$

where R_v is the gas constant for water vapor and L is the latent heat of vaporization (or sublimation). Thus

$$\frac{1}{q_s}\frac{dq_s}{dt} = \frac{L}{R_v T^2}\frac{dT}{dt} - \frac{\omega}{p}. \tag{9.9}$$

Assuming the condensation takes place as a result of saturated adiabatic expansion, the first law of thermodynamics becomes

$$\frac{-L\,dq_s}{dt} = c_p\frac{dT}{dt} - \frac{RT}{p}\,\omega. \tag{9.10}$$

Here c_p is the specific heat at constant pressure and R is the gas constant for air. Elimination of dT/dt between (9.9) and (9.10) leads to the result

$$\frac{dq_s}{dt} = \frac{q_s T}{p}\left(\frac{LR - c_p R_v T}{c_p R_v T^2 + q_s L^2}\right)\omega, \tag{9.11}$$

which is assumed to apply only during condensation and requires a negative ω (upward motion).

Expanding the left member of (9.4) in (x, y, p, t) coordinates and utilizing (9.11) for S/ρ, we obtain

$$\frac{\partial q}{\partial t} + \mathbf{V}\cdot\nabla q + \omega\frac{\partial q}{\partial p} = \frac{\delta F}{p}\,\omega, \tag{9.12}$$

where

$$F = q_s T\left(\frac{LR - c_p R_v T}{c_p R_v T^2 + q_s L^2}\right) \tag{9.13}$$

and δ may be defined as follows:

$$\begin{aligned} \delta &= 1 \quad \text{for } \omega < 0 \text{ and } q \geq q_s, \\ \delta &= 0 \quad \text{for } \omega \geq 0 \text{ or } q < q_s. \end{aligned} \tag{9.14}$$

Alternatively, the condensation may be assumed to begin at some critical relative humidity, say r_c instead of 100%, in which case condition (9.14) would contain $r_c q_s$ instead of simply q_s. Probably the most realistic hypothesis would be to make the rate of condensation dependent on the relative humidity, that is, assume δ to be a function of q/q_s.

9.2 MODIFIED THERMODYNAMIC EQUATION

In terms of the previous notation, the release of latent heat may be included in the first law of thermodynamics (9.10) in the form

$$-\frac{\delta LF\omega}{p} = c_p \frac{dT}{dt} - \frac{RT}{p}\,\omega. \tag{9.15}$$

Additional terms may be included for other heat sources, e.g., through diffusion, radiation, etc.

The individual change of temperature dT/dt may be expanded in (x, y, p, t) coordinates to

$$\frac{dT}{dt} = \frac{\partial T}{\partial t} + \mathbf{V}\cdot\nabla T + \omega\frac{\partial T}{\partial p}. \tag{9.16}$$

Hence the thermal equation (9.15) may be written as

$$\frac{\partial T}{\partial t} + \mathbf{V}\cdot\nabla T = \frac{1}{p}\left(\frac{RT}{c_p} - \frac{\partial T}{\partial \ln p} - \frac{\delta LF}{c_p}\right)\omega, \tag{9.17}$$

where δ is again defined as in (9.14). If it is desired to introduce z as a dependent variable through the geostrophic approximation, T may be replaced by means of the hydrostatic equation

$$T = -\frac{gp}{R}\frac{\partial z}{\partial p}. \tag{9.18}$$

Equations 9.12, 9.17, and 9.18, together with the horizontal equations of motion, the equation of state, and the equation of continuity, constitute a complete system governing the hydrostatic motions of air when condensation resulting from vertical motions occurs.

9.3 THE ω EQUATION AND RATE OF PRECIPITATION

If all the condensate is assumed to fall instantly as precipitation, we may write a simple equation giving the amount of precipitation P per unit area in time Δt:

$$P = \int_t^{t+\Delta t}\int_0^{\infty} -\frac{dq_s}{dt}\,\rho\,dz\,dt = \int_t^{t+\Delta t}\int_0^{p_0} -\frac{\delta F\omega}{pg}\,dp\,dt. \tag{9.19}$$

To use (9.12) and (9.17) in a numerical forecasting scheme, it is necessary to know the appropriate value of δ at every step of the iteration procedure. Initial values of q (for comparison to q_s) are obtainable by observation together with a simple computation; however the initial ω field is not directly observed and computations to obtain it are not quite so simple. The diagnostic equation for ω may be obtained by appropriate combination of the thermal equation (9.17) and the vorticity and divergence equations. The quasi-geostrophic form will be used here for illustration:

$$\frac{\partial \zeta}{\partial t} + \mathbf{V} \cdot \nabla \eta = \bar{f} \frac{\partial \omega}{\partial p}. \tag{9.20}$$

With the geostrophic approximation in the first term, (9.20) becomes

$$\nabla^2 \frac{\partial z}{\partial t} + \frac{\bar{f}}{g} \mathbf{V} \cdot \nabla \eta = \frac{\bar{f}^2}{g} \frac{\partial \omega}{\partial p}.$$

Next differentiate the previous equation with respect to p, giving

$$\frac{\partial}{\partial t} \nabla^2 \frac{\partial z}{\partial p} = -\frac{\bar{f}}{g} \frac{\partial}{\partial p} (\mathbf{V} \cdot \nabla \eta) + \frac{\bar{f}^2}{g} \frac{\partial^2 \omega}{\partial p^2}.$$

Replacing $\partial z/\partial p$ by using (9.18) leads to

$$\nabla^2 \frac{\partial T}{\partial t} = \frac{p\bar{f}}{R} \frac{\partial}{\partial p} (\mathbf{V} \cdot \nabla \eta) - \frac{p\bar{f}^2}{R} \frac{\partial^2 \omega}{\partial p^2}. \tag{9.21}$$

Now apply the operator ∇^2 to (9.17), with the result

$$\nabla^2 \frac{\partial T}{\partial t} = -\nabla^2 (\mathbf{V} \cdot \nabla T) + \frac{gp}{R} \nabla^2 (\sigma_m \omega), \tag{9.22}$$

where

$$\sigma_m = \left(\frac{RT}{c_p} - \frac{\partial T}{\partial \ln p} - \frac{\delta L F}{c_p} \right) \frac{R}{gp^2}. \tag{9.23}$$

Equating the right hand members of (9.21) and (9.22) yields the "ω equation"

$$\nabla^2 (\sigma_m \omega) + \frac{\bar{f}^2}{g} \frac{\partial^2 \omega}{\partial p^2} = \frac{R}{gp} \nabla^2 (\mathbf{V} \cdot \nabla T) + \frac{\bar{f}}{g} \frac{\partial}{\partial p} (\mathbf{V} \cdot \nabla \eta), \tag{9.24}$$

which has no time derivatives and hence is a purely diagnostic equation.

This equation can normally be solved by the relaxation methods described in Chapter 7, provided the static stability parameter (9.23) remains positive, which implies that the lapse rate must be forced to remain less than the saturated adiabatic value. Under these circumstances it is reasonable to assume that the release of latent heat results mainly from the large-scale vertical velocity ω. However, when unstable lapse rates occur, convective clouds will develop and the vertical transfers of heat and moisture are not effected simply by large-scale vertical advection. In this event, vertical flux of heat, vapor, and momentum by convective clouds and/or other "subgrid" mechanisms must be included in the equations representing the thermal and dynamical processes. It is relevant to point out here that the precipitation equation (9.19) gives reasonably good results with synoptic-scale pressure systems in middle latitudes, but has been found to be inadequate in areas of convection and especially in tropical systems which will be discussed in the next chapter.

9.4 INCLUSION OF RADIATIVE TRANSFER OF HEAT IN NUMERICAL MODELS

9.4.1 Long-wave Radiation

Although the long-wave radiation is the most important term in the *mean tropospheric heat budget*, it is usually of lesser importance in short-range forecasts of synoptic-scale disturbances, as is often the case with other diabatic processes such as sensible heat transfer, convection, latent heat release, etc. Of course, in long-term integrations of the hydrodynamical equations for purposes of simulating the general circulation and climate of the atmosphere, the inclusion of both solar and terrestrial radiations is mandatory. The general circulation models of Smagorinsky et al. (1965), Arakawa and Mintz (1965), Washington and Kasahara (1967), and Leith (1964) include such radiative processes.

When moisture is included as one of the dependent variables in a numerical weather prediction model, it is feasible to compute the long-wave radiation as a function of the water-vapor content of the atmosphere and the cloud distribution.

Danard (1969) proposed a simplified method for including long-wave radiation in numerical prediction models by assuming that the water-vapor and cloud distributions are known at and below some upper pressure level p_u. The "line-broadening" effect of pressure is taken into account by reducing the precipitable water with the factor $(p/p_0)^{0.85}$ due to Kuhn (1963). An analogous correction for variation of temperature in a vertical column leads to the following expression for the precipitable water vapor w between $p = 0$ and an arbitrary pressure level p:

$$w(p) = \frac{1}{g}\int_0^p q\left(\frac{p}{p_0}\right)^{0.85}\left(\frac{T_0}{T}\right)^{0.5} dp, \tag{9.25}$$

where p_0 and T_0 are a standard pressure and temperature respectively, and q is the specific humidity. In order to obtain the value of w at p_u, use is made of the observation that the frost point in the lower stratosphere is approximately 190 K above about 120 mb. Use of an isothermal atmosphere of 220 K leads to a value of $w = 4.6 \times 10^{-7}$ g/cm^2 at $p = 120$ mb. It is further

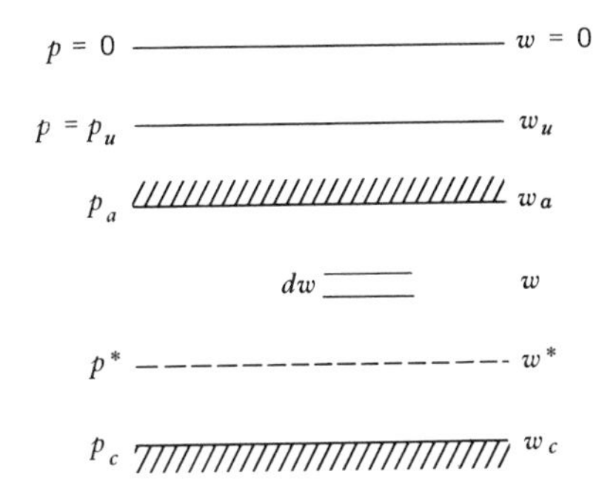

Figure 9.2. Determination of long-wave radiative flux.

assumed that q varies linearly between 120 mb and p_u. For a value of $p_u =$ 200 mb, the corrected precipitable vapor at 200 mb is

$$w_u = 4.7 \times 10^{-5} + 9.9q_{200}. \tag{9.26}$$

At all levels below p_u the value of w is computed from the distribution of q by means of (9.25), taking into account (9.26).

Clouds may be present in one or more layers in the vertical. The base and top of any cloud layer and the earth's surface are assumed to radiate as blackbodies. Consider now a level p^* between an upper cloud layer and a lower cloud layer (or the earth's surface), as shown in Figure 9.2.

It is now desired to obtain the net downward flux at p^* due to the bounding cloud layers and the water vapor contained in the intervening cloud-free layer as well. The downward water-vapor flux at p^* in the spectral band, $\lambda + d\lambda$, from the cloud at p_a may be expressed in the form (see Haltiner and Martin (1957))

$$dF^*_{a\lambda} = 2E_{a\lambda}(T, \lambda)\, d\lambda H_3[k_\lambda(w^* - w_a)]. \tag{9.27}$$

Here $E_{a\lambda}\, d\lambda$ is the Planck blackbody flux at p_a in the designated spectral band, k_λ is the absorption coefficient, and $2H_3\,[k_\lambda(w^* - w_a)]$ is a flux transmission function which determines the fraction of the initial blackbody radiation reaching p^* through the intervening water-vapor absorbing mass $w^* - w_a$. Specifically H_n is the exponential integral function of order n:

$$H_n(x) = \int_0^\infty e^{-xy} y^{-n}\, dy, \tag{9.28}$$

and

$$H_3[k_\lambda(w^* - w_a)] = \tfrac{1}{2} - k_\lambda \int_{w_a}^{w^*} H_2[k_\lambda(w^* - w)]\, dw.$$

In addition, the flux at p^* originating within the cloud-free water-vapor slab $w^* - w_a$ must be computed by summing the contributions from all the infinitesimal slabs dw making up that layer. The resulting downward flux is given by

$$dF^*_{w\lambda} = 2k_\lambda \int_{w_a}^{w^*} E_{w\lambda}\, d\lambda H_2[(k_\lambda(w^* - w)]\, dw, \tag{9.29}$$

where $E_{w\lambda}\, d\lambda$ is the blackbody flux at w in the given spectral interval. Adding (9.27) and (9.29), and integrating over all wavelengths, gives

$$F^{*\downarrow} = F_{ba} - 2\int_{w_a}^{w^*}\int_0^\infty k_\lambda E_{a\lambda} H_2[k_\lambda(w^* - w)]\, d\lambda\, dw$$
$$+ 2\int_{w_a}^{w^*}\int_0^\infty k_\lambda E_{w\lambda} H_2[k_\lambda(w^* - w)]\, d\lambda\, dw. \tag{9.30}$$

In (9.30) the second term on the right represents the flux removed from the blackbody flux F_{ba} by water-vapor absorption, while the third term is that added by emission from the cloud-free portion $w^* - w_a$ in accordance with Kirchhoff's law. The total blackbody flux is expressible in terms of temperature by the well-known Stefan-Boltzmann formula, $F_{ba} = \sigma T_a^4$. Now for an isothermal slab between w and w^*, the derivative ϵ' of the emissivity ϵ with respect to the absorbing mass evaluated at the mass $w^* - w$ is given by

$$\epsilon'(w^* - w) = \frac{2}{F_{bw}}\int_0^\infty k_\lambda E_{w\lambda} H_2[k_\lambda(w^* - w)]\, d\lambda. \tag{9.31}$$

Since the temperature dependence has been included in (9.25), the assumption is generally made that both ϵ and ϵ' are independent of temperature, and hence (9.31) applies even when the temperature varies between w and w^*. Danard made the additional approximation

$$\frac{E_{a\lambda}\, d\lambda}{F_{ba}} = \frac{E_{w\lambda}\, d\lambda}{F_{bw}}.$$

Substituting this last relation into the first integral in (9.30) and then utilizing (9.31) gives a downward flux at p^* of

$$F^{*\downarrow} = F_{ba} + \int_{w_a}^{w^*} (F_{bw} - F_{ba})\epsilon'(w^* - w)\, dw.$$

A similar expression gives the upward flux from the cloud top (or earth's surface) at p_c and the layer $(p_c - p^*)$. Thus the net flux at p^*, $F^{*\downarrow} - F^{*\uparrow}$, becomes

$$F^* = F_{ba} - F_{bc} + \int_{w_a}^{w^*} (F_{bw} - F_{ba})\epsilon'(w^* - w)\, dw$$
$$+ \int_{w^*}^{w_c} (F_{bc} - F_{bw})\epsilon'(w - w^*)\, dw. \tag{9.32}$$

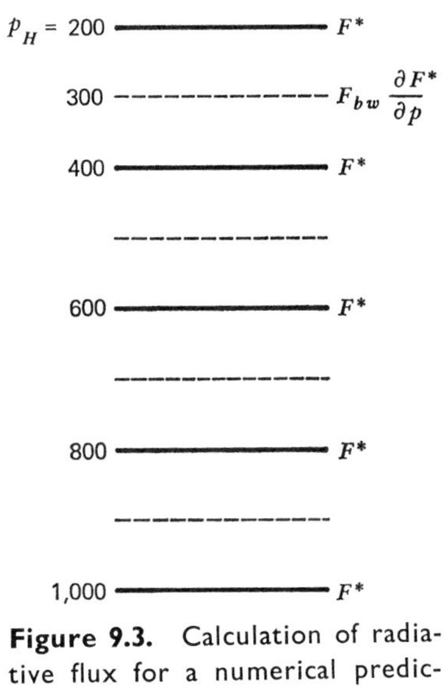

Figure 9.3. Calculation of radiative flux for a numerical prediction model.

For low elevations a water cloud acts nearly like a blackbody; however, at higher elevations with cirrus clouds, this assumption is less realistic. For cirrus clouds the assumption of a gray-body radiator presumably could be made.

The local temperature variation at level p^* resulting from the vertical flux divergence (differential flux) of long-wave radiation is given by

$$\left(\frac{\partial T}{\partial t}\right)_r = -\frac{g}{c_p}\frac{\partial F^*}{\partial p}. \tag{9.33}$$

To apply these formulations to a numerical prediction model, consider Figure 9.3. The upper pressure p_u is taken to be 200 mb and F^* is calculated at the even levels by evaluating the integrals of (9.32) over the relevant 200-mb layers in the following manner:

$$\int_{w_a}^{w^*} (F_{bw} - F_{ba})\epsilon'(w^* - w)\, dw$$
$$= \sum (\bar{F}_{bw} - F_{ba})[\epsilon(w^* - w_u) - \epsilon(w^* - w_l)]. \tag{9.34}$$

Here w_u and w_l are the values at the upper and lower levels of the contributing 200-mb layer, and $\bar{F}_{bw}$ is the blackbody flux at the midpoint. The summation is extended over all of the 200-mb layers between w_a and the particular w^* under consideration. Similarly, the second integral in (9.32) is expressible as

$$\int_{w^*}^{w_c} (F_{bc} - F_{bw})\epsilon'(w - w^*)\, dw$$
$$= \sum (F_{bc} - \bar{F}_{bw})[\epsilon(w_l - w^*) - \epsilon(w_u - w^*)], \quad (9.35)$$

where the summation extends over all layers between w^* and w_c. Clouds are assumed to exist in a 200-mb layer if the relative humidity at the midpoint exceeds a critical value r_c according to the following table:

p	900	700	500	300
r_c	0.84	0.75	0.70	0.69

For the model depicted in Figure 9.3 consider a case with 16 cloud configurations ranging from four 200-mb cloudy layers (extending from 1,000 to 200 mb) to four cloud-free layers. Kuhn's (1963) values of ϵ are used. The computer program proceeds by evaluating F^* from 200 mb down to the nearest cloud top (or the earth's surface). Then F^* is computed from the base of the highest cloud down to the top of the next cloud layer, and so forth. After the values of F^* have been computed at all of the even levels, $\partial F^*/\partial p$ is evaluated at the odd levels, 300, 500, 700 and 900 mb, by a central difference approximation. The technique can obviously be extended to give greater vertical resolution if desired. Some computations by Danard indicate that considering just long-wave radiation from water vapor overestimates the total radiative (short- and long-wave) cooling at low latitudes and low levels. The discrepancy is probably due mainly to short-wave absorption by water vapor; however long-wave absorption by carbon dioxide and ozone have also been omitted in the foregoing treatment. Nevertheless a comparison of mean cooling rates, which are about 1 °C per day, computed by other methods [e.g., London (1957), Manabe and Möller (1961)] indicate that Danard's technique is fairly satisfactory.

A comparison of long-wave cooling at cloud top with the maximum rate of latent heat release in the upper troposphere in the vicinity of an intense cyclone showed the latent heat release to be larger by a factor of five. However, the long-wave cooling can destabilize the upper part of a cloud mass and increase the vertical velocity, thus triggering other mechanisms of energy release.

In their treatment of long-wave radiation, Arakawa, Mintz, and Katayama (1968) account for the presence of CO_2 by an empirical correction to the net water-vapor radiative fluxes R at the tropopause, midtroposphere, and the ground as follows:

$$R_T' = 0.820R_T, \qquad R_M' = 0.736R_M, \qquad R_G' = 0.740R_G.$$

The reduction factors were obtained by comparing the water-vapor radiative transfer calculations to more complete evaluations including CO_2.

9.4.2 A Model for Solar Radiational Heating

Analogous to that of long-wave radiation, an exact treatment of solar radiation would be extremely involved, far too time consuming to compute, and excessive in data requirements to be practicable for a numerical prediction model at this time. Consequently, simplified models have been used. A technique designed by Katayama, used in the Mintz-Arakawa general circulation model, will serve as an illustration [see Arakawa, Katayama, and Mintz (1968)]. The solar radiation is considered to consist of two parts: (*a*) with wavelength less than 0.9 μ, in which Rayleigh scattering is significant but absorption by water vapor is negligible, and (*b*) with wavelength greater than 0.9 μ, in which absorption by water vapor is significant but scattering is slight.

The insolation incident at the top of the atmosphere is given by

$$S_0 \cos \alpha, \tag{9.36}$$

where S_0 is the solar constant and α is the zenith angle. The latter may be determined with the aid of Figure 9.4, in which φ is the latitude, δ is the declination angle of the sun, and h is the hour angle. The arcs of the spherical

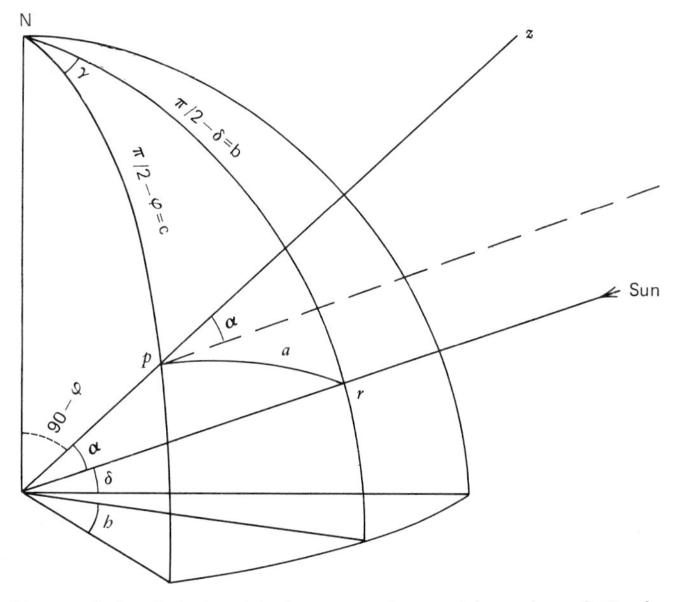

Figure 9.4. Relationship between the zenith angle α, latitude φ, sun declination, and hour angle.

triangles a, b, and c are related by

$$\cos a = \cos b \cos c + \sin b \sin c \cos h$$

or, in terms of the subtending angles,

$$\cos \alpha = \sin \varphi \sin \delta + \cos \varphi \cos \delta \cos h. \tag{9.37}$$

The division of the radiation (9.36) is as follows:

Scattered part $\qquad S_0^S = 0.651 S_0 \cos \alpha.$ $\qquad$ (9.38)

Absorbed part $\qquad S_0^A = 0.349 S_0 \cos \alpha.$

Both parts are subject to reflection from clouds and the earth's surface. However, after reflection, it is assumed that no absorption of the reflected radiation occurs.

The absorption of the absorbed parts is approximated by

$$A(w^*, \alpha) = 0.271 (w^* \sec \alpha)^{0.303}, \tag{9.39}$$

where w^* is the pressure-corrected [see (9.25)] water-vapor mass for the layer through which the beam travels.

The *albedo* for the scattered part is taken to be:

1. Clear skies $\qquad a_0 = 0.085 - 0.245 \log \left(\frac{p_s}{p_0} \cos \alpha \right),$ $\qquad$ (9.40)

2. Overcast $\qquad a_c = 1 - (1 - a_0)(1 - R_c),$ $\qquad$ (9.41)

where p_s is the surface pressure.

The albedoes of clouds and the earth's surface are:

1. Albedo of cloud: R_c.
2. Albedo of earth's surface: a_s.

The absorbed part of the insolation is partitioned with height as follows (see Figure 9.5):

1. *Above the cloud, H:*

$$S_H = S_0^A [1 - A(w_0^* - w_H^*, \alpha)]. \tag{9.42}$$

2. *At the cloud top, CT:*

$$S_{CT} = S_0^A [1 - A(w_0^* - w_{CT}^*, \alpha)]. \tag{9.43}$$

3. *In the cloud at an arbitrary level, M:*

$$S_M = S_0^A (1 - R_c) \left\{ 1 - A \left[(w_0^* - w_{CT}^*) \sec \alpha + 1.66 \frac{\Delta p_M}{\Delta p_c} w_c^* \right] \right\}. \tag{9.44}$$

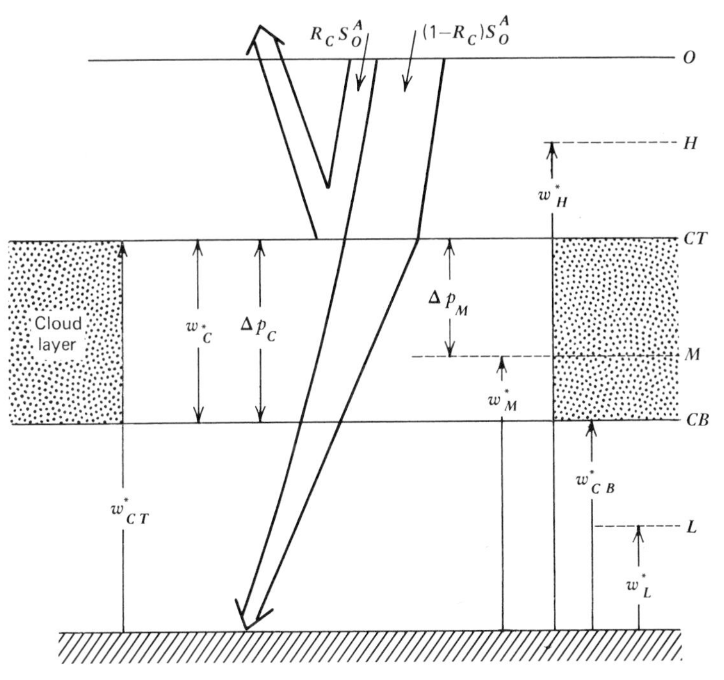

Figure 9.5. Disposition of solar radiation. [*After Arakawa, Katayama, and Mintz (1968).*]

4. *Below the cloud, L:*

$$S_L = S_0^A(1 - R_c)\{1 - A[(w_0^* - w_{CT}^*)\sec\alpha + 1.66(w_c^* + w_{CB}^* - w_L^*)]\}. \tag{9.45}$$

Here w_c^* is an equivalent water-vapor amount of cloud and

$$A[X] = 0.271X^{0.303}. \tag{9.46}$$

The factor 1.66 augments the beam length to account for the mean diffused path of the solar rays which are subject to scattering within the cloud, increasing the opportunity for further absorption. By way of further explanation, note that the quantity $1 - A$ specifies the fraction of the radiation, $S_0^A(1 - R_c)$, which is transmitted through a layer, where $S_0^A(1 - R_c)$ is the fraction of S_0^A which is not reflected at the cloud top.

The downward flux of the scattered insolation, which is constant in height, is given by

Clear skies $$S_{S0}^S = S_0^S \frac{1 - a_0}{1 - a_0 a_s}, \tag{9.47}$$

Overcast $$S_{SC}^S = S_0^S \frac{1 - a_c}{1 - a_c a_s}. \tag{9.48}$$

The denominator, e.g., $1 - a_0 a_s$, in the preceding expressions enhances the downward flux of this portion of the insolation to account for back-scattering from multiple reflections.

From the various previous formulas the absorption of solar radiation for any layer may be obtained by taking the difference between the total irradiance at the top and bottom of the layer. Moreover, the total solar irradiance at the earth's surface for clear skies is given by the sum of (9.42) and (9.47), with H taken at the surface. Similarly, for overcast skies, one employs the sum of (9.45) and (9.48):

$$\begin{aligned} &\text{Clear skies} \qquad && S_{S0}^{S} + (S_H)_{w_H^*=0} \\ &\text{Overcast} \qquad && S_{SC}^{S} + (S_L)_{w_L^*=0} \end{aligned} \tag{9.49}$$

Typical values of a_s lie in the range from 0.15 to 0.70 depending on the character of the surface; e.g., for a highly reflecting surface such as snow, a high value of a_s is appropriate. The albedo of a cloud deck also varies, with smaller values for cirrus clouds and larger values for a low overcast; a mean value of R_c would be about 0.5.

In the foregoing discussions the sky was assumed to be either clear or overcast. In the case of partly cloudy skies, the irradiance can be approximated by a linear combination of the overcast and clear values according to the fractional cloud cover, say C. Thus

$$S = (1 - C)S_{\text{clear}} + CS_{\text{overcast}}. \tag{9.50}$$

Estimates of sensible heating and evaporation require the air temperature near the earth's surface and the earth temperature as well. For these purposes the sea may be assumed to have an infinite heat capacity; hence the sea surface temperature is taken to be constant during the forecast period. On the other hand, the land surface is assumed to have zero heat capacity; thus the net flux of thermal energy across the air-land boundary must vanish.

The sensible heat flux Q_s at the air-surface boundary may be approximated by the frequently used expression

$$Q_s = \rho_s c_p C_D V_s (T_g - T_x), \tag{9.51}$$

where V_s is a surface wind speed and ρ_s is a surface density, both obtained by extrapolation from upper levels. T_g is the temperature at the earth's surface and T_x is the air's temperature at the top of the thin boundary layer. This raises the problem of determining a suitable value of T_x which will give reasonable values of sensible heat flux and evaporation. Omitting the details, which are not trivial, however, Arakawa, et al. calculated the heat flux Q_T at the top of the boundary layer as a function of the difference between a critical lapse rate γ_c and the lapse rate in the lowest layer of their model γ_M, with the latter depending on T_x. Thus

$$Q_T = \rho_s c_p K(\gamma_c - \gamma_M), \tag{9.52}$$

where K is a coefficient of turbulent heat transfer, which in turn is made to depend on the static stability. Then with the assumption of zero heat storage in the boundary layer, $Q_s = Q_T$ yields a value of T_x, the latter, in turn, permitting the calculation of the sensible heat flux and evaporation over the sea where the surface temperature is considered to be known.

Over land areas or ice-covered sea, T_g is not known and must be calculated. Since the land is assumed to have zero heat capacity, the net flux of thermal energy across the air-ground interface must vanish. The net downward long-wave flux at the ground F_g will be the difference between the radiation received from the atmosphere and the outgoing radiation from the earth's surface, the latter being assumed to radiate as a blackbody. The other energy source for the earth is the fraction of the solar radiation absorbed at the ground, say S_g. The required heat-balance equation over land is simply

$$F_g + S_g = Q_s. \tag{9.53}$$

To calculate the outgoing radiation of the land the surface temperature T_g is assumed to be a perturbation of T_x, namely,

$$T_g^4 \doteq T_x^4 + 4T_x^3(T_g - T_x).$$

The heat-balance equation (9.53) then permits the evaluation of the ground temperature.

To allow for the conduction of heat through the ice over frozen seas, another term may be added to the heat-balance equation as follows:

$$F_g + S_g + B(T_0 - T_g) = Q_s. \tag{9.54}$$

Here T_0 is the freezing point of sea water and B is the conduction coefficient for ice.

chapter ten

Tropical Forecasting, Parameterization of Convection

10.1 INTRODUCTION

The dynamics of tropical motions have not been as well understood as those of higher latitudes, primarily because the quasi-geostrophic approximation is generally not valid in the tropics because the Rossby number equals or exceeds one. The climatology of this region shows a broad band of easterly winds (trades) equatorward of the subtropical high-pressure cells which are centered at 30 to 35° latitude over the oceans. Within the friction layer equatorward motions are generated in the subtropical easterlies. The inter-tropical convergence zone (ITCZ) is a band of intense cumulus convection and heavy showers between the tradewind belts. The ITCZ is most persistent and sharply defined in the North Atlantic and the eastern and central North Pacific between 5 and 10N, but it is sometimes found between 5 and 10S in the eastern Pacific Ocean. In the western North and South Pacific the ITCZ is somewhat more diffuse and, furthermore, it is usually not well defined near large land masses. There are daily, seasonal, and other aperiodic variations in the intensity of the convection along the ITCZ.

Because of their destructive natures the hurricanes of the Atlantic and typhoons of the Pacific have been the most intensively studied of the tropical disturbances. Surprisingly, however, it is only during the last few years that some of their principal features have been simulated with dynamical models. To achieve a successful numerical simulation of a tropical storm, it has been necessary to include the convective transfer of moisture and heat. This is not unexpected since latent heat has long been believed to be the principal source of energy driving the storm. However, early investigations by Haque (1952) and Lilly (1960) on the dynamic stability of perturbations, where the release of latent heat is the direct result of the vertical velocity induced by conditional instability, show that convective scale disturbances are much more unstable than the cyclone scale. Thus this theory does not furnish a physical basis for the development or the maintenance of a tropical storm. Nevertheless, Riehl and Malkus (1948) have demonstrated the important role played by the "hot towers" (deep cumulus) in the thermal budget of a growing hurricane and further suggested that the cumulus-scale heat transfer be parameterized in the framework of the larger-scale tropical storm. Ooyama (1963) was the first to simulate a tropical storm numerically by parameterizing the cumulus convection. He assumed that the release of latent heat was proportional to the horizontal convergence of moisture in the frictional boundary layer. A linear perturbation analysis showed a dynamically unstable disturbance similar in scale and growth rate to a tropical cyclone. However, further studies by Ooyama and others have shown that the development and characteristics of a hurricane-like disturbance are very sensitive to the specific manner of formulating the frictional convergence and release of latent heat in the convection parameterization.

In addition to the intense warm-core tropical storms there are a variety of other disturbances; for example, when viewed from satellites, the ITCZ is found to be made up of cloud clusters [Reed (1970); Chang, (1970)] which are a manifestation of low-level convergence in large-scale waves propagating westward in a mean easterly current. Riehl (1945, 1948) and Palmer (1951) provided some early descriptions of these "easterly" and "equatorial" waves, although some features of these disturbances were well known to meteorologists stationed in the tropics during World War II. More recent spectral analyses of Pacific disturbances [Chang, et al., (1970)] indicate several predominant wavelengths and a range of periods from several days to a week or more. Some of these disturbances develop into tropical storms, but most maintain their intensity as moderate disturbances. There appear to be two principal types of waves, namely, one with a wavelength of 2,000 to 5,000 km, which has maximum intensity in the lower troposphere, the other with a wavelength from 6,000 to 10,000 km, which reaches its maximum intensity in the upper troposphere and has considerable vertical tilt. Each type of wave has an associated area of convection and both may exist simultaneously. However their origin, dynamics, and sources of energy are not completely understood, although they appear to be Rossby or mixed Rossby-gravity waves.

Nitta and Yanai (1969) have made analytical and observational studies of the mean easterly current for barotropic instability. They found the necessary condition for barotropic instability ($\beta - U'' = 0$) was fulfilled in summer at 600 to 700 mb in the Marshall Islands area and that the preferred scale of amplified waves was about 2,000 km. On the other hand, the easterlies were found to be barotropically stable in the vicinity of the Line Islands farther east. R. T. Williams and collaborators (1969) utilized a two-level quasi-geostrophic model including surface friction to study easterly currents with horizontal shear in the form of barotropically unstable zonal wind profiles centered at 20N. By using an initial-value approach with a hyperbolic tangent profile, an amplified disturbance was generated which is similar to observed low-level easterly waves, having rising motion upwind of the trough and little vertical tilt. With a hyperbolic secant-squared profile, the disturbance exhibits a similar structure outside the region of strong shear but with smaller vertical velocities; however, inside the region of strong shear the convergent area moves downwind of the trough. This is contrary to most easterly waves but may exist along the ITCZ.

Besides barotropic instability, several other mechanisms have been suggested for initiating waves in the tropical easterlies. Mak (1969) proposed that the large-scale tropical eddies originate from stochastic forcing by unstable baroclinic middle-latitude disturbances. According to this theory, the easterlies are dynamically stable and the tropical eddies gain their kinetic energy from pressure work at the boundaries of the subtropics, and lose it by dissipation and conversions to zonal kinetic energy and eddy available potential energy, which in turn is depleted by radiative cooling and conversion to zonal available potential energy. Mak cited some observational evidence in support of this hypothesis, particularly for the upper-level disturbances with periods of about five days and wavelengths of about 10,000 km.

A third hypothesis for the source of energy of tropical disturbances, which has already been mentioned in connection with tropical storms, is the latent heat release in the conditionally unstable atmosphere through convection forced by moisture convergence in the frictional boundary layer. This process, referred to as CISK, has been utilized by numerous investigators in connection with axially symmetric tropical storms and by Yamasaki (1970) in connection with easterly waves. Holton (1970) concluded from a theoretical study that the westward propagating waves along the ITCZ can be interpreted as Rossby waves driven by the latent heat release in precipitating cloud clusters which originate from moisture convergence in the friction layer. In addition, he found that the structure of the waves is very sensitive to the vertical shear of the zonal mean wind. It has already been noted that horizontal shear plays an important role in the structure of tropical disturbances. Spectral analyses of tropical stratospheric data together with theoretical studies indicate the presence of: (a) eastward propagating Kelvin waves with periods of about twelve days in zonal westerlies at 10–14 km having only longitudinal wind oscillations; (b) westward propagating mixed Rossby-gravity waves with periods of approximately seventeen days.

10.2 TROPICAL CYCLONES

A number of investigators, including Ooyama (1963, 1969), Charney and Eliassen (1964), Kuo (1965), Syono and Yamasaki (1966), Yamasaki (1968), and Rosenthal (1969), have simulated many important features of tropical storms. One of the principal differences between the various models lies in the precise method of parameterization of convection; other differences involve the frictional terms, number of levels, etc. No attempt will be made here to describe all of these models and only a brief summary of Ooyama's most recent results will be presented. However, the parameterization of convection will be discussed in more detail since this feature has already become a part of operational prediction models for middle latitudes as well.

Ooyama's model (1969) is axially symmetric with two principal tropospheric layers plus a frictional boundary layer. In spite of the limited vertical resolution, the numerical integrations simulate the life cycle of a typical tropical cyclone with a "remarkable degree of reality." The results showed that the supply of heat and moisture from a warm ocean is a crucial requirement for the growth and maintenance of a tropical cyclone. In a simulated landfall in which the energy supply from the ocean is cut off, the storm's intensity rapidly diminishes. Ooyama bases the parameterization of convection primarily on the premise that the deep convective clouds are the most important forms of moist convection with respect to the vertical transport of heat and moisture. The diabatic heat flux is assumed to be proportional to the vertical velocity w_F at the top of the friction layer, but not to the moisture convergence aloft nor to the evaporation as in some models. Thus

$$Q \sim \begin{cases} \eta w_F & \text{if } w_F > 0, \\ 0 & \text{if } w_F \leq 0 \end{cases} \tag{10.1}$$

where η is a nondimensional parameter determined as follows. For every unit mass from the boundary layer 0 that enters a convective cloud, $\eta - 1$ units of air in the lower layer 1 are entrained into the updraft. Hence $(\eta - 1) + 1 = \eta$ units of saturated cloud air enter the upper layer 2. The energy balance of the convective updraft is then given by

$$\eta H_2 = H_0 + (\eta - 1)H_1, \tag{10.2}$$

where

$$H \simeq c_p T + Lq + gz, \tag{10.3}$$

$$dH = \frac{c_p T}{\theta_e}\, d\theta_e = c_p\, dT + L\, dq + g\, dz = c_p\, dT + L\, dq - \alpha\, dp, \tag{10.4}$$

where θ_e is the equivalent potential temperature.

It follows from (10.2) and (10.3),

$$\eta = \frac{H_0 - H_1}{H_2 - H_1} = 1 + \frac{H_0 - H_2}{H_2 - H_1} \doteq 1 + \frac{\theta_{e0} - \theta^*_{e2}}{\theta^*_{e2} - \theta_{e1}},$$

where θ_{e0} is representative of the boundary layer, θ_{e1} of the lower main layer, and θ^*_{e2} of the upper layer, assumed saturated.

It may be desirable to recall here a few aspects of static stability. The first law of thermodynamics and (10.4) show that H is conserved by a parcel undergoing either dry or saturated (pseudo) adiabatic expansion, and thus H is constant along a moist adiabat. Thus if the air is saturated, it is statically unstable or stable according to whether $\partial H/\partial z$ is negative or positive. If unsaturated, the air is convectively unstable (also referred to as conditionally unstable) or stable according to whether $\partial H/\partial z$ is negative or positive.

Ooyama's earlier model (1963) had a fixed value of the parameter η which caused an unchecked growth of the hurricane vortex. In the present model, η is dependent on θ^*_{e2}, which is a dependent variable. Since a higher temperature in the upper layer requires a greater flux of energy from the boundary layer to enable the convective updraft to reach the upper layer, the variable η provides for a control on the maximum intensity.

Malkus and Riehl (1960) showed that to explain the central pressure of a hurricane hydrostatically, the boundary-layer value of θ_e must be about 12.5°C higher than normal; however, they also demonstrated that air traveling from about 90 km from the storm center to 30 km can pick up enough sensible heat and moisture to raise θ_e the required amount. Hence these mechanisms must be present if realistic storm development is to be expected in a numerical simulation. Ooyama combines the processes of latent and sensible heat transfer into a single expression of the form

$$C_E V_1(H_s - H_0), \tag{10.5}$$

where C_E is a nondimensional coefficient for the sea-air energy exchange and equals the drag coefficient C_D.

$$C_D = 10^{-3}(0.5 + 0.06V_1). \tag{10.6}$$

Here V_1 is the wind speed, m/sec. This formula is very similar to those of Sheppard and Miller and is probably reasonable for winds up to 50 m/sec.

Ooyama's numerical integrations yield a storm with winds of 58 m/sec in about 5.5 days, after which the maximum wind slowly decreases, but the area of hurricane and gale winds is still expanding even at the end of the ten-day computations. During the first several days, the storm develops very slowly while the central pressure drops to about 1,000 mb, which may be described as the incipient stage. Observed storms may then deepen rapidly, 20 to 30 mb in a 24-hr period, with hurricane winds in a narrow band about the center. In the numerical integration the deepening or immature stage lasts from about 60 to 130 hr, and is followed by the mature stage during which the central pressure reaches a minimum and the storm expands.

Varying the sea-surface temperature clearly demonstrated the need for relatively high sea-surface temperatures for tropical-cyclone development to hurricane intensity. For example, a value of $T_s = 27.5$ °C gave a maximum wind of almost 69 m/sec, while for 25.5 °C the wind was less than 40 m/sec. An imposed sudden cooling after four days from 27.5 to 25.5 °C resulted in a rapid weakening of the storm, while the opposite change gave rapid intensification.

In the Yamasaki-Syono (1966) model the vertical distribution of latent heat release was governed by a parameter l which remained fixed during the numerical integration. With a sufficiently large value of low-level mixing ratio and an appropriate value of l, a typhoon-scale disturbance develops, whereas when these conditions are not met, gravitational-type phenomena predominate. As in all of these studies, frictional convergence of moisture is a necessary condition for development. In a later paper Yamasaki (1968) permitted the vertical distribution of latent heat release to be a function of the difference between the temperature of the clouds and the environment. Thus the ratio l of the heat release in the lower layer to that in the upper layer varied with time. Initially l is small, but as the upper troposphere warms and stabilizes, l increases to about 0.5 and the storm intensifies rapidly. The Yamasaki model simulated the development process and mature stage very well.

In summary, it may be stated that there exist simulations of heat transfer by cumulus convection in terms of large-scale parameters that can produce cyclonic storms of hurricane intensity; however the results are quite sensitive to the specific formulation. It is reasonable to infer, therefore, that numerical prediction of the initiation, development, and dissipation of hurricanes and typhoons from observed initial data is not yet feasible. Nevertheless some attempts at numerical prediction are being conducted for tropical regions, and in some cases tropical storms are present in the forecast area. These experiments are clearly more within the purview of this text and will be described in further detail in the next section.

Mention should also be made of more or less empirical techniques that have been utilized for predicting the movement of existing hurricanes or typhoons. One such method consists of treating the storm as a point vortex being steered by the "mean" current. The vortex is first "eliminated" from the pressure field by a smoothing procedure, then the usual numerical prediction scheme is applied to the smoothed field, and the point vortex is carried along hour by hour with the predicted mean flow. Such a technique has been utilized in connection with barotropic forecasts with some degree of success.

10.3 TROPICAL PREDICTION MODELS

As of this date actual numerical weather prediction in the tropics using observed data for initialization has been quite limited indeed. One of the earliest attempts was a barotropic model by Shuman and Vanderman (1966) using a finite difference scheme devised by Shuman which will be discussed in Section 13.3. Operational numerical forecasts have sometimes consisted of a simple barotropic model with a built-in trend toward climatology, or even simply a combination of persistence and climatology. However, experiments with more sophisticated models are being conducted and several models will be described.

For the most part the details of the numerical integration techniques will be left for Chapters 12 and 13, the emphasis here being on the formulation of the differential equations and the method of simulating the relevant physical processes. Although quite successful forecasts have been made for middle and high latitudes with adiabatic, frictionless models, it is evident from the preceding discussion that tropical forecasting models presumably must include latent and sensible heat transfer, friction, convection effects, etc.

T. N. Krishnamurti (1962, 1969) has integrated both the balanced system and the primitive equations from an initial state consisting of observed data and data calculated from a balanced state. The primitive equations utilized are typical:

$$\frac{\partial \mathbf{V}}{\partial t} + \mathbf{V} \cdot \nabla \mathbf{V} = -\omega \frac{\partial \mathbf{V}}{\partial p} - f\mathbf{k} \times \mathbf{V} - mg\nabla z + \mathbf{F}, \tag{10.7}$$

$$\frac{\partial \theta}{\partial t} + \mathbf{V} \cdot \nabla \theta = -\omega \frac{\partial \theta}{\partial p} + \frac{H}{c_p}\left(\frac{p_0}{p}\right)^{R/c_p} + F_\theta, \tag{10.8}$$

$$\frac{\partial q}{\partial t} + \nabla \cdot \nabla q = -\omega \frac{\partial q}{\partial p} + M + F_q, \tag{10.9}$$

$$\left(\frac{\partial}{\partial t} + \mathbf{V}_0 \cdot \nabla\right)(z_0 - H_t) = \omega_0 \frac{R\theta_0}{gp_0}, \tag{10.10}$$

$$\frac{\partial \omega}{\partial p} = -m\left(\frac{\partial u}{\partial x} + \frac{\partial v}{\partial y}\right),$$

$$\frac{\partial z}{\partial p} = -\frac{R\theta}{gp}\left(\frac{p}{p_0}\right)^{R/c_p}.$$

Here H_t is the terrain height, $\hat{m}$ is the map factor for the Mercator projection, **F** is friction, H and M are heat and moisture sources, F_θ, F_q represent diffusion of heat and moisture from subgrid scale processes, and the subscript 0 denotes the 1,000-mb surface. Equation 10.10 is equivalent to the lower boundary condition, $w_0 = V_0 \cdot \nabla H_t$.

The atmosphere is divided vertically into 100-mb layers from 1,000 to 100 mb, and the equations are solved for u, v, and z at even levels, and for θ, q, and ω at odd levels with $\omega = 0$ at 100 mb and $\omega = \omega_0$ at 1,000 mb. Initial conditions were calculated from the balanced system (8.25), (8.26), and (8.32).

The diabatic heating is comprised of sensible heating H_S and the release of latent heat H_L. The sensible heating in $m^2 \sec^{-3}$ is given by

$$H_S = 0.00323V(T_w - T_a) \qquad \text{at 900 mb}, \tag{10.11}$$

where V is the wind velocity, m/s, and T_w and T_a are the sea and air temperatures, respectively. The latent heat release may be divided into two parts—that released under a conditionally unstable vertical stratification and that released under stable conditions by the large-scale vertical velocity. Krishnamurti did not include latent heat release from the latter process since it is probably less important in the tropics.

With regard to convection, the parameterization is based on the observation that, although the tropical regions are nearly always conditionally unstable up to 500 mb, intense cumulus activity is usually confined to limited areas and suppressed elsewhere. It appears therefore that the distribution of convection is controlled by the large-scale circulation. On this assumption, which is the same principle used in simulating tropical storms, the requirements for releasing latent heat are as follows:

1. $\partial\theta_e/\partial p > 0$ at any level above 1,000 mb.
2. There is net moisture convergence from the surface to 100 mb.

When these conditions are fulfilled, the heat released is

$$H_L = c_p a \frac{T_c - T}{\Delta t} \tag{10.12}$$

where $T_c - T$ is the temperature difference between the cloud and the environment, Δt is the time step, a is a measure of the fraction of the area covered by convective clouds ($a \leqslant 0.01$). The latent heating function results in a partial adjustment of the temperature toward the moist adiabat, unlike the Smagorinsky-Manabe (1967) procedure of adjusting the lapse rate to the neutral (saturated adiabatic) value and condensing out the excess moisture.

Another source of moisture over ocean areas only is evaporation which is accounted for by the expression

$$E_v = 0.1157(2.6 + 7.7V_0)(0.98e_w - e_0), \tag{10.13}$$

where the subscript 0 denotes 1000 mb, V_0 is in m/sec, e_w is the saturation-vapor pressure in mb over the ocean, and the units of E_v are g/(m²)(sec). The atmosphere is divided into four 200-mb layers, 10–8, 8–6, 6–4, and 4–2 decibars, in which the total moisture sources M are represented by

$$M = \frac{gE_v}{\Delta p} + \frac{a(q_c - q)}{\Delta t}. \tag{10.14}$$

Here the first term on the right is the evaporation, which is permitted only in the lowest layer of the atmosphere, and the second term represents the vertical advection of moisture on the cumulus scale. There is, of course, a loss of moisture through condensation and precipitation in amount

$$ac_p(T_c - T)/L\,\Delta t;$$

however, this is compensated by convective transfer of moisture over and above the last term (10.14).

The friction term **F** includes lateral and vertical diffusions and the surface wind stress as follows:

$$\mathbf{F} = \mu m^2\,\nabla^2\mathbf{V} + K\frac{\partial^2\mathbf{V}}{\partial p^2} - g\frac{\partial\boldsymbol{\tau}}{\partial p}, \tag{10.15}$$

where $\mu = 5 \times 10^4$ m²/sec and $K = 2 \times 10^{-2}$ mb²/sec. The surface stress $\boldsymbol{\tau}$ is defined in terms of a drag coefficient of 2.5×10^{-3} and applied at 1,000 mb, as shown in (10.16).

The initial data for the computations on 1 March 1965 covered the Pacific area from 28N to 28S and 128W to 104E. The 1,000-mb analysis showed an asymptote of convergence near 5N, a belt of easterlies over most of the northern portion of the analysis, and several cyclonic centers. At 200 mb there were two regions of westerly wind maxima and several troughs on the northern and southern edges while a broad easterly current with speeds of almost 40 knots prevailed over the equatorial latitudes.

According to Krishnamurti, the 24-hr forecast was quite satisfactory, although several of the synoptic details, which were well resolved on the $2° \times 2°$ grid, left much to be desired. There was somewhat better agreement between observed and forecast winds at 200 mb than at 1,000 mb. The forecasts after 30 hr were not very satisfactory, mainly because of boundary errors. Calculations of cloud cover, convergence asymptotes, and vertical motions appeared very realistic at 18 and 24 hr, with maximum values of the last reaching +3 and −7 cm/sec in midtroposphere and +4 cm/sec at the top of the surface friction layer at 900 mb. However, some spurious vertical advection of heat and moisture evidently took place during the adjustment from the initial data which produced rapid variations of some of the variables during the first 18 hr. An experiment was also conducted by using a constant initial mean relative humidity at each level, resulting in mass and wind fields which were very similar to the other forecasts. Some computations of energy transformation for the forecast period showed that in the vicinity of the ITCZ there was strong conversion of eddy potential to eddy kinetic energy and also smaller "barotropic" conversion from eddy to zonal kinetic energy. Away from the ITCZ, conversion rates are smaller and the barotropic process changes sign with zonal kinetic energy being converted to eddy kinetic energy, implying barotropic instability.

10.3.1 NPS Tropical Model

A ten-layer momentum-equation model with a horizontal mesh of about 150 km is under development at the Naval Postgraduate School by R. L. Elsberry and E. J. Harrison for the purposes of short-range tropical forecasting. The basic differential equations are essentially (10.7) to (10.10). To carry out the numerical integration, the equations are put in flux form. The differencing scheme is the same as that used in a middle-latitude prediction model to be described in Section 13.7; it requires virtually no smoothing and is computationally stable with a time step of six minutes. The parameterization of convection is similar to Kuo's (1965) formulation.

Only surface stress is included in the friction term which is expressed in the form

$$\boldsymbol{\tau} = \rho C_D V_0 \mathbf{V}_0, \tag{10.16}$$

where V_0 is the wind speed and the drag coefficient C_D is given by

$$C_D = \begin{cases} 0.0005 V^{1/2} & \text{for} \quad V \leq 15 \text{ m/sec}, \\ 2.6 \times 10^{-3} & \phantom{\text{for}} \quad V > 15 \text{ m/sec}. \end{cases} \tag{10.17}$$

The wind speed used in (10.17) is 80% of the 950-mb value.

Sensible heat is added only to the lowest layer with the flux assumed to be zero at 900 mb. At the sea surface,

$$H_S = \rho C_D V(T_w - T_0), \tag{10.18}$$

where T_0 is the 1,000-mb temperature and T_w the sea temperature. The flux convergence for the lowest 100-mb layer is therefore $FS = gH_S/\Delta p$. The sensible heat transfer is a significant factor in the formation of tropical storms. In some areas of high ocean temperatures the heating may increase the temperature of the lowest layer by as much as 3 °C/day.

A simple radiation scheme is employed which is a function of height only and results in a net cooling of the atmospheric column of about 1 °C/day.

Values range from -1.5 °C in the lower layers to 0.7 °C in the uppermost layer.

The criteria for the release of latent heat are as follows:

$$(a) \qquad -\frac{\partial}{\partial p}(\Phi + Lq + c_p T) < 0 \qquad (10.19)$$

at some level, which implies conditional instability.

$$(b) \qquad g^{-1}\int_{950}^{150} \nabla \cdot q\mathbf{V}\, dp + E_v > 0, \qquad (10.20)$$

which requires that the net moisture convergence including evaporation E_v be positive.

The fraction a of the grid area covered by active convection is determined by the ratio

$$a = \frac{C}{C_1 + C_2}. \qquad (10.21)$$

Here C is the surface evaporation plus the total integrated moisture convergence in the layer between the cloud base pressure B (950 mb) and the level p^* where the moist adiabat through the cloud base intersects the environment (p, T) sounding, that is,

$$C = \frac{\Delta t}{g}\int_B^{p^*} \nabla \cdot q\mathbf{V}\, dp + E_v \Delta t$$

and (10.22)

$$C_1 = \frac{-1}{g}\int_B^{p^*} (q_s - q)\, dp, \qquad C_2 = \frac{-1}{g}\int_B^{p^*} \frac{c_p}{L}(T_s - T)\, dp.$$

The subscript s denotes saturated values on the moist adiabat through the cloud base and Δt is the time step of the integration scheme. C_1 is the amount of moisture needed to saturate the environment air by increasing its specific humidity from its value q to q_s, which is the saturation value in the cloud at temperature T_s. Finally, C_2 is the amount of moisture that must be condensed to supply the latent heat necessary to raise the temperature of the environment T to the cloud temperature T_s. It follows that the average precipitation per unit mass of air, assuming all condensate precipitates is $ac_p(T_s - T)/L$; and the total precipitation in time Δt is

$$P = \frac{1}{g}\int_{p^*}^{950} \frac{ac_p}{L}(T_s - T)\, dp. \qquad (10.23)$$

Similarly, the total latent heat released is

$$\frac{1}{g}\int_{p^*}^{950} ac_p(T_s - T)\,dp \tag{10.24}$$

and the total convected moisture is

$$\frac{1}{g}\int_{p^*}^{950} a(q_s - q)\,dp. \tag{10.25}$$

To determine these quantities, the T, p values must be calculated along the moist adiabat from 950 mb to p^*, which is a rather time-consuming process. Only columns where there is net moisture convergence in the layer of 950 to 150 mb are considered.

If the criteria for latent heat release are met, the potential temperature is increased at each level up to p^* per time increment Δt by

$$\Delta\theta = a(T_s - T)\left(\frac{p_0}{p}\right)^{R/c_p}, \tag{10.26}$$

which is proportional to the fractional area covered by convective clouds. Above p^* and where $a \leqslant 0$, convection is not included. Similarly, the moisture added at each level per time increment Δt is

$$\Delta q = a(q_s - q). \tag{10.27}$$

In summary, at a gridpoint where convection is assumed to be taking place, the heat source due to convection gives the potential temperature change $\Delta\theta$ of (10.26) at each level in the layer $(950 - p^*)$. With respect to moisture at these convection gridpoints, the moisture convergence plus evaporation is spread over the layer $(950 - p^*)$, with resulting precipitation given by (10.23) plus a gain of specific humidity at each level in the cloud layer of Δq given by (10.27), which replaces the prediction equation (10.9).

The moisture gained by evaporation from the sea surface is based on the vertical flux convergence $\partial E_v/\partial p$, which is evaluated at 950 mb with $E_v = 0$ at 900 mb while at 1000 mb

$$E_v = \rho C_D V(q_w - q_0). \tag{10.28}$$

Here q_0 is linearly extrapolated to 1,000-mb from values at 850 and 950 mb. Sea-surface temperatures are obtained from climatic means. The evaporation is an important source of energy for tropical disturbances and may be overestimated in the vicinity of a hurricane in view of the observations by Landis and Leipper (1965), which showed that the ocean-surface temperatures dropped sharply as a tropical storm passed over.

The Elsberry–Harrison model is currently being tested for the southwestern Atlantic on a 32 × 15 array from about 11N to 31N. Cyclic east-west boundary conditions are imposed, with no-flux conditions at the north-south boundaries. The initialization scheme developed by Elsberry is based on a diagnostic balanced system. The principal feature of the synoptic situation is a nondeveloping tropical storm with maximum winds of about 30 knots. The 24-hr forecast handled the storm very well, with a good forecast of movement and only a slight decrease of intensity to maximum winds of 25 knots.

10.4 ARAKAWA'S PARAMETERIZATION OF CUMULUS CONVECTION

In the Arakawa-Mintz global model for simulating the general circulation and climate there are two kinds of condensation—large-scale condensation and convection. There are several types of convection as shown in Figure 10.1. Here C is the total upward mass flux from the boundary layer into the clouds; $(\eta - 1)C$ is the horizontal mass flux of surrounding air into the clouds (when $\eta < 1$, detrainment occurs), ηC is the upward mass flux in the clouds at level 2, E is the turbulent mass exchange between layer I and the boundary layer B, D is the turbulent mass exchange between B and the thin surface layer, and the M's represent large-scale mass convergence which obtains for all three cloud types.

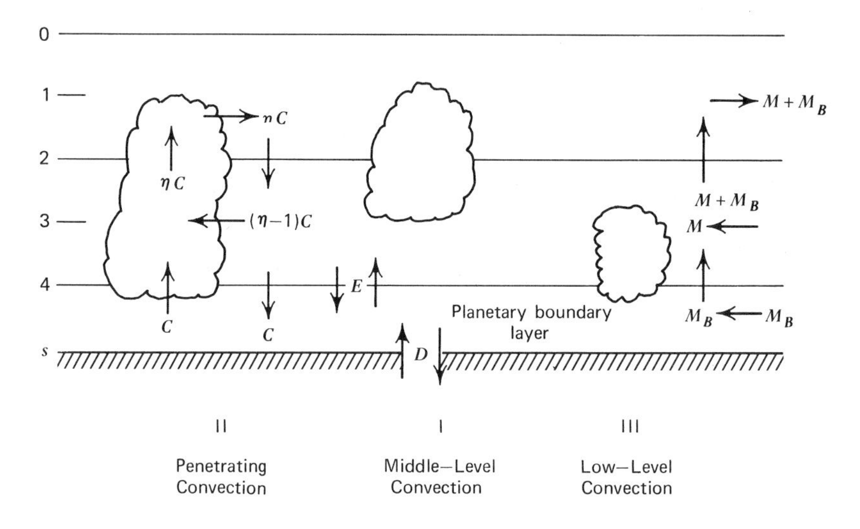

Figure 10.1. Schematic representation of the mass flux of cloud air, entrained air, boundary air, and large-scale mass convergence.

When entrainment exists, $\eta > 1$ and the total energy H_C, as defined in (10.3), is

$$H_C = \frac{CH_B + (\eta - 1)CH_3}{C + (\eta - 1)C} = H_3 + \frac{1}{\eta}(H_B - H_3). \qquad (10.29)$$

At level 1, H_C is given by

$$H_C = c_p T_{C1} + gz_1 + Lq^*_{C1},$$

where q^*_{C1} is the saturation specific humidity at the cloud temperature T_{C1}. For convenience define

$$H^*_1 = c_p T_1 + gz_1 + Lq^*_1,$$

where q^*_1 is the saturation specific humidity at temperature T_1 of the air outside the clouds at level 1. Then

$$T_{C1} - T_1 = \frac{H_C - H^*_1}{c_p(1 + \gamma_1)}, \qquad (10.30)$$

where

$$\gamma_1 = \frac{L}{c_p}\left(\frac{\partial q^*}{\partial T}\right)_1 = \frac{L}{c_p}\frac{(q^*_{C1} - q^*_1)}{(T_{C1} - T_1)}. \qquad (10.31)$$

It is apparent that for the clouds to be buoyant, that is, for $T_{C1} > T_1$, that

$$H_C > H^*_1, \qquad (10.32)$$

which is the *condition* for *convection to penetrate to the upper layer*. When there is detrainment at level 3,

$$H_C = H_B, \qquad \eta \leq 1.$$

Similarly,

$$T_{C3} - T_3 = \frac{H_B - H^*_3}{c_p(1 + \gamma_3)}.$$

Thus for convection to start in the boundary layer,

$$H_B > H^*_3. \qquad (10.33)$$

Based on these concepts, Arakawa established criteria for the three types of convection as shown in Table 10.1.

Table 10.1.

I. Middle-Level Convection	II. Penetrating Convection	III. Low-Level Convection
$H_3 > H^*_1$	$H_B > (H^*_1, H^*_3) > H_3$	$H^*_1 > H_B > H^*_3$
$\eta \to \infty$	$\eta = \dfrac{H_B - H_3}{H^*_1 - H_3}$	$\eta = 0$

The conditions on H follow from (10.32) and (10.33), and from the criterion for convectively unstable air cited earlier.

Consider first middle-level convection, type I. It is desired to have more than mere conditional instability which would correspond to $H_1^* < H_3^*$. The condition being imposed requires that there be sufficient moisture at level 3 so that H_3 exceeds H_1^*, which ensures that a parcel rising moist adiabatically will be warmer than the environment air at level 1. This requires a relative humidity at level 3 of at least

$$r_3 = \frac{q_3}{q_3^*} = 1 - \frac{1}{Lq_3^*}[(s_1 - s_3)_m - (s_1 - s_3)],$$

where $s = c_pT + gz$, and the subscript m denotes the moist adiabatic stratification. Note that the conditions $H_3^* > H_3$ and $H_1^* > H_1$ are automatically fulfilled, and that $H_3^* = H_1^*$ corresponds to a moist adiabatic lapse rate for the environment. In type I the cloud base value of H is taken to be $H_C = H_3$ so that $1/\eta$ is negligible. Formally, η may be taken to be infinite in (10.29).

Penetrating convection, type II, occurs where conditional instability exists between the boundary and level 3 but not between levels 3 and 1, so that the cloud penetrates a stable layer between levels 3 and 1. Consequently, $H_B > H_1^*$ as required by (10.32) and $H_B > H_3^*$, which is needed for convection to start in the boundary layer. However, for the stable layer (3-1), $H_3^* < H_1^*$.

For type III convection, H_B must exceed H_3^* in order to start convection from the boundary layer [see (10.33)]. Also the convection does not reach level 1; hence $H_1^* > H_B^*$. Thus the condition for low-level convection is

$$H_1^* > H_B > H_3^*.$$

The value of $\eta = 0$ represents no flux of cloud to level 1.

The prognostic equations for specific humidity with types I and II convection, as suggested by Figure 10.1 are

$$\frac{\Delta p}{g}\frac{\partial q_1}{\partial t} = \eta C(q_{C1} - q_2) + (M + M_B)q_2 - (M + M_B)q_1, \quad (10.34)$$

where, from (10.30) and (10.31),

$$q_{C1} = q_1^* + \frac{\gamma_1(H_C - H_1^*)}{L(1 + \gamma_1)},$$

where q_{C1} is the saturated cloud value. Similarly,

$$\frac{\Delta p}{g}\frac{\partial q_3}{\partial t} = \eta C\left[q_2 - q_3 + \frac{1}{\eta}(q_3 - q_4)\right] + M_B(q_4 - q_2) + M(q_3 - q_2) + E(q_B - q_3), \quad (10.35)$$

$$\frac{\Delta p_B}{g}\frac{\partial q_B}{\partial t} = C(q_4 - q_B) + M_B(q_B - q_4) + D(q_s - q_B) - E(q_B - q_3). \quad (10.36)$$

Adding these three prognostic equations gives

$$\frac{1}{g}\frac{\partial}{\partial t}[\Delta p(q_1 + q_3) + \Delta p_B q_B] = -\eta C\left[q_3 - q_{C1} + \frac{1}{\eta}(q_B - q_3)\right] + M_B(q_B - q_1) + M(q_3 - q_1) + D(q_s - q_B).$$

The first term (in brackets) on the right is the moisture loss of the environment to the clouds and gives the condensation. The next two give the net moisture supply by large-scale mass convergence, and the last term involving D represents the gain of moisture by evaporation.

Similarly, the prognostic equations for the temperature of the environment are obtained from the budgets for s as follows:

$$\frac{\Delta p}{g}\frac{\partial s_1}{\partial t} = \eta C\left[\frac{1}{1 + \gamma_1}(H_C - H_1^*) + (s_1 - s_2)\right] + (M + M_B)(s_2 - s_1), \quad (10.37)$$

$$\frac{\Delta p}{g}\frac{\partial s_3}{\partial t} = \eta C\left[(s_2 - s_3) + \frac{1}{\eta}(s_3 - s_4)\right] - M_B(s_2 - s_4) - M(s_2 - s_3) + E(s_B - s_3), \quad (10.38)$$

$$\frac{\Delta p_B}{g}\frac{\partial s_B}{\partial t} = C(s_4 - s_B) - M_B(s_4 - s_B) - E(s_B - s_3) + D(s_s - s_B). \quad (10.39)$$

The quantity s is essentially proportional to potential temperature θ, which may be substituted for s as a reasonable approximation in the foregoing equations.

For type III convection the equations are:

$$\frac{\Delta p}{g}\frac{\partial q_1}{\partial t} = (M + M_B)(q_2 - q_1),$$

$$\frac{\Delta p}{g}\frac{\partial q_3}{\partial t} = C(q_B - q_4) + M_B(q_4 - q_2) + M(q_3 - q_2) + E(q_B - q_3),$$

$$\frac{\Delta p_B}{g}\frac{\partial q_B}{\partial t} = -C(q_B - q_4) + M_B(q_B - q_4) + D(q_s - q_B) - E(q_B - q_3).$$

No precipitation is permitted in type III convection so that the condensed moisture in the cloud evaporates in layer I and the heat of vaporization is subtracted from the layer.

The corresponding equations for $\partial s/\partial t$ for each level are identical in form to the foregoing equations except that s replaces q.

To apply these formulas, an appropriate value of ηC must be determined. For example, with type I convection, (10.34), (10.35), (10.37), and (10.38) may be combined, omitting the mass convergence terms, to give an equation for $\partial(H_3 - H_1^*)/\partial t$. Assuming $H_3 - H_1^*$ decreases exponentially with time, the following expression for ηC is obtained, based on an e-folding time:

$$\eta C = \frac{1}{\tau}\frac{\Delta p}{g}\frac{1 + \gamma_1}{2 + \gamma_1}\frac{H_3 - H_1^*}{H_3 - H_1^* + \frac{1}{2}(1 + \gamma_1)(s_1 - s_2)},$$

where τ is the time scale of the cumulus convection. Values of τ of 1 hr gave satisfactory results when used in the Mintz-Arakawa general circulation model. A parameterization scheme adopted from the one described here is currently in use in the Navy's operational primitive equation model.

chapter eleven

The σ System of Coordinates

11.1 THE σ COORDINATE SYSTEM

The (x, y, p, t) coordinate system possesses certain advantages over the (x, y, z, t) system, particularly with respect to the representation of the horizontal pressure force. However, the lower boundary condition offers some difficulties and is usually approximated over level terrain by assuming $\omega \doteq 0$, and over uneven terrain by $\omega \doteq \mathbf{V} \cdot \nabla p_t$, where p_t is the pressure at terrain level.

The σ system of coordinates, devised by N. A. Phillips, which is closely related to the p system, avoids these difficulties at the lower boundary. There are several variations on this system; however, the differences are minor and the resulting equations are essentially the same. Define a new vertical coordinate σ as follows:

$$\sigma = \frac{p}{\pi}, \tag{11.1}$$

where p is the pressure at any level and π is the surface pressure. It may be noted that

$$\sigma = 0 \qquad \text{at} \qquad p = 0,$$

and

$$\dot{\sigma} = 0 \qquad \text{at} \qquad p = \pi, \qquad \text{and} \qquad p = 0.$$

It is readily verified that for an arbitrary function A (see Section 1.7 for procedure),

$$\left(\frac{\partial A}{\partial S}\right)_p = \left(\frac{\partial A}{\partial S}\right)_\sigma + \frac{\partial A}{\partial \sigma}\frac{\partial \sigma}{\partial S}, \tag{11.2a}$$

$$\frac{\partial A}{\partial p} = \frac{\partial \sigma}{\partial p}\frac{\partial A}{\partial \sigma} = \frac{1}{\pi}\frac{\partial A}{\partial \sigma}, \tag{11.2b}$$

where S represents x, y, or t. Furthermore, since σ is a function of (x, y, p, t), we can write

$$\left(\frac{\partial \sigma}{\partial S}\right)_p = -\frac{p}{\pi^2}\frac{\partial \pi}{\partial S} = -\frac{\sigma}{\pi}\frac{\partial \pi}{\partial S}. \tag{11.3}$$

By taking A in (11.2) to be the geopotential Φ, it becomes apparent that the vector equation of motion is

$$\frac{\partial \mathbf{V}}{\partial t} + (\mathbf{V}\cdot\nabla_\sigma)\mathbf{V} + \dot{\sigma}\frac{\partial \mathbf{V}}{\partial \sigma} + \nabla_\sigma\Phi - \frac{\sigma}{\pi}\frac{\partial \Phi}{\partial \sigma}\nabla\pi + f\mathbf{k}\times\mathbf{V} + \mathbf{F} = 0, \tag{11.4}$$

where the del operator is two-dimensional. Also according to (11.2b),

$$\frac{\partial \Phi}{\partial p} = \frac{1}{\pi}\frac{\partial \Phi}{\partial \sigma} = -\frac{RT}{p} = -\alpha. \tag{11.5}$$

From the preceding equation, it follows that the hydrostatic equation in the σ system is

$$\frac{\partial \Phi}{\partial \sigma} = -\frac{RT}{\sigma} = -\frac{p\alpha}{\sigma} = -\pi\alpha. \tag{11.6}$$

Differentiating (11.1) with respect to t gives

$$\omega = \frac{dp}{dt} = \pi\dot{\sigma} + \sigma\dot{\pi}. \tag{11.7}$$

Next, the continuity equation in the p system

$$\nabla_p\cdot\mathbf{V} + \frac{\partial \omega}{\partial p} = 0$$

will be transformed into the σ system by application of (11.2), giving

$$\nabla_\sigma\cdot\mathbf{V} - \frac{\sigma}{\pi}\left(\frac{\partial \pi}{\partial x}\frac{\partial u}{\partial \sigma} + \frac{\partial \pi}{\partial y}\frac{\partial v}{\partial \sigma}\right) + \frac{1}{\pi}\frac{\partial}{\partial \sigma}(\pi\dot{\sigma} + \sigma\dot{\pi}) = 0. \tag{11.8}$$

Expanding the last expression gives

$$\frac{\partial \dot{\sigma}}{\partial \sigma} + \frac{\dot{\pi}}{\pi} + \frac{\dot{\sigma}}{\pi}\frac{\partial \pi}{\partial \sigma} + \frac{\sigma}{\pi}\frac{\partial \dot{\pi}}{\partial \sigma}. \tag{11.9}$$

The last term of (11.9), in turn, may be expanded to

$$\frac{\sigma}{\pi}\frac{\partial \dot{\pi}}{\partial \sigma} = \frac{\sigma}{\pi}\frac{\partial}{\partial \sigma}\left(\frac{\partial \pi}{\partial t} + u\frac{\partial \pi}{\partial x} + v\frac{\partial \pi}{\partial y} + \dot{\sigma}\frac{\partial \pi}{\partial \sigma}\right). \tag{11.10}$$

Since $\partial\pi/\partial\sigma$ is identically zero, the substitution of (11.9) and (11.10) into (11.8) leads to

$$\nabla \cdot \mathbf{V} + \frac{\partial \dot{\sigma}}{\partial \sigma} + \frac{1}{\pi}\frac{d\pi}{dt} = 0 \tag{11.11}$$

or

$$\nabla \cdot (\pi \mathbf{V}) + \frac{\partial(\pi\dot{\sigma})}{\partial \sigma} + \frac{\partial \pi}{\partial t} = 0, \tag{11.12}$$

where the subscript σ has been omitted for simplicity. These are equivalent forms of the continuity equation in the σ system. If the latter form is integrated in the vertical (with respect to σ), there results

$$\pi\dot{\sigma} = -\int_0^\sigma \nabla \cdot (\pi \mathbf{V})\, d\sigma - \sigma\frac{\partial \pi}{\partial t}, \tag{11.13}$$

which is an equation for the vertical velocity $\dot{\sigma}$. If the upper limit of integration σ is taken to be 1, then $\dot{\sigma}$ is also zero there, and (11.13) becomes a *local tendency equation* for the *surface pressure:*

$$\frac{\partial \pi}{\partial t} = -\int_0^1 \nabla \cdot (\pi \mathbf{V})\, d\sigma. \tag{11.14}$$

Combination of (11.7) and (11.13) gives a diagnostic equation for ω:

$$\omega = \sigma \mathbf{V} \cdot \nabla \pi - \int_0^\sigma \nabla \cdot (\pi \mathbf{V})\, d\sigma. \tag{11.15}$$

Finally, the thermodynamic equation can be written in several ways. First consider the first law of thermodynamics in the conventional form

$$Q = c_p\frac{dT}{dt} - \frac{RT}{p}\frac{dp}{dt}, \tag{11.16}$$

where Q is the rate of heating. Utilizing (11.5) and substituting for T in (11.16) gives

$$Q = -\frac{c_p}{R}\frac{d}{dt}\left(\sigma\frac{\partial \Phi}{\partial \sigma}\right) - \frac{RT}{p}\omega. \tag{11.17}$$

An alternative form in terms of potential temperature is

$$\frac{\partial \theta}{\partial t} + \mathbf{V} \cdot \nabla \theta + \dot{\sigma}\frac{\partial \theta}{\partial \sigma} = \frac{\theta}{c_p T} Q. \tag{11.18}$$

The equation of horizontal motion, (11.4), the hydrostatic equation (11.6), the equation of continuity (11.12), and the thermodynamic equation (11.17) or (11.18) constitute a complete system of equations for dry air in terms of the σ system. In addition, the equation for local surface pressure tendency, (11.14), and an equation for "vertical velocity" $\dot{\sigma}$, namely, (11.13), are useful for purposes of numerical prediction.

If moisture is to be included, (9.4) is easily expressed in the σ coordinate system as follows:

$$\frac{\partial q}{\partial t} + \mathbf{V} \cdot \nabla q + \dot{\sigma} \frac{\partial q}{\partial \sigma} = \frac{S}{\rho}.$$

Multiplying this equation by π, the continuity equation (11.12) by q, and adding the two resulting equations gives the flux form

$$\frac{\partial \pi q}{\partial t} + \nabla \cdot (\pi q \mathbf{V}) + \frac{\partial}{\partial \sigma} \pi q \dot{\sigma} = \frac{\pi S}{\rho}.$$

If the only sources of moisture are evaporation E_v and precipitation P (expressed in total grams of water per unit area and time), then vertical integration of the foregoing equation yields a form which clearly shows the local sources of moisture to be evaporation and horizontal flux convergence while precipitation is a sink.

$$\frac{\partial}{\partial t} \int_0^1 \pi q \, d\sigma + \nabla \cdot \int_0^1 \pi q \mathbf{V} \, d\sigma = g(E_v - P).$$

With these processes, the diabatic term Q in (11.18) must include the transfer of latent heat as discussed in Chapters 9 and 10.

11.2 ENERGY RELATIONS

Combination of the adiabatic form of the thermal equation (11.18) and the equation of continuity, (11.12), leads directly to the result

$$\frac{\partial (\pi \theta)}{\partial t} + \nabla \cdot (\pi \theta \mathbf{V}) + \frac{\partial}{\partial \sigma} \pi \theta \dot{\sigma} = 0. \tag{11.19}$$

If this equation is integrated over the entire atmosphere both laterally and vertically (from $\sigma = 1$ to $\sigma = 0$), assuming no flux across the boundaries, it follows from the divergence theorem that

$$\frac{\partial}{\partial t} \overline{\pi \theta} = 0, \tag{11.20}$$

where the bar represents the integral mean over the entire atmosphere.

An equation for the kinetic-energy change is obtainable by taking the dot product of the vector velocity $\mathbf{V}$ with each term of (11.4), giving, neglecting friction

$$\frac{\partial(\frac{1}{2}V^2)}{\partial t} + \mathbf{V}\cdot\nabla(\tfrac{1}{2}V^2) + \dot\sigma\,\frac{\partial(\frac{1}{2}V^2)}{\partial\sigma} + \mathbf{V}\cdot\nabla\Phi - \frac{\sigma}{\pi}\frac{\partial\Phi}{\partial\sigma}\,\mathbf{V}\cdot\nabla\pi = 0. \quad (11.21)$$

Multiplying this equation by π and adding the continuity equation (11.12) after multiplication by $\frac{1}{2}V^2$ leads to the result,

$$\frac{\partial}{\partial t}(\tfrac{1}{2}\pi V^2) + \nabla\cdot(\tfrac{1}{2}\pi V^2\mathbf{V} + \pi\Phi\mathbf{V}) + \frac{\partial}{\partial\sigma}(\tfrac{1}{2}\pi V^2\dot\sigma + \pi\Phi\dot\sigma)$$

$$+ \frac{\partial\pi}{\partial t}\frac{\partial}{\partial\sigma}\Phi\sigma - \left(\sigma\frac{\partial\pi}{\partial t} + \sigma\mathbf{V}\cdot\nabla\pi + \pi\dot\sigma\right)\frac{\partial\Phi}{\partial\sigma} = 0 \quad (11.22)$$

The first law of thermodynamics, (11.16), may be rewritten by expanding dT/dt and dp/dt;

$$\frac{\partial(c_pT)}{\partial t} + \mathbf{V}\cdot\nabla(c_pT) + \dot\sigma\,\frac{\partial(c_pT)}{\partial\sigma}$$

$$- \alpha\left[\frac{\partial(\pi\sigma)}{\partial t} + \mathbf{V}\cdot\nabla(\pi\sigma) + \dot\sigma\,\frac{\partial(\pi\sigma)}{\partial\sigma}\right] = 0. \quad (11.23)$$

Here adiabatic flow has been assumed and p is replaced by $\pi\sigma$ from (11.1). Multiplying (11.23) by π and adding it to the continuity equation (11.12) gives

$$\frac{\partial(c_p\pi T)}{\partial t} + \nabla\cdot(c_p\pi T\mathbf{V}) + \frac{\partial}{\partial\sigma}(c_p\pi T\dot\sigma)$$

$$- \pi\alpha\left[\frac{\partial(\pi\sigma)}{\partial t} + \mathbf{V}\cdot\nabla(\pi\sigma) + \dot\sigma\,\frac{\partial(\pi\sigma)}{\partial\sigma}\right] = 0. \quad (11.24)$$

By virtue of (11.5) the last terms of (11.22) and (11.24) are equal except for sign; hence combination of these two equations yields

$$\frac{\partial}{\partial t}[\pi(c_pT + \tfrac{1}{2}V^2)] + \nabla\cdot[\pi(c_pT + \tfrac{1}{2}V^2 + \Phi)\mathbf{V}]$$

$$+ \frac{\partial}{\partial\sigma}[\dot\sigma\pi(c_pT + \tfrac{1}{2}V^2 + \Phi)] + \frac{\partial\pi}{\partial t}\frac{\partial(\Phi\sigma)}{\partial\sigma} = 0. \quad (11.25)$$

Application of the divergence theorem shows that if there is *no flux* across the *boundaries* of the atmosphere, the sum of the *potential* and *kinetic* energies is *conserved* for an *adiabatic, frictionless* atmosphere, that is,

$$\frac{\partial}{\partial t}\int \pi(c_pT + \tfrac{1}{2}V^2)\,dx\,dy\,d\sigma = 0. \qquad (11.26)$$

Note that $\rho\,dx\,dy\,dz = g^{-1}\pi\,dx\,dy\,d\sigma$.

When (11.25) is integrated vertically, a term $\Phi_G\,\partial\pi/\partial t$ results from the last term. However, as shown in Section 4.3, when the ground is not at sea level, the potential energy of a column of air also includes the term $(\Phi p)_G$, which should appear in the local change in potential energy as $(\Phi\,\partial p/\partial t)_G$ on the left side of (11.25). Thus the two terms exactly cancel one another when (11.25) is integrated vertically, leaving (11.26) as correct in the mean.

An important implication of theorem (11.26) will now be discussed. The differential energy relations (11.22), (11.24), (11.25), and (11.26) were derived directly from the hydrodynamical system of equations (motion, thermodynamics, and continuity), namely, (11.4), (11.12), and (11.16), and thus constitute constraints on this system with respect to energy. If the system (11.4), (11.12), (11.16) is utilized for numerical prediction, care should be taken that the finite difference analogues of this system fulfill the same constraints, as pointed out by Lorenz and Arakawa. This does not happen automatically even if the finite difference analogues for the derivatives are mutually consistent with respect to truncation error, computational instability, etc. Finally, note that, when (11.25) was integrated to obtain (11.26), the vertical derivative vanished independently of the horizontal divergence term.

chapter twelve

Advection Schemes and Nonlinear Instability

12.1 INTRODUCTION, NONLINEAR INSTABILITY

In Chapter 5 the problem of computational instability was discussed in connection with the numerical solution of linear partial differential equations. The Lax theorem, together with the condition Eq. (5.55) for avoiding *linear computational instability*, will assure convergence of a difference scheme. As might be expected, experience has shown that such a condition is also required in the numerical solution of nonlinear equations. However, in dealing with nonlinear equations there is another problem called *aliasing* which can lead to erroneous amplification, especially of very short waves. The *existence* of this *nonlinear instability*, which was first noted and discussed by Norman Phillips (1959), does not depend on the relationship between the time increment and grid length, and consequently is not removed by simply decreasing either of these quantities.

To provide some insight into the origin of this phenomenon, consider a simplified barotropic model

$$\frac{\partial \zeta}{\partial t} = -\mathbf{V} \cdot \boldsymbol{\nabla} \zeta \tag{12.1}$$

or

$$\frac{\partial \nabla^2 \psi}{\partial t} = -J(\psi, \nabla^2 \psi) \tag{12.2}$$

where

$$\mathbf{V} = \mathbf{k} \times \boldsymbol{\nabla} \psi \qquad \text{and} \qquad \zeta = \nabla^2 \psi. \tag{12.3}$$

Now suppose that the stream function ψ at any specific time has been expressed as the double Fourier series

$$\psi = \sum_{\substack{m=0 \\ n=1}}^{\infty} \left(a_{mn} \cos \frac{2\pi m x}{L_x} + b_{mn} \sin \frac{2\pi m x}{L_x} \right) \sin \frac{\pi n y}{L_y}, \tag{12.4}$$

where L_x and L_y are the dimensions of the region. Substitution of (12.4) into the right side of (12.2), together with use of trigonometric identities, shows that the interaction of two waves with numbers m_1, n_1 and m_2, n_2 in the Jacobian term will produce a time rate of change of vorticity in four waves with numbers

$$\begin{aligned} &m_1 + m_2, \quad && n_1 + n_2, \\ &m_1 + m_2, && n_1 - n_2, \\ &m_1 - m_2, && n_1 + n_2, \\ &m_1 - m_2, && n_1 - n_2. \end{aligned} \tag{12.5}$$

Thus nonlinear interaction results in the transfer of vorticity and energy between different parts of the spectrum. In fact, it is apparent that wavelengths which may not have been present initially can eventually appear.

When the region is discretized into a finite mesh of points, as is done in the finite difference method of numerical weather prediction, only a finite number of waves can be represented. For example, consider a one-dimensional set of $(M + 1)$ points $0, d, 2d, \ldots, id, \ldots, Md$, for which the function ψ is known. Then it is normally possible to determine $(M + 1)$ coefficients A_p such that a series of cosine (and/or sine) functions represents the function over the given set of points, e.g.,

$$\psi_i = \sum_{p=0}^{M} A_p \cos \frac{pi\pi}{M} = \sum_{p=0}^{M} A_p \cos \frac{2p\pi x_i}{2Md}.$$

Thus the function ψ has been expressed as a sum of M harmonics, one having the *maximum wavelength* $2Md$ corresponding to $p = 1$, and one with the *minimum wavelength* of $2d$ corresponding to $p = M$. Hence there is a maximum and a minimum wavelength that can be resolved from known values of ψ over the finite sequence of points, and *wavelengths outside the range of these maximum and minimum values will be misrepresented in terms of the permissible harmonics.*

Return now to (12.2) and suppose it is desired to solve this equation on the following finite set of points covering a two-dimensional region of length L_x and width L_y:

$$\begin{aligned} &x = id, \quad && i = 0, \ldots, I, \quad && Id = L_x, \\ &y = jd, && j = 0, \ldots, J, && Jd = L_y, \end{aligned}$$

where d is the grid distance. For simplicity assume that the streamfunction ψ and vorticity ζ vanish at the northern and southern boundaries, and furthermore that ψ is periodic in x. Thus

$$\begin{aligned} \psi(x, 0) &= \psi(x, Jd) = 0, \\ \psi(0, y) &= \psi(Id, y). \end{aligned} \tag{12.6}$$

This leaves $I(J - 1)$ arbitrary points or, stating it another way, there are $I(J - 1)$ degrees of freedom with respect to gridpoint values of ψ.

The stream function can now be represented by a finite series of the form (12.4), with $I(J - 1)$ degrees of freedom in the coefficients a_{mn} and b_{mn} as follows:

$$\psi_{ij} = \sum_{m=0}^{I/2} \sum_{n=1}^{J-1} \left(a_{mn} \cos \frac{2\pi m i}{I} + b_{mn} \sin \frac{2\pi m i}{I} \right) \sin \frac{\pi n j}{J}. \tag{12.7}$$

This series appears to have $(I + 2)(J - 1)$ arbitrary coefficients. However, the b_{0n} may be taken as zero since the sine term in the parentheses vanishes at $x = 0$, which reduces the number of arbitrary coefficients by $J - 1$. The cyclic requirement (12.6) further diminishes the number of degrees of freedom in the coefficients by another $J - 1$. These two conditions reduce the number of degrees of freedom in the coefficients to $I(J - 1)$, equaling that of the gridpoints.

It is clear from the finite series (12.7) that the maximum resolvable wave number in the x direction is $I/2$, which corresponds to a wavelength of $2d$. Hence if two waves with numbers m_1 and m_2 interact and the sum $(m_1 + m_2)$ exceeds $I/2$, that particular resulting wave cannot be resolved and will be misrepresented in terms of permissible waves. This misrepresentation is known as *aliasing*. Suppose that $m_1 + m_2 = I - s$, then the cosine term of the series (12.7) for the wave resulting from the interaction is

$$\cos \frac{2\pi(I - s)i}{I} = \cos 2\pi i \cos \frac{2\pi i s}{I} + \sin 2\pi i \sin \frac{2\pi i s}{I} = \cos \frac{2\pi i s}{I}.$$

A similar result is obtained for the sine term. It is therefore clear that, if the nonlinear interaction produces a wave number that exceeds the maximum resolvable value, it will falsely appear as the number

$$s = I - (m_1 + m_2).$$

Now suppose, for example, that $m_1 = I/2$ and $m_2 = I/4$, then $s = I/4$ and there has been a feedback of energy into the four-grid-length wave. Repetition of this process can eventually cause a rapid increase in the kinetic energy of such short wavelengths, which is unrelated to any real physical process. This spurious computational phenomenon is known as *nonlinear instability*. It is evident that for such a feedback to occur for a particular wave, say to m_1,

$$I - (m_1 + m_2) = m_1,$$

or

$$2m_1 = I - m_2.$$

Furthermore, since $m_2 \leq I/2$, then $m_1 \geq I/4$. Hence for such spurious amplification to occur, the wavelengths must be between $2d$ and $4d$.

The preceding discussion indicates how nonlinear instability may develop unless some appropriate measures are taken to prevent it. Phillips (1959) gave a simple example showing that the ordinary centered difference scheme applied to a barotropic model does indeed lead to nonlinear instability. Obviously a mere reduction in the time or space increments will not eliminate this difficulty. Phillips further showed that the instability could be controlled in his general circulation model by periodical spatial smoothing of the prognostic fields, which was accomplished by removing wavelengths smaller than $4d$ by using a Fourier analysis. In fact, small-scale irregularities from various sources can be eliminated or reduced in some instances by one of several techniques: (*a*) by spatial or temporal smoothing; (*b*) by the inclusion of diffusion terms of the form $\nu\nabla^2 A$ in the predictive equation for A [see (10.15)]; or (*c*) by the implicit smoothing or selective damping inherent in certain finite difference schemes.

A simple type of explicit smoothing can be effected by applying the operator $(1 + k\nabla^2)$ to a field as follows:

$$\bar{A} = A + k\nabla^2 A, \qquad 0 < k \leq \tfrac{1}{4},$$

where ∇^2 is the finite difference Laplacian defined in Section 5.7. Repeated applications of this operator, which influences larger and larger scales, may be expressed in the form

$$\bar{A}^n = (1 + k\nabla^2)^n A.$$

An analogous formula for simple time smoothing would be

$$\tilde{A}(t) = A(t) + k_1[A(t + \Delta t) - 2A(t) + A(t - \Delta t)].$$

12.2 INTERCHANGE OF ENERGY BETWEEN WAVES

Some insight about the process of energy exchange between different wave components can be obtained by further consideration of the simplified barotropic vorticity equation (12.1). Assume the solution of (12.1) is expressible as a series of orthogonal functions

$$\psi = \sum_{i=1}^{n} \psi_i, \tag{12.8}$$

where the ψ_i are characteristic solutions of the following equation:

$$\nabla^2 \psi_i = -\mu_i^2 \psi_i, \tag{12.9}$$

and the eigenvalues μ_i are wave numbers. (An example of such a representation is the trigonometric series (12.4).) As shown earlier in Chapter 4, the mean kinetic energy over a global pressure surface, $\bar{K} = \frac{1}{2}\overline{(\nabla \psi)^2}$, and the mean squared vorticity, $\overline{\zeta^2} = \overline{(\nabla^2 \psi)^2}$, are conservative with respect to time for the barotropic vorticity equation (12.2). Because of the orthogonality of the functions (12.8), it may be shown that

$$\bar{K} = \tfrac{1}{2} \sum \overline{(\nabla \psi_i)^2} \equiv \sum \bar{K}_i, \tag{12.10a}$$

$$\overline{\zeta^2} = \sum \overline{(\nabla^2 \psi_i)^2} = 2 \sum \mu_i^2 \bar{K}_i. \tag{12.10b}$$

Since both $\overline{\zeta^2}$ and $\bar{K}$ are conserved, their ratio, which is essentially a kinetic-energy *weighted-average wave number is also conserved.*

Now suppose there are only two waves present, then (12.10*a*) and (12.10*b*) imply that there can be no transfer of energy, since the mean kinetic energy of each component must remain constant. Next consider the case of three wave components with decreasing wave numbers μ_1, μ_2, and μ_3 (hence increasing wavelengths). Then from (12.9) and (12.10), it follows that (omitting bars)

$$K_1 + K_2 + K_3 = C_1, \tag{12.11}$$

$$\mu_1^2 K_1 + \mu_2^2 K_2 + \mu_3^2 K_3 = C_2, \tag{12.12}$$

where C_1 and C_2 are constants. Elimination of K_1 and K_3 from the previous pair of equations gives

$$\begin{aligned} (\mu_1^2 - \mu_2^2) K_2 + (\mu_1^2 - \mu_3^2) K_3 &= \mu_1^2 C_1 - C_2, \\ (\mu_1^2 - \mu_3^2) K_1 + (\mu_2^2 - \mu_3^2) K_2 &= C_2 - \mu_3^2 C_1. \end{aligned} \tag{12.13}$$

Now since $\mu_1 > \mu_2 > \mu_3$, the coefficients of the K's and the right-hand members of (12.13) are positive. It follows that, if K_2 decreases, both K_1 and K_3 will increase, and vice versa. This result may be represented symbolically as

$$K_1 \leftarrow K_2 \rightarrow K_3 \qquad \text{or} \qquad K_1 \rightarrow K_2 \leftarrow K_3. \tag{12.14}$$

Thus with three waves there cannot be transport of energy in just one direction, that is, consistently toward the shortest or longest wavelengths. When there are more than three waves present, the interchange of energy between waves is more complex. However, in a general way, decreasing kinetic energy in intermediate wavelengths tends toward increases in the energy of the shorter and longer wavelengths and vice versa. Moreover, as shown by R. Fjortoft (1963), the *energy cannot flow consistently in one direction, say toward higher wave numbers; but rather such flow of energy is clearly limited and the higher the wave number, the more it is limited.* This was implied by the earlier statement that the average wave number is conserved.

It may be concluded, therefore, that a continual transfer of energy toward short wavelengths is not characteristic of the atmosphere and hence should certainly be avoided in a numerical prediction scheme. Thus the spurious transfer of energy to the very short waves by aliasing must be controlled. Methods of treating this problem will be the subject of subsequent sections.

12.3 FINITE DIFFERENCE APPROXIMATIONS FOR JACOBIANS

When only the rotational component is considered, the wind is expressible in terms of a stream function and the nonlinear advective terms can be written as Jacobians, for example, as in (12.2). It has also been noted that on a finite grid of points aliasing could lead to nonlinear computational instability. Arakawa (1966) has devised a number of finite difference schemes for nonlinear terms which prevent this false cascade of energy. His technique as applied to Jacobians will now be described.

It has been shown that both the mean squared vorticity and the mean kinetic energy are conserved with respect to the differential barotropic vorticity equation. However, this is no guarantee that these quantities will be conserved when an arbitrary finite difference scheme is imposed on the differential equation. Recall that these conservative properties were established for the barotropic model by multiplying (12.1) by ζ and ψ and integrating over a global isobaric surface. The corresponding conservation principles for a time-continuous space-difference scheme would be established by multiplying the difference approximation of the Jacobian in (12.2) for an arbitrary point (i, j) by ζ_{ij} or ψ_{ij} and summing over the entire grid. It is apparent, therefore, that a finite difference Jacobian J for which

$$\overline{\psi \mathbb{J}(\psi, \zeta)} = \overline{\zeta \mathbb{J}(\psi, \zeta)} = 0 \tag{12.15}$$

would preserve the mean kinetic energy and the mean square vorticity in the absence of time-truncation error.

As a guide for designing various finite difference Jacobians, note that the analytic Jacobian may be expressed alternatively as

$$\begin{aligned} J(A, B) = \frac{\partial A}{\partial x}\frac{\partial B}{\partial y} - \frac{\partial A}{\partial y}\frac{\partial B}{\partial x} &= \frac{\partial}{\partial x}\left(A\frac{\partial B}{\partial y}\right) - \frac{\partial}{\partial y}\left(A\frac{\partial B}{\partial x}\right) \\ &= \frac{\partial}{\partial y}\left(B\frac{\partial A}{\partial x}\right) - \frac{\partial}{\partial x}\left(B\frac{\partial A}{\partial y}\right). \end{aligned} \tag{12.16}$$

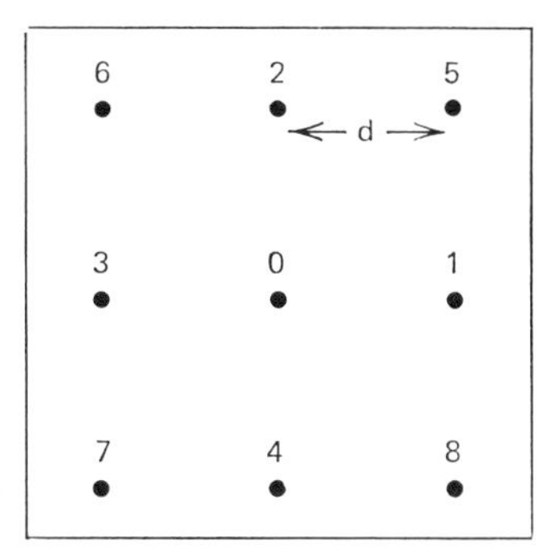

Figure 12.1. Nine-point grid block for various finite difference approximations of the Jacobian.

Each of the foregoing formulations may be placed in finite difference form based on the adjacent nine-point block of grid points, in the previous order, as follows:

$$J^{++} = \frac{1}{4d^2}[(A_1 - A_3)(B_2 - B_4) - (A_2 - A_4)(B_1 - B_3)], \tag{12.17a}$$

$$J^{+\times} = \frac{1}{4d^2}[A_1(B_5 - B_8) - A_3(B_6 - B_7) - A_2(B_5 - B_6) + A_4(B_8 - B_7)], \tag{12.17b}$$

$$J^{\times +} = \frac{1}{4d^2}[B_2(A_5 - A_6) - B_4(A_8 - A_7) - B_1(A_5 - A_8) + B_3(A_6 - A_7)]. \tag{12.17c}$$

The superscript + or × denotes the points from which the finite difference approximations for the derivatives of A and B (in that order) are formed. A + symbol indicates the use of the points 1, 2, 3, and 4, and the × symbol indicates the points 5, 6, 7, and 8. The first form (12.17*a*) has been widely used in short-range numerical weather prediction, but, as shown by Phillips (1959), it can give rise to nonlinear instability if the integration period is prolonged.

All of the Jacobians (12.17) are of the general bilinear form,

$$J_{ij}(\psi, \zeta) = \sum_{k,l,m,n}^{-1,0,+1} C_{i,j,k,l,m,n}\psi_{i+k,j+l}\zeta_{i+m,j+n}. \tag{12.18}$$

The coefficients C may be determined so that the finite difference approximation possesses as many of the properties of the analytic Jacobian as possible. Now if (12.18) is multiplied by ζ_{ij}, the left side represents the local rate of change of ζ_{ij}^2. In this case the right side may then be written as

$$\sum_{m,n}^{-1,0,+1} a_{i,j,i+m,j+n}\zeta_{i+m,j+n}\zeta_{ij}, \tag{12.19}$$

which shows that the rate of change of ζ^2 at the point (i, j) may be considered to be the consequence of interaction between the given point and adjacent points. Similarly, the change of ζ^2 at the point $(i + m, j + n)$ will receive a corresponding contribution from the point ζ_{ij}. It is evident that a net change of mean squared vorticity may be avoided if the gain at the point (i, j) due to interaction with the point $(i + m, j + n)$ is exactly balanced by a corresponding loss at $(i + m, j + n)$ due to point (i, j). This will occur if the coefficients fulfill the condition

$$a_{i,j,i+m,j+n} = -a_{i+m,j+n,i,j}. \tag{12.20}$$

The various Jacobians of (12.17) will now be examined with respect to the condition (12.20). Consider first the gain of square vorticity at the point 0 due to the point 1, and vice versa. With the usual form of the Jacobian (12.17*a*) the results are

$$\zeta_0 J_0^{++} \sim -\zeta_0\zeta_1(\psi_2 - \psi_4) + \text{other terms},$$
$$\zeta_1 J_1^{++} \sim \zeta_1\zeta_0(\psi_5 - \psi_8) + \text{other terms},$$

where the subscript on J denotes the point where the evaluation is made. These terms obviously *do not cancel*, showing that the ordinary central difference approximation does not conserve mean square vorticity. On the other hand, the following typical terms show that $J^{\times+}$ does conserve $\overline{\zeta^2}$ since the contributions at adjacent points exactly cancel one another:

$$\zeta_0 J_0^{+\times} \sim \zeta_0\zeta_5(\psi_1 - \psi_2) + \text{other terms},$$
$$\zeta_5 J_5^{+\times} \sim \zeta_5\zeta_0(\psi_2 - \psi_1) + \text{other terms}.$$

A similar examination shows that $J^{\times+}$ does not conserve $\overline{\zeta^2}$, namely,

$$\zeta_0 J_0^{\times+} \sim -\zeta_0\zeta_1(\psi_5 - \psi_8) + \text{other terms},$$
$$\zeta_1 J_1^{\times+} \sim \zeta_1\zeta_0(\psi_2 - \psi_4) + \text{other terms}.$$

However, a comparison of the terms comprising ζJ^{++} and $\zeta J^{\times+}$ shows that they are merely opposite in sign. It is evident, therefore, that their sum will conserve $\overline{\zeta^2}$. Hence the appropriate forms of the finite difference Jacobian which fulfill this integral property are

$$J^{+\times} \text{ and } \tfrac{1}{2}(J^{++} + J^{\times+}) \qquad (\text{conserve } \overline{\zeta^2}).$$

A similar treatment with multiplication of each of the Jacobians (12.17) by ψ (or from symmetry) shows that the mean kinetic energy is conserved by

$$J^{\times+} \text{ and } \tfrac{1}{2}(J^{++} + J^{+\times}) \qquad (\text{conserve } \overline{V^2}).$$

It follows from these two results, as discovered by A. Arakawa (1966), that both mean square vorticity and mean square kinetic energy will be conserved by the average of the three Jacobians of (12.17), namely,

$$J = \tfrac{1}{3}(J^{++} + J^{+\times} + J^{\times+}) \qquad \text{(Arakawa Jacobian).} \tag{12.21}$$

In addition, it has been shown that when the mean vorticity and mean kinetic energy are preserved, so is the average wave number. As a result of these conservation properties, the Arakawa Jacobian prevents the continued growth of very short waves, which is characteristic of nonlinear instability. Although aliasing is still present in the form of a phase error, this can be tolerated. In fact, the linear terms may also produce phase and amplitude errors, as described in Chapter 5. Lilly (1965) has given a rigorous proof by spectral methods that Arakawa's quadratic conserving scheme eliminates the type of nonlinear instability demonstrated by Phillips.

In Chapter 13 the numerical integration of the primitive equations of motion will be discussed including an application of Arakawa's technique for avoiding nonlinear instability in this system. Specifically, finite difference approximations will be given for treating general advective terms of the type $\mathbf{V} \cdot \nabla A$, where $\mathbf{V}$ is the total velocity and A is an arbitrary scalar.

The conservative properties of the Arakawa Jacobian have only been established here for the time continuous case. In other words, the product $-\zeta_0 J_0(\psi, \zeta)$ corresponds to $\frac{1}{2}\,\partial\zeta_0^2/\partial t$, which is conserved in the mean for the barotropic model. However, when the time derivative is approximated by finite differences, this property may not hold; for example, with centered time differences and the Arakawa Jacobian, it follows that

$$\frac{\overline{\zeta_{0n}(\zeta_{0n+1} - \zeta_{0n-1})}}{2\Delta t} = \overline{\zeta_{0n} J_{0n}} = 0.$$

Thus

$$\overline{\zeta_{0n}\zeta_{0n+1}} = \overline{\zeta_{0n}\zeta_{0n-1}}. \tag{12.22}$$

But this is not the correct conservation property for mean square vorticity, which for this case could be written as

$$\overline{\zeta^2_{0n+1}} = \overline{\zeta^2_{0n-1}}.$$

On the other hand, if $\zeta_{0n} = \frac{1}{2}(\zeta_{0n+1} + \zeta_{0n-1})$ in (12.22), the correct result would follow. However, it would take a rather complicated implicit scheme to make $\overline{\frac{1}{2}(\zeta_{0n+1} + \zeta_{0n-1})J_{0n}}$ vanish. Nevertheless the desired result would be achieved in practice with the central time difference scheme if ζ varies smoothly between successive time steps. Unfortunately this is not always the case and a sawtooth variation occurs when there is a *separation* of the solutions at the even and odd time steps. The possibility of such a separation is perhaps not too surprising if one examines, for example, the difference scheme, 5.12. Here it is seen that the value of F_{mn+1} (at time $n + 1$ and grid-point m) depends on F_{mn-1}, F_{m+1n} and F_{m-1n} but not on F_{mn}. Thus the solutions at adjacent time steps may become decoupled and may eventually lose coherence to the point where instability develops. Such decoupling may be counteracted by occasionally smoothing the solutions over several time steps and then restarting with a forward time step.

12.4 LAGRANGIAN METHODS

12.4.1 Introduction

It has been shown that the finite difference technique may lead to nonlinear instability when applied to the advective terms of the prediction equations. Simple spatial smoothing or special filters can sometimes control the erroneous amplification of very short waves; however a more effective method is Arakawa's scheme which prevents nonlinear instability by conserving the mean square of critical quantities. Another way of preventing the spurious growth of the meteorological variables by incorrectly evaluating the advective terms is to use a Lagrangian approach. The essence of this method is to determine at time t the position of the air particle which will arrive at a particular gridpoint at the future time $t + \Delta t$. The predicted value of a conservative meteorological variable at that gridpoint at $t + \Delta t$ is then simply taken to be the value at the previous position of the particle, as obtained by interpolation. If there are nonconservative physical processes acting on the particle while enroute, the effects of such sources or sinks must be added.

12.4.2 A Quasi-Lagrangian Scheme

Crowley (1968) has discussed several second- and fourth-order quasi-Lagrangian schemes, with the latter demonstrating excellent properties when applied to the advection of a circular pattern in a purely rotational wind field. Consider again the simple advective equation

$$\frac{\partial F}{\partial t} + u\frac{\partial F}{\partial x} = 0.$$

Recall that the analytic solution of this equation is $F(x - ut)$, where F is an arbitrary function. Hence the value of F at a distance $\delta x = u\,\Delta t$ upstream from the point $x = m\,\Delta x$ should give the value of F at x at time $(n + 1)\,\Delta t$. This is illustrated in Figure 12.2 for $u > 0$. Thus

$$F[m\,\Delta x, (n + 1)\,\Delta t] = F[(m + r)\,\Delta x, n\,\Delta t], \tag{12.23}$$

where

$$r = -\frac{u\,\Delta t}{\Delta x}.$$

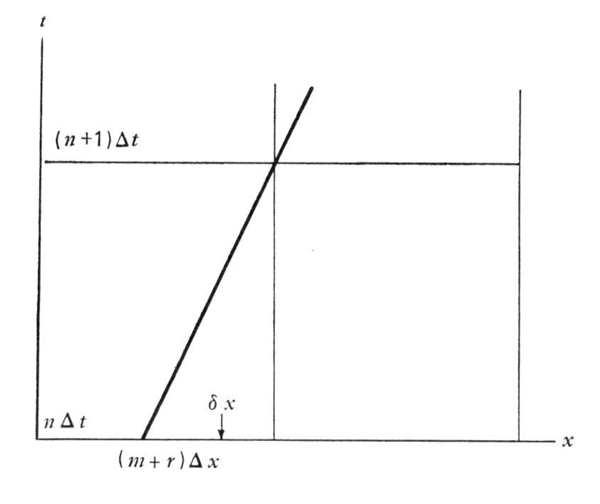

Figure 12.2. A quasi-Lagrangian scheme for horizontal advection.

The line through $[m\,\Delta x, (n+1)\,\Delta t]$ in Figure 12.2 is referred to as a *characteristic*. Next the value of F at the point $(m+r)\,\Delta x$ may be obtained by interpolation as follows:

$$F[(m+r)\,\Delta x, n\,\Delta t] = F_m^n - u\,\Delta t\,\frac{(F_{m+1}^n - F_{m-1}^n)}{2\,\Delta x} + \frac{(u\,\Delta t)^2}{2}\frac{(F_{m+1}^n - 2F_m^n + F_{m-1}^n)}{\Delta x^2}. \quad (12.24)$$

Equation 12-24 may also be looked upon as a truncated Taylor series with the derivatives approximated by centered differences. The foregoing scheme has errors of first order with respect to t and second order with respect to x. A more compact representation of (12.23) and (12.24) is

$$F_{mn+1} = (I - A)F_{mn}, \quad (12.25)$$

where I is the identity matrix (of order 1) and

$$AF_{mn} = u\,\Delta t\,\frac{(F_{m+1n} - F_{m-1n})}{2\,\Delta x} - \frac{(u\,\Delta t)^2}{2}\frac{(F_{m+1n} - 2F_{mn} + F_{m-1n})}{\Delta x^2}.$$

This scheme is stable for $|u\,\Delta t/\Delta x| \leq 1$, which implies that interpolation takes place between the points $m\,\Delta x$ and $(m-1)\,\Delta x$ (for $u > 0$), rather than extrapolation to a point beyond $(m-1)\,\Delta x$.

A similar treatment gives a fourth-order scheme with respect to x but a first-order scheme with respect to time, which is known as the Lax-Wendroff method. A two-step scheme, which is second order with respect to x and t, is

$$F_{mn+1} = (I - C)F_{mn}, \quad (12.26)$$

where

$$C = B(I - A), \qquad BF_m = \frac{u\,\Delta t}{2\,\Delta x}(F_{m+1} - F_{m-1}),$$

$$AF_m = \tfrac{1}{2}BF_m - \frac{(u\,\Delta t)^2}{8\,\Delta x^2}(F_{m+1} - 2F_m + F_{m-1}).$$

This two-step scheme is stable for $u\,\Delta t/\Delta x \leq 2$, but it does tend to damp disturbances. However, the damping is quite limited if $u\,\Delta t/\Delta x < 1$.

Two-dimensional calculations

A simple two-dimensional advection equation is

$$\frac{\partial F}{\partial t} + \mathbf{V} \cdot \nabla F = 0. \tag{12.27}$$

Marchuk's method of fractional time steps has been used for two-dimensional advection by C. E. Leith (1965) in his general circulation studies, namely

$$F_{lmn+1} = (I - A)(I - B)F_{lmn}. \tag{12.28}$$

Here A and B represent advection in the x- and y-directions, which is carried out in two steps as follows:

$$\begin{aligned} F^*_{lm} &= (I - B)F_{lmn} \qquad \text{(step 1)}, \\ F_{lmn+1} &= (I - A)F^*_{lm} \qquad \text{(step 2)}. \end{aligned} \tag{12.29}$$

This scheme is stable if each one-dimensional scheme is stable. It gave excellent results with a circular pattern being advected with a rotational wind field. It is an interesting fact that, if both operators A and B are applied simultaneously, computational instability occurs, that is,

$$F_{lmn+1} = (I - A - B)F_{lmn}. \qquad \text{(unstable)}.$$

Crowley also presented a fourth-order conservation scheme which was based on a flux form of (12.27), namely,

$$\frac{\partial F}{\partial t} + \nabla \cdot F\mathbf{V} - F\nabla \cdot \mathbf{V} = 0. \tag{12.30}$$

The difference equation for (12.30) may be written in the symbolic form

$$F_{lmn+1} = (I + D)(I - A)(I - B)F_{lmn}, \tag{12.31}$$

where D represents the divergence term, which is treated as a source superimposed on the advective process.

Both the second-order and fourth-order advection and conservation schemes were tested on a barotropic model at the Naval Postgraduate School.

The fourth-order schemes gave much smoother forecasts and retained the intensity of the synoptic features better than the ordinary central difference Jacobian; the latter, however, had slightly smaller rms errors at 500 mb. The apparent reason for this discrepancy is that the forecasts with the central difference Jacobian required smoothing which reduced the intensity of the centers and as a result also usually decreased the rms errors. This simply points up the fact that the rms error is not an adequate measure of forecast accuracy and must be augmented by other verification techniques.

12.4.3 Lagrangian Advective Schemes

Krishnamurti (1969) has successfully used a quasi-Lagrangian method in a multilevel primitive (Newtonian) equation model for tropical forecasting. Recall that the basic principle of the Lagrangian technique is to locate the position P of a parcel which will arrive at a particular gridpoint Q at the end of a time step Δt, as shown schematically in Figure 12.3. Then if the behavior of any dependent variable is governed by the simple advective equation (12.27), which implies that $dF/dt = 0$, the value of F at Q at time $(t + \Delta t)$ will be given by

$$F(Q, t + \Delta t) = F(P, t). \tag{12.32}$$

On the other hand, if the property F is nonconservative, that is, if

$$\frac{dF}{dt} = A,$$

the change in F enroute from P to Q due to the source A must be taken into account.

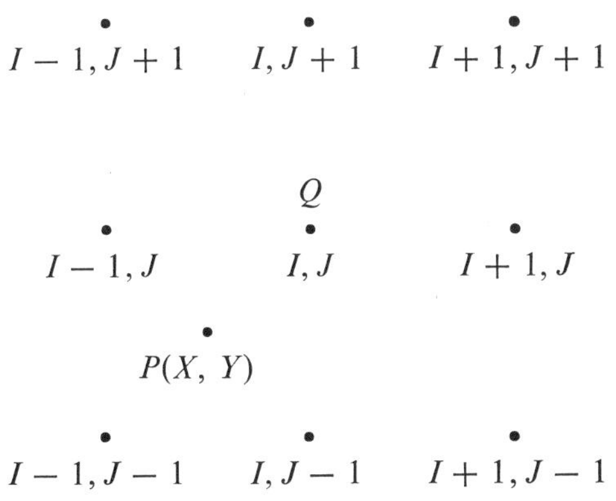

$I-1, J+1$ $\quad$ $I, J+1$ $\quad$ $I+1, J+1$

Q

$I-1, J$ $\quad$ I, J $\quad$ $I+1, J$

$P(X, Y)$

$I-1, J-1$ $\quad$ $I, J-1$ $\quad$ $I+1, J-1$

Figure 12.3. Interpolation grid.

Since advective terms need not be evaluated directly in this method, the nonlinear instability described earlier can be avoided. There are, however, several obvious difficulties involved in applying the method. First, since the dependent variables are carried only at gridpoints, an interpolation scheme must be used to determine their values between gridpoints, for example, at point P. In this connection, the point should lie within the interpolation mesh or the advective scheme is likely to become unstable. Second, the trajectory point P must be accurately determined. Finally, sources or sinks must be calculated and interpolated. Mather (1969) improved Krishnamurti's interpolation method through the use of a nine-point interpolation formula of the form

$$F(P) = \sum_{\substack{i=I-1\\ j=J-1}}^{\substack{j=J+1\\ i=I+1}} W_{ij} F_{ij}, \qquad (12.33)$$

where W_{ij} is the nine-point Lagrange polynomial

$$W_{ij} = \prod_{\substack{k=I-1\\ k\neq i}}^{k=I+1} \frac{x - x_k}{x_i - x_k} \prod_{\substack{l=J-1\\ l\neq j}}^{l=J+1} \frac{y - y_l}{y_j - y_l}. \qquad (12.34)$$

Using the elementary relations between velocity, constant acceleration a, and distance S, namely,

$$u(t + \Delta t) = u(t) + a\,\Delta t,$$
$$S = u\,\Delta t + \tfrac{1}{2}a\,\Delta t^2 = \Delta t\,\frac{u(t + \Delta t) + u(t)}{2}. \qquad (12.35)$$

Krishnamurti determined the coordinates (X, Y) of the point P by successive approximations as follows:

Step 1 $\quad X_P^0 = -\Delta t u(Q, t), \qquad Y_P^0 = -\Delta t v(Q, t).$

Step 2

$$u(Q, t + \Delta t) = u(P, t) + \Delta t\left(\frac{du}{dt}\right)_{(P,t)} \qquad (12.36)$$
$$v(Q, t + \Delta t) = v(P, t) + \Delta t\left(\frac{dv}{dt}\right)_{(P,t)}.$$

The accelerations are calculated from the forces in the equations of motion and interpolated to the point P.

Step 3. The final estimate for locating the point P then uses the relations

$$X_P^1 = -\frac{\Delta t}{2}[u(P, t) + u(Q, t + \Delta t)],$$
$$Y_P^1 = -\frac{\Delta t}{2}[v(P, t) + v(Q, t + \Delta t)], \qquad (12.37)$$

after which the variables may be interpolated to the new position (X_P^1, Y_P^1). Finally, the prediction for the dependent variables is made, for example,

$$u(Q, t + \Delta t) = u(P, t) + \Delta t \left(\frac{du}{dt}\right)_{Q, t+\Delta t}, \text{ etc.,} \tag{12.38}$$

where the acceleration has been now calculated at Q at time $t + \Delta t$. All of the dependent variables are similarly calculated from the dynamical equations. This two-step predictor-corrector method is similar to the Matsuno time-differencing scheme. Krishnamurti used a time step of ten minutes in his primitive equation model, which was discussed in Chapter 10.

The interpolation equation (12.33) is a more accurate form than actually used by Krishnamurti, and was used by Mather (1969), who elaborated further on Krishnamurti's scheme in the barotropic model:

$$\begin{aligned} \frac{d\mathbf{V}}{dt} &= \nabla\Phi - f\mathbf{k} \times \mathbf{V} = A\mathbf{i} + B\mathbf{j}, \\ \frac{dZ}{dt} &= -\nabla \cdot \mathbf{V} = C, \qquad Z = \ln \Phi. \end{aligned} \tag{12.39}$$

The first guess of the x-coordinate of P is given by

$$XP^{01} = -u^{00}\,\Delta t - \tfrac{1}{2}A^{00}\,(\Delta t)^2, \tag{12.40}$$

where the superscript denotes the time step and the guess number in that order and all quantities are evaluated at Q. Similar equations apply for the YP^{01} and ZP^{01}. Next, all quantities may be interpolated to the point P. A forward-difference extrapolation may now be carried out to estimate the variables at Q at time $t + \Delta t$, for example,

$$u^{11} = uP^{01} + AP^{01}\,\Delta t. \tag{12.41}$$

The second guess for the position becomes

$$XP^{0n+1} = -\tfrac{1}{2}(uP^{01} + u^{1n})\,\Delta t, \tag{12.42}$$

where $n = 1$, and similarly for the y-coordinate. Next the dependent variables may again be interpolated to the new location, from which another estimate of the velocity at Q is made as follows:

$$u^{n+1} = uP^{0n+1} + \tfrac{1}{2}(AP^{0n+1} + A^{1n})\,\Delta t. \tag{12.43}$$

Obviously (12.42) and (12.43) can be iterated as often as desired.

Mather solved the barotropic equations for a sinusoidal wave pattern with cyclic east-west boundaries and rigid walls at the northern and southern boundaries ($v = 0$), where a one-sided interpolation formula was used. After eight days some small-scale irregularities appeared for both 1 and 3 iterations. Note, however, that no diffusion terms were included which are frequently used to suppress such small-scale instabilities.

12.5 SPECTRAL METHODS

The methods of numerical integration discussed thus far are based on the discrete representation of the data on a grid or mesh of points covering the space over which a prediction of the variables is desired. Then the local time derivatives of the quantities to be predicted are determined by expressing the horizontal and vertical advection terms, sources of vorticity, etc., in finite difference form. Finally, the time extrapolation is achieved by one of many possible formulas, for example, (6.18). The finite difference technique, which is also applicable to the momentum (primitive) equations (Chapter 13), has a number of associated problems such as truncation error, linear and nonlinear instability. Despite these difficulties, the finite difference method has been the most practical method of producing forecasts numerically from the dynamical equations.

There is another approach called the *spectral* method which avoids some of the difficulties cited previously, in particular, nonlinear instability; however the required computations are comparatively time consuming. As a result its use has been limited to research studies and some experimental forecasts. Nevertheless, more advanced computers may ultimately render the method practicable, and it is desirable to provide at least a brief description here. In a general sense, the mode of representation of data depends on the nature of the data and the shape of the region over which the representation is desired. An alternative to depiction on a mesh or grid of discrete points is a representation in the form of a series of orthogonal functions. This requires the determination of the coefficients of these functions, and the representation is said to be a *spectral* representation or a series expansion in *wave number space*. When such functions are used, the *space derivatives can be evaluated analytically, eliminating the need for approximating them with finite differences.*

As indicated earlier the choice of orthogonal functions depends in part on the geometry of the region to be represented, and for meteorological data a natural choice is a series of spherical harmonics. The first published work on the application of this technique to meteorological prediction is that of Silberman (1954). He considered the barotropic vorticity equation in spherical coordinates

$$\frac{\partial \zeta}{\partial t} = -\frac{1}{a}\left(v_\theta \frac{\partial}{\partial \theta} + \frac{v_\lambda}{\sin\theta}\frac{\partial}{\partial \lambda}\right)(\zeta + 2\Omega\cos\theta), \tag{12.44}$$

$$\zeta = \frac{1}{a\sin\theta}\left[\frac{\partial}{\partial\theta}(v_\lambda \sin\theta) - \frac{\partial v_\theta}{\partial \lambda}\right], \tag{12.45}$$

where a is the earth's radius, θ is the colatitude, λ is the longitude, and v_λ and v_θ are the vorticity components in the direction of increasing λ and θ. In terms of a stream function ψ, the velocity components are

$$v_\lambda = \frac{1}{a}\frac{\partial \psi}{\partial \theta}, \qquad v_\theta = -\frac{1}{a\sin\theta}\frac{\partial \psi}{\partial \lambda},$$

and the vorticity equation becomes

$$\nabla_s^2 \frac{\partial \psi}{\partial t} = \frac{1}{a^2\sin\theta}\left(\frac{\partial\psi}{\partial\lambda}\frac{\partial}{\partial\theta} - \frac{\partial\psi}{\partial\theta}\frac{\partial}{\partial\lambda}\right)(\nabla_s^2\,\psi + 2\Omega\cos\theta), \tag{12.46}$$

where

$$\nabla_s^2 = \frac{1}{a^2\sin\theta}\left[\frac{\partial}{\partial\theta}\left(\sin\theta\frac{\partial}{\partial\theta}\right) + \frac{1}{\sin\theta}\frac{\partial^2}{\partial\lambda^2}\right].$$

The stream function is represented in terms of spherical harmonics which are the solutions Y_n^m of the equation

$$a^2\nabla_s^2 Y_n^m + n(n+1)Y_n^m = 0. \tag{12.47}$$

These functions are expressible in the form

$$Y_n^m = e^{im\lambda}P_n^m. \tag{12.48}$$

If (12.48) is substituted into (12.47), the result is the ordinary differential equation

$$\frac{d^2P_n^m}{d\theta^2} + \cot\theta\,\frac{dP_n^m}{dt} + \left[n(n+1) - \frac{m^2}{\sin^2\theta}\right]P_n^m = 0. \tag{12.49}$$

This is a form of the Legendre equation, and its solutions P_n^m are known as *Legendre* functions of order m and degree n. The characteristic solutions of (12.49) are expressible as a set of orthonormal functions such that

$$\int_0^\pi P_n^m P_s^m \sin\theta\, d\theta = \delta_{ns}, \tag{12.50}$$

where δ_{ns} is the Kronecker delta,

$$\delta_{ns} = 1 \qquad \text{if } n = s$$

and

$$\delta_{ns} = 0 \qquad \text{if } n \neq s.$$

The order m may take on negative values for which the Legendre function is

$$P_n^{-m} = (-1)^m P_n^m.$$

For integral values of n and m, the Legendre functions are simply polynomials, the orders of which increase with n.

At any particular time t' the stream function may be expressed as the finite sum

$$\psi_{t=t'} = a^2\Omega \sum_{n=|m|}^{n'} \sum_{m=-m'}^{m'} K_n^m(t') Y_n^m. \tag{12.51}$$

Since the series is finite, disturbances of sufficiently small scale are not represented, which constitutes a truncation error; however, this is not necessarily undesirable. The harmonic coefficients K_n^m are complex, and the condition for a real ψ is that

$$K_n^{-m} = (-1)^m K_n^{m*}$$

where the * denotes the complex conjugate. The stream function tendency is given by

$$\left(\frac{\partial \psi}{\partial t}\right)_{t=t'} = a^2\Omega \sum_{n=|m|}^{n''} \sum_{m=-m''}^{m''} \left(\frac{dK_n^m}{dt}\right)_{t=t'} Y_n^m. \tag{12.52}$$

When the expressions for ψ and $\partial\psi/\partial t$ are substituted into the vorticity equation (12.46) and then this equation is multiplied by $Y_n^{-m} \sin\theta$ and integrated from 0 to 2π with respect to λ and from 0 to π with respect to θ, the result is

$$\left(\frac{dK_n^m}{dt}\right)_{t=t'} = \frac{2i\Omega m K_n(t')}{n(n+1)} + \frac{i\Omega}{2} \sum_{s=|r|}^{n'} \sum_{r=-m'}^{m'} \sum_{k=|j|}^{n'} \sum_{j=-m'}^{m'} K_k^j(t') K_s^r(t') H_{kns}^{jmr}, \tag{12.53}$$

where H_{kns}^{jmr} is zero unless $j + r = m$, in which case,

$$H_{kns}^{jmr} = \frac{s(s+1) - k(k+1)}{n(n+1)} \int_0^\pi P_n^m \left(jP_k^j \frac{dP_s^r}{d\theta} - r\frac{dP_k^j}{d\theta} P_s^r\right) d\theta. \tag{12.54}$$

The quantities H_{kns}^{jmr} are called *interaction coefficients*, which are zero unless

$$k + n + s = \text{odd integer} \qquad \text{and} \qquad |k - s| < n < k + s.$$

Also $m'' \leq 2m'$ and $n'' \leq 2n' - 1$.

After the right side of the tendency equation (12.53) has been calculated, the future values of the expansion coefficients may be determined by extrapolating forward in time as in the finite difference technique; for example,

$$K_n^m(t + \Delta t) = K_n^m(t - \Delta t) + 2\Delta t \frac{dK_n^m(t)}{dt}. \tag{12.55}$$

The calculation of the interaction coefficients is a lengthy task, especially when the technique is extended to multilevel baroclinic models. An advantage of the method, however, is that nonlinear instability is avoided completely because all nonlinear interactions are computed analytically and all contributions to wave numbers outside the truncated series are automatically eliminated.

Following Silberman's initial paper on the spectral method, there were contributions by Platzman (1960), Platzman and Baer (1961), Baer (1964), Robert (1966), Simons (1968).

Robert (1966) proposed a modification to Silberman's method for numerical integration of the primitive equations in which some simpler functions are substituted for the spherical harmonics. These functions are in fact the basic elements required to generate spherical harmonics, namely,

$$G_n^m(\lambda, \varphi) = e^{im\lambda} \cos^M \varphi \sin^n \varphi.$$

Here λ and φ are the longitude and latitude respectively, M is the absolute value of m and gives the number of waves along a latitude circle, and both M and n are either positive integers or zero.

Robert applied this technique in a general circulation experiment with the atmosphere initially at rest and isothermal, using the momentum (primitive) equations in spherical curvilinear form, with p as the vertical coordinate.

Centered differences were used for the vertical derivatives when possible. Except for an initial forward step, the time integration was performed according to the two-step procedure:

$$F^*(t + \Delta t) = F(t - \Delta t) + 2\,\Delta t\left(\frac{\partial F}{\partial t}\right)_t^*,$$

$$F(t) = F^*(t) + 0.01[F^*(t + \Delta t) + F(t - \Delta t) - 2F^*(t)],$$

where the * represents a preliminary value. The second step acts as a weak filter and stabilizes the central difference scheme, which would otherwise be unstable in this case because of dissipative terms in the prediction equations. A time step of 20 minutes was sufficient to maintain stability. At first only the coefficients in the range $-2 \leq n \leq 2$; $-2 \leq m \leq 2$, were used, giving 15 coefficients for each variable. After 200 days, the limit on n and m was increased from 2 to 4. Another 22 days of integration gave weather maps that appeared fairly realistic in spite of the relatively low resolution. Although no comparisons were made to the finite difference method, Robert concluded that the method could not compete with the finite difference method for operational models, but it could be advantageous for general circulation experiments and, perhaps, for extended-range forecasts where only certain statistical averages would have meaning in any event.

Simons (1968) proposed a generalization of the spectral method wherein the atmospheric flow field is represented by orthogonal functions in both the horizontal and vertical directions, which removes the need for vertical finite differences as well as horizontal ones.

12.5.1 An Example of the Spectral Method

As an illustration of the spectral method, we present a simple example due to Lorenz (1960). Consider the vorticity equation (12.2) applied to a plane region over which the stream function is doubly periodic, that is,

$$\psi\left(x + \frac{2\pi}{k}, y + \frac{2\pi}{l}, t\right) = \psi(x, y, t), \tag{12.56}$$

where k and l are constants. Thus the area is finite but unbounded, and in that respect it is analogous to the spherical earth. Note also that (12.2) applies with a constant coriolis parameter so that rotation is not excluded. Next assume that the stream function is representable in terms of the characteristic solutions of the equation

$$\nabla^2\psi - c\psi = 0, \tag{12.57}$$

which is the analogue to (12.47). The solutions are trigonometric functions; thus ψ is expressible as a double Fourier series, which for convenience may be written as

$$\psi = \sum_{\substack{m=0\\n=n_0}}^{\infty} - \frac{1}{m^2k^2 + n^2l^2}\,[A_{mn}\cos(mkx + nly) + B_{mn}\sin(mkx + nly)]. \tag{12.58}$$

The coefficients are as yet unknown functions of time, except for the initial values which are assumed known. It is apparent from (12.58) that the characteristic values of (12.57) are $c_{mn} = -(m^2k^2 + n^2l^2)$.

For actual prediction purposes a finite series of the form (12.58) is used and the time derivatives of the coefficients must be determined from the vorticity equation. Consider a simple case where m and n take on only the values of 0, 1, −1.

Then after combining like terms, (12.58) is expressible in the form

$$\begin{aligned}\psi = &-\frac{A_{10}}{k^2}\cos kx - \frac{A_{01}}{l^2}\cos ly - \frac{A_{11}}{k^2 + l^2}\cos(kx + ly)\\ &-\frac{A_{1-1}}{k^2 + l^2}\cos(kx - ly) - \frac{B_{10}}{k^2}\sin kx - \frac{B_{01}}{l^2}\sin ly\\ &-\frac{B_{11}}{k^2 + l^2}\sin(kx + ly) - \frac{B_{1-1}}{k^2 + l^2}\sin(kx - ly).\end{aligned}$$

It turns out that, if the B's are zero initially, they will remain so; also if $A_{1-1} = -A_{11}$ initially, it will remain so. With these assumptions, Lorenz obtains the "maximum simplification" of the stream function for use with (12.2), namely,

$$\psi = -\frac{A}{l^2}\cos ly - \frac{F}{k^2}\cos kx - \frac{2G}{k^2 + l^2}\sin ly \sin kx. \tag{12.59}$$

Substituting this streamfunction into the vorticity equation (12.2), utilizing trigonometric identities, and finally equating coefficients of like terms leads to the following differential equations for the coefficients:

$$\begin{aligned}
\frac{dA}{dt} &= -\left(\frac{1}{k^2} - \frac{1}{k^2 + l^2}\right)klFG \equiv K_1FG, \\
\frac{dF}{dt} &= \left(\frac{1}{l^2} - \frac{1}{k^2 + l^2}\right)klAG \equiv K_2AG, \\
\frac{dG}{dt} &= -\frac{1}{2}\left(\frac{1}{l^2} - \frac{1}{k^2}\right)klAF \equiv K_3AF.
\end{aligned} \tag{12.60}$$

Note that the interaction coefficients K_1, K_2, and K_3 are constants and, hence, remain the same throughout the period of integration. The set (12.60), which is analogous to (12.53), can be solved analytically; however when the spectral technique is applied, say to a hemisphere with real data, the resulting system of equations is much more complex and must be solved by numerical methods. If the leapfrog scheme is used here, the numerical integration of (12.60) would be analogous to (12.55), that is,

$$A_{n+1} = A_{n-1} + 2\Delta t K_1 F_n G_n, \tag{12.61}$$

etc., where the subscript denotes the time step. As usual, a forward step must be used for the first time step from the initial values of the coefficients A_0, F_0, and G_0. To avoid linear instability, Δt must be a fairly small fraction of the period of the most rapidly oscillating variable. Equation 12.61 and similar ones for F and G permit the calculation of future values of the coefficient of the series (12.59); thus the prediction of stream function is achieved.

chapter thirteen
Primitive Equation Integrations

13.1 INTRODUCTION

In Chapter 8 it was noted that as fewer approximations are made in order to achieve more accuracy with the filtered equation, they become increasingly more complicated until any advantages gained by filtering out the gravity waves are offset by other complexities. As a result, research in the late fifties turned toward utilization of the momentum equations, usually referred to as the *primitive equations*. Although sound waves are filtered out by the hydrostatic approximation (except for the Lamb wave), gravity waves are permitted. Linear instability can be avoided by merely reducing the time step to 5 to 10 minutes for the typical grid sizes of several hundred kilometers. A more crucial problem is nonlinear instability, and to obtain satisfactory results, the finite difference equations, boundary conditions, and initial conditions must be carefully formulated.

Only in the second half of the sixties did computers become fast enough to permit forecasts with the primitive equations on an operational basis. As a consequence, much of the early work in this field was carried out by research groups with an interest in simulating the general circulation of the atmosphere where the integration period is months to years. In such lengthy calculations it is vitally important to maintain the constraints on energy, etc., for even a slight error per time step can accumulate and eventually lead to catastrophic failures. As a result, a number of the finite difference schemes in use were based on conservation principles. Some of the early investigators who developed successful primitive equation models were K. Hinkelmann, C. E. Leith, Y. Mintz and A. Arakawa, F. G. Shuman, and J. Smagorinsky and collaborators.

The general circulation experiments are usually begun with an atmosphere at rest, which presents no particular problem with respect to initialization. However, prediction with real data requires the specification of the initial state, which is a difficult problem because observational data are incomplete, both with respect to kind and density. Moreover, besides ordinary instrumentation errors, the data are cluttered with "noise," that is, the effects of micro- and mesoscale phenomena. Consequently, the data cannot be merely interpolated to the gridpoints and the forecast commenced. In the filtered models, an objective analysis of the geopotential field and rotational winds computed geostrophically or from the balance equation are quite satisfactory initial conditions. Although such initial conditions have been used with the primitive equations, they are not entirely satisfactory because of the sizable inertial-gravity oscillations that are generated during the adjustment process. N. A. Phillips (1960) has shown that this "noise" is suppressed if the initial wind field is assumed to have divergence equivalent to that implied by the quasi-geostrophic theory of numerical weather prediction. If the geopotential field is considered reliable, the stream function and the potential function for the rotational and divergent wind components are derivable, respectively, from the balance and continuity equation. The latter, $\nabla^2\chi = -\partial\omega/\partial p$, first requires the solution of the ω equation. On the other hand, if the winds are more reliable, the vorticity ζ_0 may be calculated directly from the winds, after which the equation $\nabla^2\psi = \zeta_0$ may be relaxed to give the stream function ψ. With this ψ, the Φ field can be calculated from the balance equation. Then the ω equation must be solved, presumably through an iterative process with the vorticity equation (for consistency with respect to the energetics). However for initialization, this lengthy procedure can probably be shortened somewhat, although there has been little experimentation of this kind. In any event, the final step would be the calculation of χ by relaxing the continuity equation as previously.

Such procedures certainly do not solve the initialization problem completely but do help to suppress the "noise." A more complete treatment of various methods of objective analysis and initialization will be given in Chapter 14.

In general, the treatment of the various physical processes is simpler and more direct with the primitive system than with the filtered system. Hence the remainder of this Chapter will be devoted primarily to methods of numerical integration for the hyperbolic primitive equations, including both linear and nonlinear systems.

13.2 FINITE DIFFERENCE SCHEMES

R. D. Richtmyer in an NCAR technical note has summarized some difference schemes used on a simple hyperbolic system of the form

$$\frac{\partial U}{\partial t} + A\frac{\partial U}{\partial x} = 0. \tag{13.1}$$

Here U may be interpreted as a vector of the dependent variables, and A as the matrix of coefficients of the spatial derivatives, which generally depends on U, x, and t. However for simplicity, a single equation with a constant coefficient A will be considered. The rationale for this approach, which was also used in Chapter 5, is that when instabilities develop in the numerical solution of partial-differential equations, they tend to appear first in a small region in the form of small-amplitude short-wavelength oscillations superimposed on an otherwise smooth solution. Since for a small region the coefficients are nearly constant and, furthermore, the initial amplitudes of the instabilities are small, it may be reasoned that a linear system with constant coefficients is sufficient to study some aspects of computational instability. There is, of course, the more subtle nonlinear instability which was discussed in Chapter 12.

Several common stable differencing schemes are as follows:

Leapfrog

$$U_i^{n+1} = U_i^{n-1} - \frac{A\,\Delta t}{\Delta x}(U_{i+1}^n - U_{i-1}^n) \qquad 0(\Delta x^3),$$

Diffusing

$$U_i^{n+1} = \tfrac{1}{2}(U_{i+1}^n + U_{i-1}^n) - \left(\frac{A\,\Delta t}{2\,\Delta x}\right)(U_{i+1}^n - U_{i-1}^n), \qquad 0(\Delta x^2), \tag{13.2}$$

Lax-Wendroff

$$U_i^{n+1} = U_i^n - \frac{A\,\Delta t}{2\,\Delta x}(U_{i+1}^n - U_{i-1}^n) + 2\left(\frac{A\,\Delta t}{2\,\Delta x}\right)^2 \times (U_{i+1}^n - 2U_i^n + U_{i-1}^n) \qquad 0(\Delta x)^3,$$

Implicit

$$U_i^{n+1} = U_i^n - \frac{A\,\Delta t}{4\,\Delta x}[(U_{i+1}^{n+1} - U_{i-1}^{n+1}) + (U_{i+1}^n - U_{i-1}^n)] \qquad 0(\Delta x)^2.$$

The leapfrog and implicit differencing schemes have been discussed to some extent in Chapter 5. The former is quite accurate and has some very desirable characteristics; however, it does not conserve kinetic energy, and nonlinear instability may develop during extended integration if not controlled by smoothing, which may be damaging. Another fault of the leapfrog method is that the solutions at adjacent time steps tend to become decoupled, primarily because of the computational mode which changes phase at every time step. Hence it is necessary to average, in some fashion, every 50 or so time steps and begin again with a forward time step.

The diffusing scheme, though stable for $A\,\Delta t/\Delta x < 1$ has strong smoothing or damping effects which may be undesirable. The Lax-Wendroff scheme includes a second-order difference term which may be established as follows. Differentiating (13.1) with respect to t and substituting from (13.1) gives

$$\frac{\partial^2 U}{\partial t^2} = -A\,\frac{\partial^2 U}{\partial x\,\partial t} \equiv -A\,\frac{\partial}{\partial x}\frac{\partial U}{\partial t} = A^2\,\frac{\partial^2 U}{\partial x^2}\,. \tag{13.3}$$

Next expand U as a power series:

$$U_i^{n+1} = U_i^n + \frac{\partial U}{\partial t}\,\Delta t + \frac{\partial^2 U}{\partial t^2}\frac{\Delta t^2}{2}\cdots, \tag{13.4}$$

and replace the time derivatives by space derivatives by means of (13.1) and (13.3), giving

$$U_i^{n+1} = U_i^n - A\,\Delta t\left(\frac{\partial U}{\partial x}\right)_i + 2\left(\frac{A\,\Delta t}{2}\right)^2\left(\frac{\partial^2 U}{\partial x^2}\right)_i. \tag{13.5}$$

Finally replace the space derivatve by central difference equivalents which yields the Lax-Wendroff form of (13.2). A similar scheme was independently used by Leith in his studies of the general circulation.

If the leapfrog and diffusing schemes are used alternately, the two-step Lax-Wendroff scheme results, namely,

$$U_i^{n+1} = \tfrac{1}{2}(U_{i+1}^n + U_{i-1}^n) - \frac{A\,\Delta t}{2\,\Delta x}(U_{i+1}^n - U_{i-1}^n), \tag{13.6}$$

$$U_i^{n+2} = U_i^n - \frac{A\,\Delta t}{\Delta x}(U_{i+1}^{n+1} - U_{i-1}^{n+1}). \tag{13.7}$$

Substitution of (13.6) into (13.7) for a constant A yields precisely the Lax-Wendroff form of (13.2) for a time interval of $2\,\Delta t$ and a grid spacing of $2\,\Delta x$.

In order to determine the accuracy of any numerical integration technique, it is necessary to compare the numerical result to the analytic solution. However, an analytical solution can be obtained only if the differential system is quite simple in structure, usually linear with constant coefficients. But a numerical scheme devised for a simple system may not work for the meteorological equations which are nonlinear and rather complicated. A reasonable inference, nevertheless, is that a numerical integration scheme which is unsatisfactory for a simple linear system would most likely also be unsatisfactory for a more complex set of nonlinear differential equations.

Table 13.1

Summary of the Properties of the Methods Studied*

Method	Difference Equation	Number of Time Levels	Computational Stability	Physical Mode		Computational Mode
				Amplitude	Phase	Amplitude
Implicit:						
(*A*) backward	$h^{\tau+1} - h^{\tau} = \Delta t F^{\tau+1}$	2	Absolutely stable	Highly selective damping	Retardation	None
(*B*) trapezoidal	$h^{\tau+1} - h^{\tau} = \Delta t(F^{\tau+1} + F^{\tau})$	2	Absolutely stable	No change	Little retardation	None
(*C*) partly	$h^{\tau+1} - h^{\tau} = \Delta t F_1^{\tau} + \frac{\Delta t}{2}(F_2^{\tau+1} + F_2^{\tau})$	2	Unstable for meteorological wave and one gravity wave			None
(*D*) partly	$h^{\tau+1} - h^{\tau-1} = 2\,\Delta t F_1^{\tau} + 2\,\Delta t F_2^{\tau+1}$	3	(Very weak) unstable for meteorological wave	Damping of gravity wave and weak amplifying of meteorological wave		Damping

Explicit:						
(*O*) forward	$h^{\tau+1} - h^{\tau} = \Delta t F^{\tau}$	2	Unstable			None
(1) leapfrog (centered)	$h^{\tau+1} - h^{\tau-1} = 2\Delta t F^{\tau}$	3	Conditionally stable ($b < 1$)	No change	Moderate acceleration	No change
Iterative:						
(2) Euler-backward	$h^{*} - h^{\tau} = \Delta t F^{\tau}$ $h^{\tau+1} - h^{\tau} = \Delta t F^{*}$	2	Conditionally stable ($b < 1$)	Moderately selective damping	Large acceleration	None
(3) modified Euler-backward	$h^{*} - h^{\tau} = \frac{\Delta t}{2} F^{\tau}$ $h^{**} - h^{\tau} = \Delta t F^{*}$ $h^{\tau+1} - h^{\tau} = \Delta t F^{**}$	2	Conditionally stable ($b < \sqrt{2}$)	Highly selective damping	Moderate acceleration	None
(4) leapfrog-trapezoidal	$h^{*} - h^{\tau-1} = 2\Delta t F^{\tau}$ $h^{\tau+1} - h^{\tau} = \frac{\Delta t}{2}(F^{*} + F^{\tau})$	3	Conditionally stable ($b < \sqrt{2}$)	Little damping	Little error	Very effective damping (in particular of meteorological wave)
(5) leapfrog-backward	$h^{*} - h^{\tau-1} = 2\Delta t F^{\tau}$ $h^{\tau+1} - h^{\tau} = \Delta t F^{*}$	3	Conditionally stable ($b < 0.8$)	Moderately selective damping	Moderate acceleration	Damping

* In "difference equation," F_1 and F_2 represent nonlinear and linear terms, respectively, and $F = F_1 + F_2$. "Number of time levels" means what is associated with each marching step. In "computational stability," $b = \mu c \Delta t$, where c is an appropriate phase velocity. Under "physical mode," retardation or acceleration means a fictitious change of phase velocity resulting only from finite differencing in time. (*From Kurihara, 1965.*)

A number of integration schemes has been investigated and described in a series of articles in the *Monthly Weather Review*. Y. Kurihara (1965) studied the linear differential system

$$\frac{\partial u}{\partial t} + U\frac{\partial u}{\partial x} = fv - \frac{\partial \Phi}{\partial x},$$

$$\frac{\partial v}{\partial t} + U\frac{\partial v}{\partial x} = -fu, \tag{13.8}$$

$$\frac{\partial \Phi}{\partial t} + U\frac{\partial \Phi}{\partial x} = fUv - gH\frac{\partial u}{\partial x}.$$

If h represents u, v, or Φ, the system (13.8) may be written as

$$\frac{\partial h}{\partial t} = F_1 + F_2 = F, \tag{13.9}$$

where $F_1 = -U\,\partial h/\partial x$ and F_2 is the right side of (13.8). Table 13.1 gives a qualitative summary of the characteristics of ten different integration schemes.

The separation of F into the parts, F_1 and F_2, is prompted by the fact that the slow, low-frequency meteorological waves are primarily associated with the nonlinear advective terms while the linear terms represented by F_2 are responsible for the high-frequency inertial-gravitational oscillations. The latter fast-moving waves are more likely to be the cause of computational instability and also should usually be suppressed as undesirable noise in a meteorological forecast, except as necessary to effect "geostrophic adjustment" by removing noncyclostrophic imbalances between the wind and pressure. The implicit methods can successfully eliminate the computational instability but only at the expense of time-consuming inversion of large matrices. However the two- or three-step iterative methods listed in Table 13.1 are also quite useful in suppressing noise and are simpler to apply. Since (13.9) is a linear equation, the use of more than two time levels results in the appearance of a computational mode.

It should be pointed out that a combination of several of the schemes may be more effective than a single scheme. For example, after applying the leapfrog method 1 for a sizable number of time steps because of its simplicity, method 4 could be used for a few steps to dampen the computational mode, and then method 3 for a few steps to suppress the noise before finally returning to the leapfrog method. The Euler-backward scheme (method 2) has also been used for a few steps to restart after many leapfrog steps in order to eliminate the computational mode and suppress noise.

Lilly (1965) examined a number of time integration schemes as applied to a spectral form of the simplified barotropic vorticity equation. The time variations of the amplitudes of the Fourier components are expressible as a system of nonlinear ordinary differential equations as follows:

$$\frac{dA_i}{dt} = F_i(A_1, A_2, \ldots, t).$$

Of the eight schemes applied to a four-component system, the best overall—considering simplicity, efficiency, and accuracy—appeared to be the Adams-Bashford scheme as follows:

$$A_i^{n+1} = A_i^n + \Delta t(\tfrac{3}{2}F_i^n - \tfrac{1}{2}F_i^{n-1})$$

Since there are three time levels present, this scheme has a computational mode; however, it tends to damp which is favorable. On the other hand, the physical mode tends to amplify; but the rate of erroneous amplification is small if Δt is relatively small especially for the longer wavelengths. Thus the Adams-Bashford scheme is suitable unless the period of integration is very lengthy.

13.3 UNITED STATES WEATHER BUREAU BAROTROPIC MODEL

The previous section dealt with a linear system which is, of course, not suitable for actual forecasting. For the purpose of examining a more complicated numerical integration scheme applicable to nonlinear systems, the following notation will be introduced. Let $t = n\,\Delta t$; $x = i\,\Delta x$; $y = j\,\Delta y$; $n, i, j = 0, \pm\frac{1}{2}, \pm 1, \pm\frac{3}{2} \ldots.$

$$\bar{F}^x = \tfrac{1}{2}(F_{i+1/2,j} + F_{i-1/2,j}); \qquad F_x = \frac{\delta_x F}{\Delta x} = \frac{F_{i+1/2,j} - F_{i-1/2,j}}{\Delta x}, \tag{13.10}$$

$$\bar{F}^{xx} = \overline{\bar{F}^x}^x = \tfrac{1}{4}(F_{i+1} + 2F_i + F_{i-1}). \tag{13.11}$$

It is readily verified that

$$\bar{F}_x^x = \frac{F_{i+1,j} - F_{i-1,j}}{2\Delta x}, \tag{13.12}$$

$$F_{xx} = \frac{F_{i+1,j} - 2F_{ij} + F_{i-1,j}}{\Delta x^2}. \tag{13.13}$$

F. Shuman (1960) tested a number of finite difference forms on a barotropic model for long periods of integration. Of these schemes, which he termed momentum, semimomentum, advective, and filtered factored, a version of the semi-momentum form proved most satisfactory.

This method was applied to the following barotropic system which represents a one-layer fluid with a free surface; a map factor m for the Mercator projection is included.

$$\begin{aligned} \frac{\partial u}{\partial t} - fv + m\left(u\,\frac{\partial u}{\partial x} + v\,\frac{\partial u}{\partial y} + \frac{\partial gh}{\partial x}\right) &= 0, \\ \frac{\partial v}{\partial t} + fu + m\left(u\,\frac{\partial v}{\partial x} + v\,\frac{\partial v}{\partial y} + \frac{\partial gh}{\partial y}\right) &= 0, \\ \frac{\partial h}{\partial t} + m\left[u\,\frac{\partial h}{\partial x} + v\,\frac{\partial h}{\partial y} + h\left(\frac{\partial u}{\partial x} + \frac{\partial v}{\partial y}\right)\right] - hv\,\frac{\partial m}{\partial y} &= 0. \end{aligned} \tag{13.14}$$

The difference equations are:

$$\begin{aligned} &\bar{u}_t^t + \overline{[-\bar{f}^{xy}\bar{v}^{xy} + \bar{m}^{xy}(\bar{u}^{xy}\bar{u}_x^y + \bar{v}^{xy}\bar{u}_y^x + g\bar{h}_x^y)]}^{xy} = 0, \\ &\bar{v}_t^t + \overline{[+\bar{f}^{xy}\bar{u}^{xy} + \bar{m}^{xy}(\bar{u}^{xy}\bar{v}_x^y + \bar{v}^{xy}\bar{v}_y^x + g\bar{h}_y^x)]}^{xy} = 0, \\ &\bar{h}_t^t + \overline{[-\bar{h}^{xy}\bar{v}^{xy}\bar{m}_y^x]}^{xy} \\ &\qquad + \overline{\{\bar{m}^{xy}[\bar{u}^{xy}\bar{h}_x^y + \bar{v}^{xy}\bar{h}_y^x + \bar{h}^{xy}(\bar{u}_x^y + \bar{v}_y^x)]\}}^{xy} = 0. \end{aligned} \tag{13.15}$$

This system has been applied operationally at 500 mb in the middle latitudes and tropics. The u and v wind components constitute initial data from which the h field is obtained by means of the balance equation derived from the first two equations in (13.14). The first boundary conditions tried at the northern and southern walls ($\sim$46°N and S) were

$$u_y = \bar{v}^y = h_y = 0. \tag{13.16}$$

However this forecast "blew up" in about two weeks due to contamination by gravity waves generated at the boundaries. The correct boundary conditions as obtained from the first two equations of motion (13.15), with $\bar{v}^y = 0$, are

$$\begin{aligned} \bar{u}_t^{ty} + \overline{\bar{m}^y(\bar{u}^{xy}\bar{u}_x^y + gh_x^y)}^x &= 0, \\ \bar{f}^y\bar{u}^y + \bar{m}^y gh_y &= 0. \end{aligned} \tag{13.17}$$

The x-averaging is not needed for f and m since these parameters are constant along latitude circles. Using a time interval of 10 minutes, the forecast was continued for 100 days without smoothing by using a special initial field of wave number 3. The fields were well behaved in a meteorological sense, indicating the stability of the calculation. While no separate smoothing was performed, it should be noted that the difference equations involve considerable averaging, as represented by the superior bars.

13.4 UNITED STATES WEATHER BUREAU (USWB) BAROCLINIC MODEL

A six-layer baroclinic model using the modified version of Phillips' σ coordinate system has also been devised by Shuman and is being run operationally by the USWB. The equations for a polar stereographic projection are:

$$\frac{\partial u}{\partial t} + \dot{\sigma}\frac{\partial u}{\partial \sigma} - fv + m\left(u\frac{\partial u}{\partial x} + v\frac{\partial u}{\partial y} + \frac{\partial gz}{\partial x} + c_p\theta\frac{\partial \pi}{\partial x}\right) + F_x = 0,$$

$$\frac{\partial v}{\partial t} + \dot{\sigma}\frac{\partial v}{\partial \sigma} + fu + m\left(u\frac{\partial v}{\partial x} + v\frac{\partial v}{\partial y} + \frac{\partial gz}{\partial y} + c_p\theta\frac{\partial \pi}{\partial y}\right) + F_y = 0,$$

$$\frac{\partial gz}{\partial \sigma} + c_p\theta\frac{\partial \pi}{\partial \sigma} = 0, \quad (13.18)$$

$$\frac{\partial \theta}{\partial t} + \dot{\sigma}\frac{\partial \theta}{\partial \sigma} + m\left(u\frac{\partial \theta}{\partial x} + v\frac{\partial \theta}{\partial y}\right) - H = 0,$$

$$\frac{\partial}{\partial t}\frac{\partial p}{\partial \sigma} + \frac{\partial}{\partial \sigma}\left(\frac{\partial p}{\partial \sigma}\dot{\sigma}\right) + m\left[\frac{\partial}{\partial x}\left(\frac{\partial p}{\partial \sigma}u\right) + \frac{\partial}{\partial y}\left(\frac{\partial p}{\partial \sigma}v\right)\right] - \frac{\partial p}{\partial \sigma}\left(u\frac{\partial m}{\partial x} + v\frac{\partial m}{\partial y}\right) = 0,$$

$$\pi = \left(\frac{p}{P}\right)^{R/c_p}, \qquad P = 1{,}000 \text{ mb}.$$

The frictional force and diabatic heating are represented by F and H. The vertical structure is shown in Figure 13.1.

The isentropic cap is quiescent leaving six active layers. The finite difference equations, which are a generalization of the barotropic form, are as follows:

$$\bar{u}^{\sigma t}_{t} + \overline{(\bar{\dot{\sigma}}^{\sigma}\bar{u}^{xy}_{\sigma} - \bar{f}^{xy}\bar{v}^{\sigma xy} + F_x)}^{xy}$$
$$+ \overline{\{\bar{m}^{xy}[\bar{u}^{\sigma xy}\bar{u}^{\sigma y}_{x} + \bar{v}^{\sigma xy}\bar{u}^{\sigma x}_{y} + \overline{(g\bar{z}^{y}_{x} + c_p\bar{\theta}^{xy}\bar{\pi}^{y}_{x})}^{\sigma}]\}}^{xy} = 0,$$

$$\bar{v}^{\sigma t}_{t} + \overline{(\bar{\dot{\sigma}}^{\sigma}\bar{v}^{xy}_{\sigma} + \bar{f}^{xy}\bar{u}^{\sigma xy} + F_y)}^{xy}$$
$$+ \overline{\{\bar{m}^{xy}[\bar{u}^{\sigma xy}\bar{v}^{\sigma y}_{x} + \bar{v}^{\sigma xy}\bar{v}^{\sigma x}_{y} + \overline{(g\bar{z}^{x}_{y} + c_p\bar{\theta}^{xy}\bar{\pi}^{x}_{y})}^{\sigma}]\}}^{xy} = 0,$$

$$gz_\sigma + c_p\bar{\theta}^{\sigma}\pi_\sigma = 0, \quad (13.19)$$

$$\bar{\theta}^{\sigma t}_{t} + \overline{[\bar{\dot{\sigma}}^{\sigma}\bar{\theta}^{xy}_{\sigma} + \bar{m}^{xy}(\bar{u}^{\sigma xy}\bar{\theta}^{\sigma y}_{x} + \bar{v}^{\sigma xy}\bar{\theta}^{\sigma x}_{y}) - H]}^{xy} = 0,$$

$$\bar{p}^{t}_{\sigma t} + \overline{[\bar{p}^{xy}_{\sigma}\dot{\sigma}_{\sigma} + \bar{m}^{xy}(\bar{u}^{\sigma xy}\bar{p}^{y}_{\sigma x} + \bar{v}^{\sigma xy}\bar{p}^{x}_{\sigma y})]}^{xy}$$
$$+ \overline{\bar{p}^{xy}_{\sigma}[\bar{m}^{xy}(\bar{u}^{\sigma y}_{x} + \bar{v}^{\sigma x}_{y}) - \bar{u}^{\sigma xy}\bar{m}^{y}_{x} - \bar{v}^{\sigma xy}\bar{m}^{x}_{y}]}^{xy} = 0.$$

It is apparent that the first term in each equation represents the ordinary leapfrog or central difference scheme. Note that $\bar{u}^{\sigma}$, $\bar{v}^{\sigma}$, and $\bar{\theta}^{\sigma}$ appear essentially as dependent variables; however u_σ, v_σ, and θ_σ are also needed so that derivatives must be constructed from the averaged quantities.

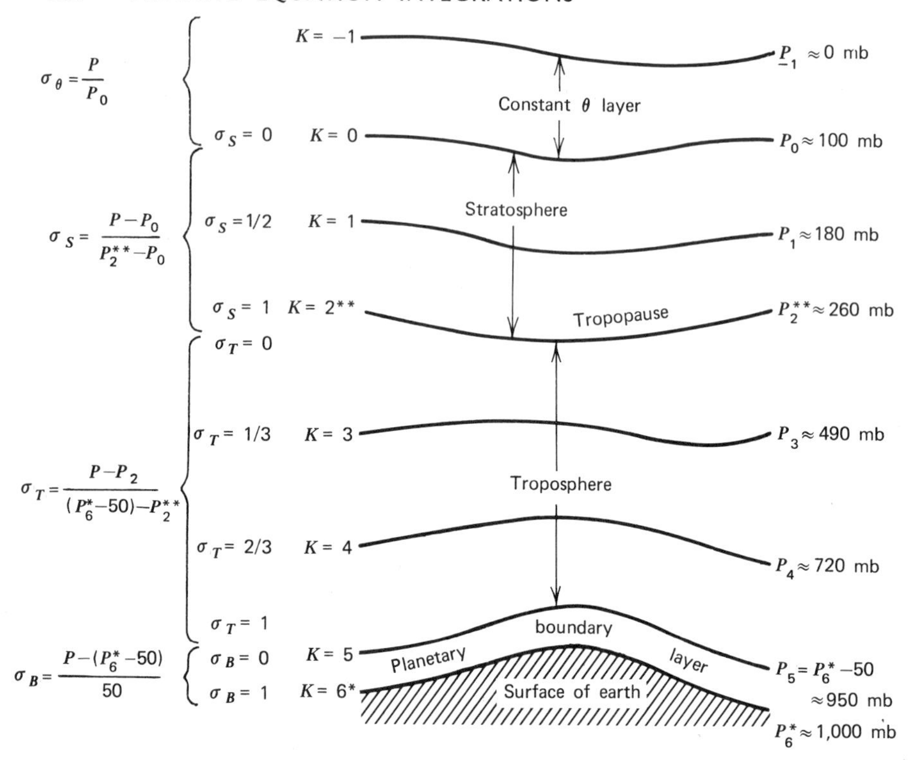

Figure 13.1. The vertical structure of the USWB six-layer baroclinic model. [*After Shuman and Hovermale (1968).*]

These may be obtained, say for θ, by minimizing

$$\sum (\theta - \bar{\theta}^{\sigma\sigma})^2 \tag{13.20}$$

A "vertical velocity" equation for $\dot{\sigma}$ may be obtained by differencing the last equation of (13.19) with respect to σ, which eliminates the time derivative and gives the following diagnostic equation:

$$\bar{p}_\sigma^{xy}\dot{\sigma}_{\sigma\sigma} + \bar{m}^{xy}[\bar{u}_\sigma^{\sigma xy}\bar{p}_{\sigma x}^{y} + \bar{v}_\sigma^{xy}\bar{p}_{\sigma y}^{x} + \bar{p}_\sigma^{xy}(\bar{u}_{\sigma x}^{\sigma y} + \bar{v}_{\sigma y}^{\sigma x})] \\ - \bar{p}_\sigma^{xy}(\bar{u}_\sigma^{\sigma xy}\bar{m}_x^{y} + \bar{v}_\sigma^{\sigma xy}\bar{m}_y^{x}) = 0. \tag{13.21}$$

The vertical velocity $\dot{\sigma}$ is defined at the center of each horizontal square mesh. The tropospheric boundary conditions are $\dot{\sigma}_2 = \dot{\sigma}_6 = 0$, and for the stratosphere, $\dot{\sigma}_0 = \dot{\sigma}_2 = 0$, which is based on the assumption that the stratosphere is bounded by material surfaces through which there is no mass transport.

The numerical integration consists of calculating σ and π from the last equation of (13.18), z by integrating from the third equation of (13.18), then u, v, and θ from $\bar{u}^\sigma$, $\bar{v}^\sigma$, and $\bar{\theta}^\sigma$ by (13.20), vertical velocity from (13.21), the tendencies from (13.19), and finally time extrapolation with the computed tendencies.

The friction terms are applied only in the surface friction layer and are formulated as follows:

$$F_x = \frac{gp}{\Delta p} C_D[(\bar{u}^{\sigma xy})^2 + (\bar{v}^{\sigma xy})^2]^{1/2} \bar{u}^{\sigma xy}.$$

The heating term involves only the turbulent transfer of heat from warm water to cold air:

$$H = K(\bar{\theta}^\sigma - T_w),$$

where T_w is the water temperature and $K = 10^{-4}$ sec^{-1}. In addition, latent heat and radiational cooling are included by parameterization in terms of the large-scale variables. The boundary conditions at the lateral walls (which are assumed to lie between the last two rows of the grid) are

$$\bar{u}^y = \bar{v}^y = \theta_y = p_{\sigma y} = 0, \qquad y = \text{const.}$$

Similar conditions hold for the x-constant walls.

Shuman's primitive equation model has been found to be generally superior to the USWB three-level baroclinic model; it makes a better prognosis of development, although displacements were not as good, initially at least. In addition to the previously described features, the model includes radiation from snow surfaces and moisture analyses, and prognoses for the latent heat release and precipitation forecasts. The problem of proper initialization for integration of the primitive equations is not fully resolved. Usually the horizontal velocity field has been initially related to the pressure field by means of the balance equation, while the vertical winds are initially taken to be zero. Some sizable oscillations of the pressure field occur which are not characteristic of the actual atmosphere, but they tend to diminish considerably after several hours.

13.5 STAGGERED GRID SYSTEMS

When computing space derivatives of the physical variable at a given gridpoint by means of finite difference approximations, it may be noted [see, for example (13.12) and (13.15)] that much of the data come from adjacent gridpoints. This suggests that there could be a considerable savings in computing time and even storage if each variable is retained at alternate gridpoints. Such a staggering of data with respect to the grid system, which may also be alternated with each successive time step, was first introduced into meteorological forecasting by A. Eliassen. His system consisted of retaining u at alternate gridpoints between adjacent values of Φ in the y-direction, v between adjacent values of Φ in the x-direction, and ω at the central point of two-grid squares with Φ at the corners. In addition, the values of the dependent variables at odd time steps are computed at gridpoints diagonally adjacent to the points at which the variables are known at the even time steps, as illustrated in Figure 13.2.

Eliassen's complete finite difference system will not be shown here; however a simpler system used by W. L. Gates (1967) to study Rossby waves in an ocean with a steady zonal wind stress on its surface will be presented. The vertically integrated equations of motion for this case are:

$$\frac{\partial u}{\partial t} + \nabla \cdot u\mathbf{V} = fv - gH\frac{\partial h}{\partial x} + A\nabla^2 u + \frac{1}{\rho_0}(\tau_x^w - \tau_x^b), \tag{13.22}$$

$$\frac{\partial v}{\partial t} + \nabla \cdot v\mathbf{V} = -fu - gH\frac{\partial h}{\partial y} + A\nabla^2 v + \frac{1}{\rho_0}(\tau_y^w - \tau_y^b),$$

$$\frac{\partial h}{\partial t} = -\nabla \cdot \mathbf{V}, \tag{13.23}$$

where $\mathbf{V}$ is the vertically integrated horizontal velocity, $z = h$ is the height of the free surface, $z = -H$ is the depth of the horizontal bottom, and the superscripts w and b refer to the surface wind stress and bottom stress. To simplify the analysis, take

$$\tau_x^b = \tau_y^b = \tau_y^w = 0.$$

The boundary conditions are

$$u = v = 0 \quad \text{at} \quad x = 0, L$$

and

$$y = 0, W. \tag{13.24}$$

This leads to $\partial h/\partial x = \partial h/\partial y = 0$ at these locations. An alternative boundary condition to the foregoing is to apply (13.24) to the continuity equation (13.23), which leads to

$$\frac{\partial h}{\partial t} = -\frac{\partial V_n}{\partial n}, \tag{13.25}$$

where n is normal to the boundary. The last condition insures that the total mass is conserved within the basin prescribed by (13.24).

The main purpose for presenting this particular model is merely to illustrate a staggered grid system which is shown in Figure 13.3.

The solid dots denote the positions where the velocity components are known (row and column index sum is odd) while h is prescribed at the

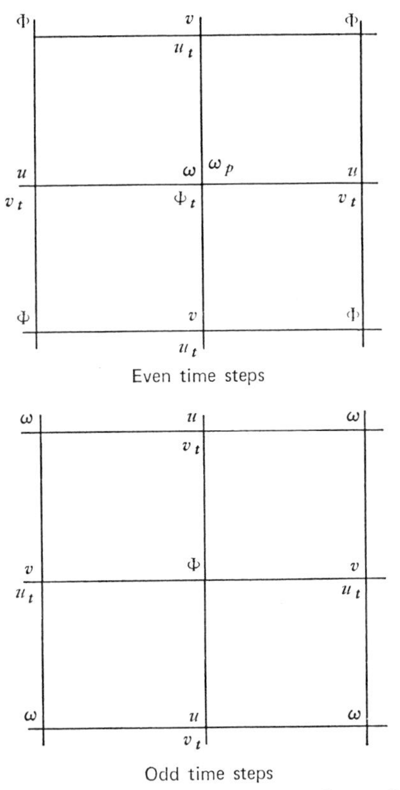

Figure 13.2. A staggered grid system due to A. Elliassen.

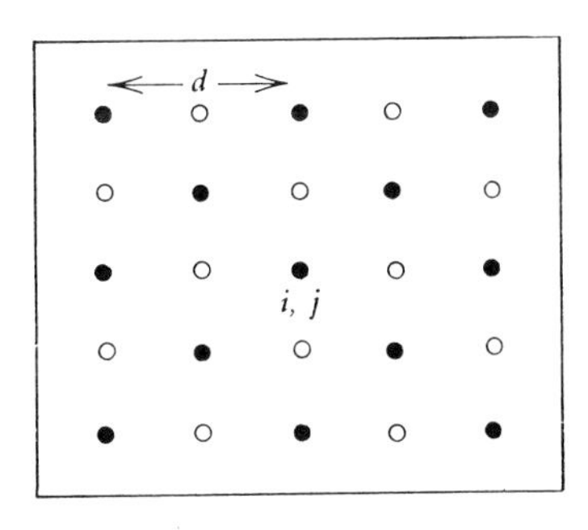

Figure 13.3. Gates' (1967) staggered grid.

positions denoted by open circles. Centered differencing is used as follows:

$$x = \frac{id}{2}, \qquad y = \frac{jd}{2}, \qquad \begin{cases} i = 1, 2, \ldots, N + 1, \\ j = 1, 2, \ldots, M + 1, \end{cases}$$

$$t = n\,\Delta t, \qquad n = 1, 2, \ldots.$$

$$\begin{aligned} u_{ij}^{n+1} &= u_{ij}^{n-1} + 2\,\Delta t f_j v_{ij}^n - \frac{2\,\Delta t}{d}\, gH(h_{i+1,j}^n - h_{i-1,j}^n) \\ &+ \frac{4A\,\Delta t}{d^2}(u_{i+1,j+1}^{n-1} + u_{i-1,j+1}^{n-1} + u_{i+1,j-1}^{n-1} + u_{i-1,j-1}^{n-1} - 4u_{ij}^{n-1}) \\ &- \frac{\Delta t}{2hd}\,[(u_{i+1,j+1}^n + u_{i+1,j-1}^n)^2 - (u_{i-1,j+1}^n + u_{i-1,j-1}^n)^2 \\ &+ (u_{i+1,j+1}^n + u_{i-1,j+1}^n)(v_{i+1,j+1}^n + v_{i-1,j+1}^n) \\ &- (u_{i+1,j-1}^n + u_{i-1,j-1}^n)(v_{i+1,j-1}^n + v_{i-1,j-1}^n)] + \frac{2\,\Delta t}{\rho_0}\tau_{x,i,j}^w, \end{aligned} \tag{13.26}$$

$$\begin{aligned} v_{ij}^{n+1} &= v_{ij}^{n-1} - 2\,\Delta t f_j u_{ij}^n - \frac{2\,\Delta t}{d}\, gH(h_{i,j+1}^n - h_{i,j-1}^n) \\ &+ \frac{4A\,\Delta t}{d^2}(v_{i+1,j+1}^{n-1} + v_{i-1,j+1}^{n-1} + v_{i+1,j-1}^{n-1} + v_{i-1,j-1}^{n-1} - 4v_{ij}^{n-1}) \\ &- \frac{\Delta t}{2hd}\,[(v_{i+1,j+1}^n + v_{i-1,j+1}^n)^2 - (v_{i+1,j-1}^n + v_{i-1,j-1}^n)^2 \\ &+ (u_{i+1,j+1}^n + u_{i+1,j-1}^n)(v_{i+1,j+1}^n + v_{i+1,j-1}^n) \\ &- (u_{i-1,j+1}^n + u_{i-1,j-1}^n)(v_{i-1,j+1}^n + v_{i-1,j-1}^n)], \end{aligned} \tag{13.27}$$

$$h_{ij}^{n+1} = h_{ij}^{n-1} - \frac{2\,\Delta t}{d}(u_{i+1,j}^n - u_{i-1,j}^n + v_{i,j+1}^n - v_{i,j-1}^n) \tag{13.28}$$

Some features of these difference equations are as follows: (*a*) the "leap-frog" time differencing technique gives no damping of the solutions; (*b*) the computational mode was controlled by appropriate formulation of the difference equations and boundary conditions; (*c*) *the frictional diffusion terms are evaluated at the previous time step*, $n - 1$, *in order to insure linear computational stability;* (*d*) the Laplacian for these terms is obtained from the corners of the square surrounding the point where the time extrapolation is being effected, which amounts to a rotation of 45° of the usual centered difference formula for the Laplacian over a slightly longer grid distance

$\sqrt{2}d/2$; (*e*) the advective or inertial terms require the use of averaged values of u and v because of the staggered grid. The boundary condition (13.25) is approximated by a centered time difference and a one-sided space difference. Initially a forward difference is used instead of the centered difference system [(13.26) to (13.28)]. The results of this integration will not be discussed except to mention that the finite differencing scheme was well behaved for integration periods of 60 days.

13.6 GENERAL CIRCULATION MODELS

The first successful attempt to simulate the general circulation of the atmosphere by numerically integrating the hydrodynamical equations was conducted by N. Phillips (1956). He utilized a two-level geostrophic model which included simplified forms of heating and friction. From an initial condition of rest, a circulation evolved which possessed some of the principal features of the observed mean circulation.

Subsequent models utilizing the momentum equations were developed by C. E. Leith at the Lawrence Radiation Laboratory, by J. Smagorinsky at ESSA's Geophysical Fluid Dynamics Laboratory, by Y. Mintz and A. Arakawa at UCLA, and somewhat later by A. Kasahara and W. Washington at NCAR. All of these later efforts have one or more improved features such as (*a*) increased horizontal coverage, either hemispheric or global, (*b*) increased vertical resolution, and (*c*) more complete simulation of the physical processes. The general circulation experiments usually begin with an atmosphere at rest; however several of the models have been utilized for experimental forecasts up to two weeks from observed initial data. Features of some of these models have already been described earlier and only some brief additional remarks are made.

Leith's (1965) six-level global model had radiation, evaporation, condensation, surface friction, and horizontal diffusion. A partially implicit method was used for the time extrapolation as follows:

$$u_V^{n+3/2} - u_V^{n+1/2} = \Delta_a u_V^{n+1/2} + F v_V^{n+3/2} - \left(\frac{\Delta t}{2\,\Delta x}\right)(\Phi_V + \Phi_Q - \Phi_{V-1} - \Phi_{Q-1})$$

$$v_V^{n+3/2} - v_V^{n+1/2} = \Delta_a v_V^{n+1/2} - F u_V^{n+3/2} - \left(\frac{\Delta t}{2\,\Delta y}\right)(\Phi_{V-1} + \Phi_V - \Phi_{Q-1} - \Phi_Q),$$

where Δ_a represents the change due to advection and diffusion.

$$F = \left(f + \frac{u^{n+1/2}\tan\varphi}{a}\right)\Delta t.$$

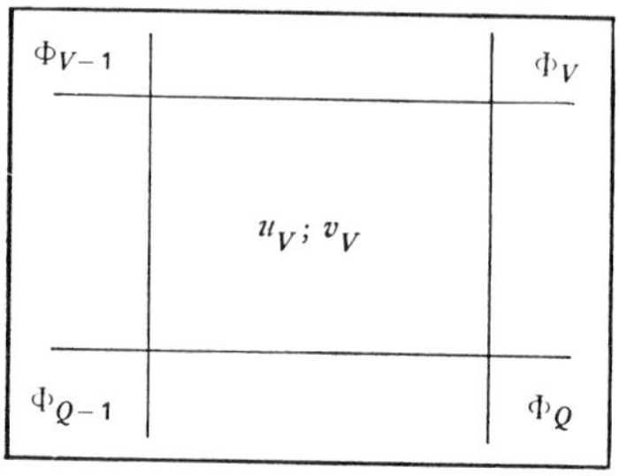

Figure 13.4. Partial grid for Leith's (1965) finite differencing scheme.

The advection scheme consists of two steps [see (12.29)] as follows:

$$\psi_{i,j}^{n+\frac{1}{2}} = \psi_{i,j}^{n} - \frac{\beta}{2}(\psi_{i,j+1}^{n} - \psi_{i,j-1}^{n}) + \frac{\beta^2}{2}(\psi_{i,j+1}^{n} - 2\psi_{i,j}^{n} + \psi_{i,j-1}^{n}),$$

$$\psi_{i,j}^{n+1} = \psi_{i,j}^{n+\frac{1}{2}} - \frac{\alpha}{2}(\psi_{i+1,j}^{n+\frac{1}{2}} - \psi_{i-1,j}^{n+\frac{1}{2}}) + \frac{\alpha^2}{2}(\psi_{i+1,j}^{n+\frac{1}{2}} - 2\psi_{i,j}^{n+\frac{1}{2}} + \psi_{i-1,j}^{n+\frac{1}{2}}),$$

where

$$\alpha = \frac{u\,\Delta t}{\Delta x}, \qquad \beta = \frac{v\,\Delta t}{\Delta y}.$$

The Mintz-Arakawa model is global with two output levels in the vertical plus a boundary layer. It includes the seasonal and diurnal variation of solar radiation, orography, land-sea-ice surface effects, surface friction, terrestrial radiation, clouds, and large-scale and convective precipitation with concomitant thermal effects. Time extrapolation is achieved with the Euler two-step scheme, which may be illustrated by the equation $\partial u/\partial t = F(u)$, as follows:

$$\begin{aligned} u^1(t + \Delta t) &= u(t) + \Delta t F(u, t) \qquad \text{(first approximation)},\\ u(t + \Delta t) &= u(t) + \Delta t F[u^1(t + \Delta t)] \qquad \text{(forecast)}. \end{aligned} \tag{13.29}$$

As indicated in Table 13.1, this integration scheme results in damping of high-frequency waves, for example, gravity waves, but negligible damping of synoptic disturbances. A time step of 12 minutes is used with a grid of 5° longitude and 4° latitude. Some features of this model have already been described earlier in Chapters 9 and 10. A complicated space-differencing technique devised by Arakawa prevents nonlinear instability, conserves the first and second moments, and provides for appropriate constraints regarding the conservation of energy. A less sophisticated version of Arakawa's scheme was partially described in Section 12.3 and will be further illustrated in Section 13.7, which describes the Navy primitive equation model. A complete description of the Mintz-Arakawa model has been given by Langlois and Kwok (1969). The treatment of the thin boundary layer allows no storage of heat with equal fluxes through the top and bottom. The oceans are regarded as infinite heat reservoirs while the land surface is taken to be a perfect insulator with zero heat capacity. Thus the net flux of thermal energy across the air-land boundary is zero, that is, the sum of the flux of short and long radiation plus sensible heat transfer must vanish. On the other hand, heat transfer is allowed to take place through ice on the ocean as follows:

$$F_4 + Q_s - S_4 = B(T_0 - T_g),$$

where F_4 is the net upward long-wave radiation (at $\sigma = 1$) which depends on the surface temperature T_g, S_4 is the rate at which the ice absorbs solar radiation, Q_s is the sensible heat transfer from the surface to the air, T_0 is the freezing point of water, and B is the thermal conductivity of ice. Substituting in the appropriate expressions for F_4, Q_s, and S_4 permits a determination of the ground temperature and, subsequently, sensible heat transfer and evaporation. Moisture is removed from the atmosphere through condensation in the model.

The GFDL hemispheric model, devised by Smagorinsky with the collaboration of Manabe and others (1963, 1965), has the greatest vertical resolution with nine σ levels. The physics include solar and terrestrial radiation as functions of latitude, orography, and boundary-layer fluxes of heat, vapor and momentum. Integrations for the northern hemisphere on a polar stereographic projection starting from an isothermal atmosphere at rest showed good correspondence to the mean overall structure of the atmosphere below 30 km. The dominant wave number of the meridional component is 5 to 6 in the troposphere and 3 in the stratosphere. The level of the jet stream as well as the maximum northward transport of momentum agrees with observation, but the jet stream is much stronger than the observed annual mean. Computation of energy transformations shows that the ratios of eddy to zonal kinetic energy and of eddy to zonal available potential energy are much smaller than those in the actual atmosphere. The kinetic energy of the stratosphere is maintained against its conversion into potential energy and dissipation through interaction with the troposphere, in qualitative agreement with observation.

Extended predictions of two weeks have been made by Miyakoda, Smagorinsky, Strickler, and Hembree (1969) using the "moist" general circulation model. The quality of the forecasts improved with the inclusion of all of the physical processes mentioned previously, as compared to experiments omitting some. The experiment succeeded in forecasting the genesis of second- and third-generation extratropical cyclones and their subsequent behavior. The correlation coefficients between predicted and observed 500-mb height change remained above 0.5 for 13 days in one winter case and 9 days in another. Some weaknesses were (*a*) forecast temperatures were too low; (*b*) a general weakening of the intensities of pressure systems after about one week reflecting an underestimation of eddy kinetic energy; (*c*) over-forecasting the intensity of the tropospheric westerlies.

The NCAR global general circulation model developed by Kasahara and Washington (1967) utilizes spherical coordinates [see (1.23)] with height as the vertical coordinate. Hydrostatic equilibrium is maintained as in the other models, and energy transfer due to solar and terrestrial radiation, small-scale turbulence and convection, orography, and release of latent heat are simulated. The classical tendency equation derived from the hydrostatic and continuity equations is used to calculate the local pressure change, namely,

$$\frac{\partial p}{\partial t} - \left(\frac{\partial p}{\partial t}\right)_{Z_T} = g\rho w - g\int_Z^{Z_T} \nabla \cdot \rho \mathbf{V}\, dz,$$

where Z_T is the top level of the model at which it is assumed that $w_T = 0$. Some experiments with the model showed the importance of orography in determining certain features of the mean winter circulation such as the Asiatic and North American highs and the Aleutian and Icelandic lows. Baumhefner (1970) made some global forecasts with this two-level model which showed reasonable skill out to two days at the surface and four days in the middle troposphere. The best initialization scheme of those attempted was the complete balance equation, although simpler schemes were nearly equal in forecasting skill.

13.7 NAVY MODEL

As a vehicle for research on numerical weather prediction, but with the potentiality for operational forecasting, a primitive equation model was initiated at the Naval Postgraduate School. The vertical coordinate ζ is a variation of the σ system described in Chapter 11 as follows:

$$\sigma = F(\zeta).$$

Therefore

$$\frac{\partial}{\partial \sigma} = -S\frac{\partial}{\partial \zeta},$$

where

$$S = -\left(\frac{\partial F}{\partial \zeta}\right)^{-1}. \tag{13.30}$$

After multiplying the equation of motion (11.4) by the surface pressure π, the continuity equation by the velocity components u and v, and adding, the results may be written, with appropriate modifications for the new vertical coordinate and the map scale [see (1.45 and 11.4)], as

$$\frac{\partial(\pi u)}{\partial t} = -m^2\left[\frac{\partial}{\partial x}\left(\frac{\pi u^2}{m}\right) + \frac{\partial}{\partial y}\left(\frac{\pi uv}{m}\right)\right] - \pi S\frac{\partial wu}{\partial \zeta}$$

$$+\left[2\Omega \sin\varphi - \frac{1}{2a^2}(vx - uy)\right]\pi v - m\pi\frac{\partial\Phi}{\partial x} - mRT\frac{\partial\pi}{\partial x} + \pi F_x, \qquad (13.31)$$

$$\frac{\partial(\pi v)}{\partial t} = -m^2\left[\frac{\partial}{\partial x}\left(\frac{\pi uv}{m}\right) + \frac{\partial}{\partial y}\left(\frac{\pi v^2}{m}\right)\right] - \pi S\frac{\partial wv}{\partial \zeta}$$

$$-\left[2\Omega \sin\varphi - \frac{1}{2a^2}(vx - uy)\right]\pi u - m\pi\frac{\partial\Phi}{\partial y} - mRT\frac{\partial\pi}{\partial y} + \pi F_y. \qquad (13.32)$$

The modified forms of the hydrostatic equation (11.6), the continuity equation (11.12), and the thermodynamic equation (11.16) are respectively,

$$\frac{\partial\Phi}{\partial\zeta} = \frac{RT}{S\sigma}, \qquad \sigma = \frac{p}{\pi},$$

$$\frac{\partial\pi}{\partial t} = -m^2\left[\frac{\partial}{\partial x}\left(\frac{\pi u}{m}\right) + \frac{\partial}{\partial y}\left(\frac{\pi v}{m}\right)\right] - \pi S\frac{\partial w}{\partial\zeta}, \qquad (13.33)$$

$$\frac{\partial(\pi T)}{\partial t} = -m^2\left[\frac{\partial}{\partial x}\left(\frac{\pi uT}{m}\right) + \frac{\partial}{\partial y}\left(\frac{\pi vT}{m}\right)\right] - \pi S\frac{\partial(wT)}{\partial\zeta}$$

$$+\frac{RT}{c_p\sigma}\left[-\pi w + \sigma\frac{\partial\pi}{\partial t} + m\sigma\left(u\frac{\partial\pi}{\partial x} + v\frac{\partial\pi}{\partial y}\right)\right] + \frac{Q}{c_p}. \qquad (13.34)$$

The corresponding difference equations proposed by R. T. Williams are based on the Arakawa technique and are similar to those used by Smagorinsky (1965). This system, which is centered in time, avoids nonlinear instability by requiring the advective terms to conserve the squared quantities, such as πV^2, in the mean, that is, when summed over the entire grid. In addition, the total energy, potential plus kinetic, is conserved through requirements placed on the vertical differencing scheme [see (11.22) through (11.25)].

The adiabatic frictionless difference equations are:

$$(\pi u)_{ijk}^{n+1} = (\pi u)_{ijk}^{n-1} + 2\,\Delta t\Big\{-\mathscr{L}(u)_{ijk} + \Big[2\Omega(\sin\varphi)_{ij} - \frac{1}{2a^2}(v_{ijk}x_i - u_{ijk}y_j)\Big]\pi_{ij}v_{ijk} - \frac{m_{ij}\pi_{ij}(\Phi_{i+1jk} - \Phi_{i-1jk})}{2d} - \frac{m_{ij}RT_{ijk}(\pi_{i+1j} - \pi_{i-1j})}{2d}\Big\}^n, \quad (13.35a)$$

$$(\pi v)_{ijk}^{n+1} = (\pi v)_{ijk}^{n-1} + 2\,\Delta t\Big\{-\mathscr{L}(v)_{ijk} - \Big[2\Omega(\sin\varphi)_{ij} - \frac{1}{2a^2}(v_{ijk}x_i - u_{ijk}y_j)\Big]\pi_{ij}u_{ijk} - \frac{m_{ij}\pi_{ij}(\Phi_{ij+1k} - \Phi_{ij-1k})}{2d} - \frac{m_{ij}RT_{ijk}(\pi_{ij+1} - \pi_{ij-1})}{2d}\Big\}^n \quad (13.35b)$$

$$\pi_{ij}^{n+1} = \pi_{ij}^{n-1} + 2\,\Delta t\left(\frac{\partial\pi}{\partial t}\right)_{ij}^n \quad (13.36)$$

where [see (1.48) and (11.14)]

$$\left(\frac{\partial\pi}{\partial t}\right)_{ij}^n = -\sum_{k=1}^{K}\frac{\Delta\zeta m_{ij}^2}{S_k 2d} \times \left[\left(\frac{\pi u}{m}\right)_{i+1jk} - \left(\frac{\pi u}{m}\right)_{i-1jk} + \left(\frac{\pi v}{m}\right)_{ij+1k} - \left(\frac{\pi v}{m}\right)_{ij-1k}\right]^n, \quad (13.37)$$

$$(\pi T)_{ijk}^{n+1} = (\pi T)_{ijk}^{n-1} + 2\,\Delta t\Big\{-\mathscr{L}(T)_{ijk} + \frac{\kappa T_{ijk}}{\sigma_k} \times \Big[-\pi_{ij}\frac{w_{ijk+1} + w_{ijk}}{2} + \sigma_k\left(\frac{\partial\pi}{\partial t}\right)_{ij}\Big] + \frac{\kappa T_{ijk}m_{ij}^2}{2d}[u_{ijk}(\pi_{i+1j} - \pi_{i-1j}) + v_{ijk}(\pi_{ij+1} - \pi_{ij-1})]\Big\}^n.$$

$$(13.38)$$

The difference operator $\mathscr{L}$ is defined as

$$\mathscr{L}(\psi)_{ijk} = \frac{m_{ij}^2}{4d}\left\{\left[\left(\frac{u\pi}{m}\right)_{i+1jk} + \left(\frac{u\pi}{m}\right)_{ijk}\right](\psi_{i+1jk} + \psi_{ijk})\right.$$

$$- \left[\left(\frac{u\pi}{m}\right)_{ijk} + \left(\frac{u\pi}{m}\right)_{i-1jk}\right](\psi_{ijk} + \psi_{i-1jk}) + \left[\left(\frac{v\pi}{m}\right)_{ij+1k} + \left(\frac{v\pi}{m}\right)_{ijk}\right]$$

$$\left.\times (\psi_{ij+1k} + \psi_{ijk}) - \left[\left(\frac{v\pi}{m}\right)_{ijk} + \left(\frac{v\pi}{m}\right)_{ij-1k}\right](\psi_{ijk} + \psi_{ij-1k})\right\}$$

$$+ \frac{\pi_{ij}S_k}{\Delta\zeta}\left(w_{ijk+1}\frac{\psi_{ijk+1} + \psi_{ijk}}{2} - w_{ijk}\frac{\psi_{ijk} + \psi_{ijk-1}}{2}\right), \tag{13.39}$$

$$w_{ijk+1} = w_{ijk} - \frac{\Delta\zeta}{\pi_{ij}S_k}\left\{\left(\frac{\partial\pi}{\partial t}\right)_{ij} + \frac{m_{ij}^2}{2d}\left[\left(\frac{\pi u}{m}\right)_{i+1jk} - \left(\frac{\pi u}{m}\right)_{i-1jk}\right.\right.$$

$$\left.\left. + \left(\frac{\pi v}{m}\right)_{ij+1k} - \left(\frac{\pi v}{m}\right)_{ij-1k}\right]\right\}, \tag{13.40}$$

$$w_{ij1} = 0. \tag{13.41}$$

The geopotential field is computed hydrostatically from the temperature field as follows:

$$\Phi_{ijk+1} = \Phi_{ijk} + \frac{R\,\Delta\zeta}{2}\left[\left(\frac{T}{S\sigma}\right)_{ijk+1} + \left(\frac{T}{S\sigma}\right)_{ijk}\right], \tag{13.42}$$

$$\Phi_{ij1} = (\Phi_{\text{surf}})_{ij} + \frac{\Delta\zeta}{2}\frac{RT_1}{\sigma_1 S_1}. \tag{13.43}$$

The surface geopotential is calculated from the height of the earth's surface above mean sea level.

13.7.1 Boundary Conditions

At the horizontal boundaries (see Figure 13.5), which lie between the outer two rows of grid points, the normal velocity components are assumed to vanish as represented by the following equations. At the earth's surface and the top of the atmosphere, the vertical velocity w vanishes. Note that the values of u, v, T, Φ are carried at a level $\frac{1}{2}\,\Delta\zeta$ above the w levels; however the same vertical index is used for the two corresponding levels.

$$\left(\frac{u\pi}{m}\right)_{1jk} = -\left(\frac{u\pi}{m}\right)_{2jk}, \qquad \left(\frac{u\pi}{m}\right)_{I+1jk} = -\left(\frac{u\pi}{m}\right)_{Ijk},$$

$$\left(\frac{v\pi}{m}\right)_{i1k} = -\left(\frac{v\pi}{m}\right)_{i2k}, \qquad \left(\frac{v\pi}{m}\right)_{iJ+1k} = -\left(\frac{v\pi}{m}\right)_{iJk}.$$

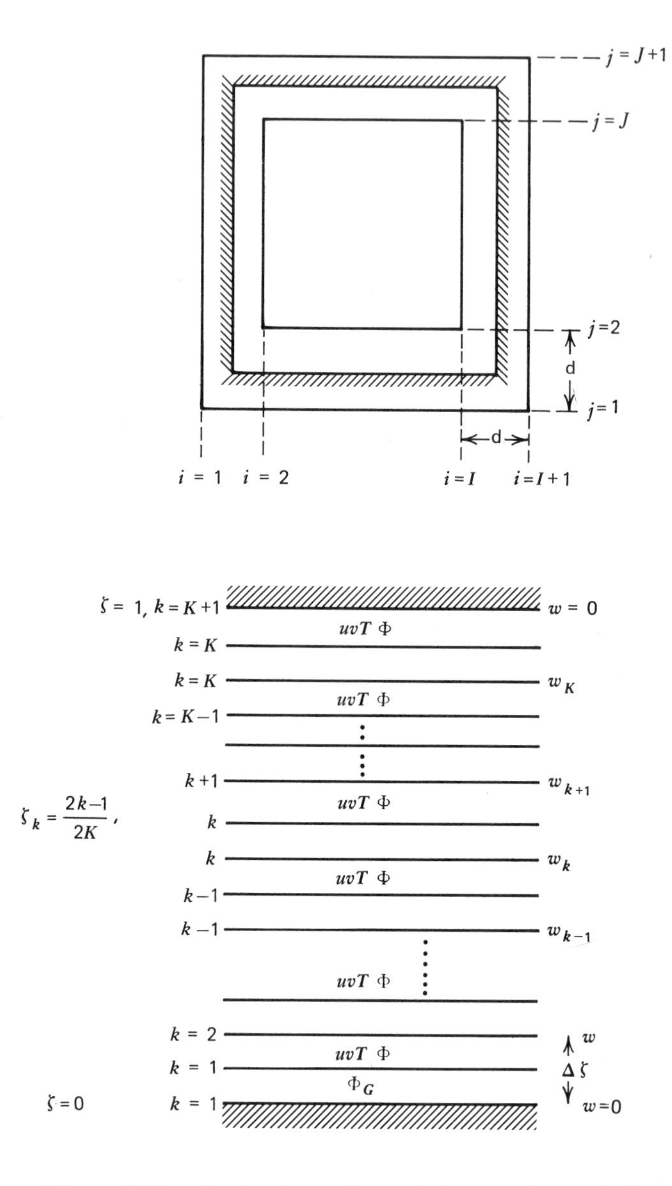

Figure 13.5. The horizontal boundaries and the vertical indexing for the Navy primitive equation model.

13.7.2 Energy Conservation for the Modified σ Coordinate System

The energy conservation properties of these finite difference equations will now be considered for which the time-continuous space-difference equations in one dimension are:

$$\left(\frac{\partial \pi u}{\partial t}\right)_{ik} = -\mathscr{L}(u)_{ik} + f\pi_i v_{ik} - m\pi\frac{\Phi_{i+1k} - \Phi_{i-1k}}{2d} - m_i RT_{ik}\frac{\pi_{i+1} - \pi_{i-1}}{2d}, \tag{13.44}$$

$$\left(\frac{\partial \pi v}{\partial t}\right)_{ik} = -\mathscr{L}(v)_{ik} - f\pi_i u_{ik}, \tag{13.45}$$

$$\left(\frac{\partial \pi T}{\partial t}\right)_{ik} = -\mathscr{L}(T)_{ik} + \frac{\kappa T_{ik}}{\sigma_k} \times \left\{-\pi_i\frac{w_{ik+1} + w_{ik}}{2} + \sigma_k\left[\frac{\partial \pi_i}{\partial t} + \frac{m_i u_{ik}(\pi_{i+1} - \pi_{i-1})}{2d}\right]\right\}, \tag{13.46}$$

$$\mathscr{L}(\psi)_{ik} = \frac{m_i^2}{4d}\left\{\left[\left(\frac{u\pi}{m}\right)_{i+1k} + \left(\frac{u\pi}{m}\right)_{ik}\right](\psi_{i+1k} + \psi_{ik}) - \left[\left(\frac{u\pi}{m}\right)_{ik} + \left(\frac{u\pi}{m}\right)_{i-1k}\right] \times (\psi_{ik} + \psi_{i-1k})\right\} + \frac{\pi S_k}{2\,\Delta\zeta}[w_{ik+1}(\psi_{ik+1} + \psi_{ik}) - w_{ik}(\psi_{ik} + \psi_{ik-1})]. \tag{13.47}$$

First it will be shown that squared quantities are conserved by the advective terms. In this connection note that for any parameter ψ

$$\frac{\partial}{\partial t}\left(\frac{\pi\psi^2}{2}\right) = \psi\frac{\partial(\pi\psi)}{\partial t} - \frac{\psi^2}{2}\frac{\partial \pi}{\partial t}. \tag{13.48}$$

Now consider the sum in one horizontal dimension and substitute into the foregoing expression from (13.47) and (13.37), with $v = 0$:

$$\begin{aligned}\sum_{i,k}\frac{1}{m_i^2 S_k}\left[-\psi_{ik}\mathscr{L}(\psi)_{ik} - \frac{\psi_{ik}^2}{2}\frac{\partial \pi_i}{\partial t}\right] &= \sum_{i,k} -\frac{\psi_{ik}}{4\,dS_k}[(\alpha_{i+1k} + \alpha_{ik})(\psi_{i+1k} + \psi_{ik}) - (\alpha_{ik} + \alpha_{i-1k})(\psi_{ik} + \psi_{i-1k})] \\ &\quad - \frac{\pi_i\psi_{ik}}{2m_i^2\,\Delta\zeta}[w_{ik+1}(\psi_{ik+1} + \psi_{ik}) - w_{ik}(\psi_{ik} + \psi_{ik-1})] \\ &\quad + \frac{\psi_{ik}^2}{4\,dS}(\alpha_{i+1k} - \alpha_{i-1k}) + \frac{\pi_i\psi_{ik}^2}{2m_i^2\,\Delta\zeta}(w_{ik+1} - w_{ik}) \\ &= \sum_{i,k}\frac{-1}{4\,dS_k}[(\alpha_{i+1k} + \alpha_{ik})\psi_{ik}\psi_{i+1k} - (\alpha_{ik} + \alpha_{i-1k})\psi_{ik}\psi_{i-1k}] \\ &\quad - \frac{\pi_i}{2m_i^2\,\Delta\zeta}(w_{ik+1}\psi_{ik}\psi_{ik+1} - w_{ik}\psi_{ik}\psi_{ik-1}),\end{aligned} \tag{13.49}$$

where

$$\alpha = \frac{\pi u}{m}.$$

The quantity in brackets in the last equality vanishes when summed horizontally (with respect to i), provided that $\alpha = 0$ at lateral boundaries between the last two columns of gridpoints. Similarly, the last term in the foregoing equation vanishes when summed vertically if w vanishes at the top and bottom of the atmosphere. Note that the total mount of any property A, that is, kinetic or potential energy, for the entire atmosphere is given by

$$\begin{aligned}\int A\rho\, dx\, dy\, dz &= \frac{1}{g}\int A\, dx\, dy\, dp \\ &= \frac{1}{g}\int A\frac{dX}{m}\frac{dY}{m}\pi\, d\sigma \\ &= \frac{1}{g}\int \frac{\pi}{Sm^2} A\, dX\, dY\, d\zeta\end{aligned}$$

where $dX = m\, dx$, $dY = m\, dy$, $dp = \pi\, d\sigma$, and $d\zeta = -S\, d\sigma$.

These results will now be used to show that the mean kinetic energy is conserved. Since it has been shown by (13.49) that the advective terms conserve the mean square of any quantity, in this case u or v, and the Coriolis force makes no contribution to kinetic-energy production, only the pressure force terms in (13.44) need be considered. Thus the mean change in kinetic energy is given by

$$\begin{aligned}&\sum_{ik}\frac{1}{m_i^2S_k}\frac{\partial}{\partial t}\left(\pi\frac{u^2+v^2}{2}\right)_{ik} \\ &= \sum_{ik}\frac{1}{m_i^2S_k}\left[u_{ik}\left(\frac{\partial\pi u}{\partial t}\right)_{ik} + v_{ik}\left(\frac{\partial\pi v}{\partial t}\right)_{ik} - \frac{u_{ik}^2+v_{ik}^2}{2}\frac{\partial\pi_i}{\partial t}\right] \\ &= \sum_{ik}\left(\frac{\pi_i u_{ik}}{m_iS_k}\frac{\Phi_{i+1k}-\Phi_{i-1k}}{2d} - \frac{Ru_{ik}T_{ik}}{m_iS_k}\frac{\pi_{i+1}-\pi_{i-1}}{2d}\right) \\ &= \sum_{ik}\left\{\frac{\Phi_{ik}[(\pi u/m)_{i+1k} - (\pi u/m)_{i-1k}]}{2\,dS_k}\right. \\ &\quad - \frac{[(\pi u/m)_{i+1k}\Phi_{ik} + (\pi u/m)_{ik}\Phi_{i+1k} - (\pi u/m)_{ik}\Phi_{i-1k} - (\pi u/m)_{i-1k}\Phi_{ik}}{2\,dS_k} \\ &\quad \left. - \frac{R}{m_iS_k}\frac{u_{ik}T_{ik}(\pi_{i+1}-\pi_{i-1})}{2d}\right\} \\ &= \sum_{ik}\left[-\frac{\Phi_{ik}\pi_i}{m_i^2\,\Delta\zeta}(w_{ik+1}-w_{ik}) - \frac{\Phi_{ik}}{m_i^2S_k}\frac{\partial\pi_i}{\partial t} - \frac{Ru_{ik}T_{ik}}{m_iS_k}\frac{\pi_{i+1}-\pi_{i-1}}{2d}\right. \\ &\quad \left. - \frac{(\pi u/m)_{i+1k}\Phi_{ik} + (\pi u/m)_{ik}\Phi_{i+1k} - (\pi u/m)_{ik}\Phi_{i-1k} - (\pi u/m)_{i-1k}\Phi_{ik}}{2\,dS_k}\right]\end{aligned} \tag{13.50}$$

The continuity equation (13.37) has been used to obtain the last form. The last term vanishes for periodic boundaries.

The potential energy change is given by

$$\sum_{ik} \frac{1}{m_i^2 S_k} \frac{\partial}{\partial t}(c_p \pi_i T_{ik}) = R \sum_{ik} \left[- \frac{T_{ik}\pi_i(w_{ik+1} + w_{ik})}{m_i^2 S_k 2\sigma_k} + \frac{T_{ik}}{m_i^2 S_k} \frac{\partial \pi_i}{\partial t} + \frac{T_{ik} u_{ik}}{m_i S_k} \frac{\pi_{i+1} - \pi_{i-1}}{2d} \right], \quad (13.51)$$

which follows from (13.46), wherein the term $\mathscr{L}(T)_{ik}$ cancels when summed over the grid. Recall that $\kappa = R/c_p$.

Comparison of the kinetic and potential energy equations, (13.50) and (13.51), shows that the total energy will be conserved if

$$\sum_k \frac{\Phi_{ik}(w_{ik+1} - w_{ik})}{\Delta\zeta} + \sum_k \frac{RT_{ik}}{S_k \sigma_k} \frac{w_{ik+1} + w_{ik}}{2} = 0 \quad (13.52)$$

and

$$\sum_k \frac{\Phi_{ik}}{S_k} = \sum_k \frac{RT_{ik}}{S_k}. \quad (13.53)$$

When (13.52) is expanded from $k = 1$ to $K + 1$ and the vertical boundary conditions are imposed, it is easily seen that the sum will be zero, provided

$$\frac{\Phi_{k+1} - \Phi_k}{\Delta\zeta} = \frac{R}{2}\left(\frac{T_{k+1}}{S_{k+1}\sigma_{k+1}} + \frac{T_k}{S_k \sigma_k}\right), \quad (13.54)$$

which is a suitable approximation for the hydrostatic equation $\partial\Phi/\partial\zeta = RT/S\sigma$ for the layer from k to $k + 1$. Expansion of (13.53) gives

$$\frac{\Phi_1}{S_1} + \ldots + \frac{\Phi_K}{S_K} = R\left(\frac{T_1}{S_1} + \ldots + \frac{T_K}{S_K}\right) \quad (13.55)$$

Since the thickness of each of the layers $(\Phi_{k+1} - \Phi_k)$ is known for all $1 \leq k \leq K - 1$ from (13.54), (13.55) represents a condition on Φ_1. If (13.54) is used to eliminate progressively the Φ's, Φ_1 can be expressed in the form

$$\Phi_1 = \Phi_G + R\,\Delta\zeta \sum_{k=1}^{K} \lambda_k T_k, \quad (13.56)$$

where Φ_G is the geopotential at the ground and the λ's are coefficients that depend on the temperature, vertical "map factors," $S_k = \Delta\zeta/\Delta\sigma_k$, and the equality,

$$\Delta\zeta \sum_{k=1}^{K} S_k^{-1} = 1.$$

When the vertical-grid spacing is the same in both σ and ζ coordinates, that is, $\Delta\sigma \equiv \Delta\zeta$, then $S_k \equiv 1$ for all k. In this case (13.56) reduces to simply

$$\Phi_1 = \Phi_G + \frac{R\,\Delta\zeta}{2}\frac{T_1}{\sigma_1}, \tag{13.57}$$

which is a one-sided difference approximation for the hydrostatic equation. This relationship may be established directly from (13.55), which reduces to simply $\sum \Phi_k = \sum RT_k$. The last equation may be rewritten for the present purpose as

$$K\Phi_1 + (K-1)(\Phi_2 - \Phi_1) + \ldots + 2(\Phi_{K-1} - \Phi_{K-2}) + (\Phi_K - \Phi_{K-1}) = \sum_{k=1}^{K} RT_k.$$

Next, each of the geopotential differences may be replaced with temperatures by means of (13.54) (with all S's equal to unity). The resulting equation reduces to (13.57) after some manipulation and use of the following relations:

$$K\,\Delta\zeta = 1, \qquad \sigma_k = \frac{K - k + \frac{1}{2}}{K}.$$

The model described here has been programmed by LCDR P. G. Kesel, USN, while a graduate student at the Naval Postgraduate School (1966–1968). During his subsequent assignment to the Fleet Numerical Weather Central Kesel and F. J. Winninghoff of FNWC and NPS adapted and developed the model for operational use. The operational version includes orography, solar and terrestrial radiation, a moisture equation, condensation, evaporation, parameterized convection, latent and sensible heat transfer, surface friction, and diffusion of momentum. The treatment of the diabatic processes is similar to that described in Chapters 9 and 10. The one-sided approximation of the hydrostatic equation (13.57) proved to be unsatisfactory and a better approximation for the mean temperature of the lowest layer is presently in use. The integration scheme is stable although some small-scale noise is generated in the mountainous areas. For output purposes this noise is removed by a filter. Following Kurihara, the pressure force is computed on pressure surfaces in the mountain areas and then interpolated to the σ surfaces. Thus far the results with the model for 36-hr forecasts have been very favorable in comparison with the previous filtered operational models. The inclusion of the diabatic processes has been crucial in several instances of spectacular development of cyclones which have deepened 30 mb or more in 24 hr.

The British Meteorological Office ten-level primitive equation designed by Bushby and Timpson (1967) has simpler treatments [see Gadd and Keers

(1970)] of solar and terrestrial radiation which appear to give quite satisfactory results for short-range forecasting. A heat-balance equation is used to determine surface temperature from which surface fluxes of sensible and latent heat are calculated over land. Over the oceans monthly mean sea-surface temperatures are used for surface temperature, which is a more or less standard procedure.

chapter fourteen
Objective Analysis

14.1 GENERAL OBJECTIVE ANALYSIS

To carry out a numerical forecast it is necessary to have the initial fields of the meteorological parameters. The discussion of this topic has been left to the last so that the reader is already familiar with the specific initial requirements of the various numerical models.

Perhaps the most obvious way of obtaining an objective analysis of a two-dimensional scalar field, such as temperature, pressure, geopotential, etc., would be to fit a two-dimensional surface to the observed data by the method of least squares. The first attempt of this kind for objective synoptic analysis was made by Panofsky (1949), who used a cubic polynomial of the following form in a prescribed area:

$$Z(x, y) = \sum_{i,j=0}^{3} a_{ij} x^i y^j, \qquad i + j \leq 3. \tag{14.1}$$

It may be desirable to review briefly the method of least squares at this point. Obviously the problem is to determine the coefficients a_{ij}. Since there are ten constants a_{ij} in (14.1), at least ten sets of observations (x, y, Z) are needed to determine unique values of the a_{ij}; any number less than ten sets makes the problem indeterminate. On the other hand, more than ten sets of observations overdetermines the problem in general. In the latter event, the coefficients may be determined by requiring the sum of the squares of the differences between computed and observed values of Z to be a minimum. Thus it is desired to determine coefficients a_{ij} which will minimize the sum

$$S = \sum_{n=1}^{N} \left(Z_{0n} - \sum_{ij} a_{ij} x_n^i \, y_n^j \right)^2,$$

where Z_{0n} is an observed value of Z corresponding to the position (x_n, y_n) and N is the number of observations. A necessary condition for S to be a minimum in this respect is that the partial derivatives of S with respect to the a_{ij} vanish, that is,

$$\frac{\partial S}{\partial a_{ij}} = 0, \qquad i + j \leq 3.$$

This requirement gives a set of ten equations in the ten unknown values a_{ij} which can be solved for the latter.

A difficulty with the Panofsky method is that at the boundaries of continuous areas, which were each fitted with a different polynomial, discontinuities occurred; however these could be removed by smoothing the resulting values.

To avoid this difficulty and also to obtain the values at a set of gridpoints as needed for numerical prediction, the procedure may be altered by locally fitting a quadratic surface of the form (14.1) in a specified neighborhood of each gridpoint in order to determine the gridpoint value of the scalar being analyzed.

In the case of pressure-height analyses, wind observations may also be introduced through the geostrophic approximation

$$\frac{\partial Z}{\partial x} = \frac{fv}{g}, \qquad \frac{\partial Z}{\partial y} = -\frac{fu}{g}.$$

In this event the quantity to be minimized in applying the least-squares technique would be

$$\sum_i (Z_i - Z_{oi})^2 + W\left\{\sum_j \left[\left(\frac{\partial Z}{\partial x}\right)_j - \left(\frac{fv_o}{g}\right)_j\right]^2 + \left[\left(\frac{\partial Z}{\partial y}\right)_j + \left(\frac{fu_o}{g}\right)_j\right]^2\right\}. \quad (14.2)$$

Here the subscript o denotes an observed value of height or wind, W is a weight factor which specifies the importance placed on each type of observation and the indices i, j range over all the observations. By taking the gridpoint as the origin, the value of a_{oo} in (14.1) is the gridpoint value desired. If there are insufficient observations in the neighborhood of any gridpoint, it may be omitted on a first scan and then picked up later by using values subsequently calculated at adjacent gridpoints.

Nevertheless, in regions of sparse data, as are often encountered in hemispheric or global analysis, the foregoing interpolation method may not lead to a reasonable analysis. In manual analysis this difficulty is generally overcome by using a judicious combination of the previous analysis (say

6 or 12 hr ago), the prognosis, and perhaps the climatological mean. This principle was soon adapted to the objective analysis schemes when numerical weather prediction became operational in the mid-fifties. For example, the guess field for pressure-height might be

$$Z^{(0)} = \frac{\sum W_i Z_i}{\sum W_i} \tag{14.3}$$

where the Z_i include the value 12 hr earlier, the forecast value for the present time, and the normal value. The W_i are weight factors which depend on the location, season, mean accuracy of the forecast technique, etc.

Modifications of the initial guess field are based on observations. For example, suppose there is a height observation Z_o at point 0 in the neighborhood of a gridpoint G. A new estimate of the height at the gridpoint $Z_G^{(1)}$ could then be obtained from the relation

$$Z_{G1}^{(1)} - Z_G^{(0)} = Z_o - Z_o^{(0)}, \tag{14.4}$$

where $Z_o^{(0)}$ is the initial estimate of the value of Z at the observation point obtained by interpolation from the initial guess field, which might be a forecast, climatology, persistence, etc.

If there is also a wind observation at point O, another estimate of the height value at the gridpoint may be obtained by calculating the gradient of height geostrophically, as illustrated in Figure 14.1. The point G represents a gridpoint, O is an observation point, $\mathbf{V}$ is the wind at O, $\mathbf{r}$ is the position vector

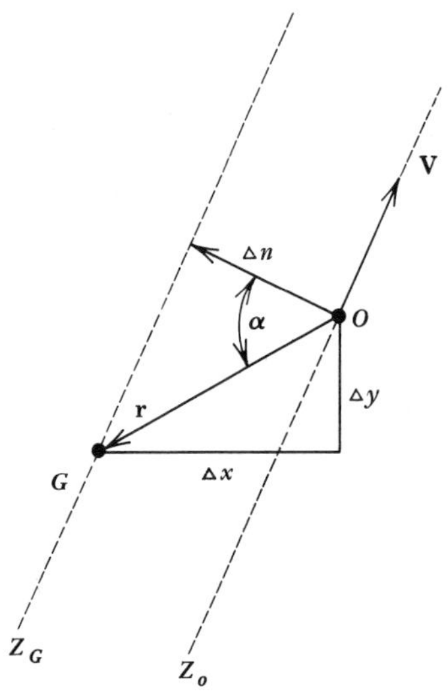

Figure 14.1. Estimation of the pressure height at a gridpoint G from the observed wind vector at point O.

of G relative to O, ∇Z is the gradient of height, and Δn is the normal distance between the height contours Z_G and Z_o. Then

$$\Delta Z = Z_G - Z_o = r\frac{\Delta Z}{\Delta n}\cos\alpha = \mathbf{r}\cdot\nabla Z.$$

The geostrophic relation gives $g\nabla Z = \mathbf{V} \times f\mathbf{k}$, thus

$$Z_G = Z_o + \frac{f}{g}\mathbf{r}\cdot(\mathbf{V}\times\mathbf{k}) \tag{14.5}$$

or, since $\mathbf{r} = -(\Delta x\mathbf{i} + \Delta y\mathbf{j})$,

$$Z_{G2}^{(1)} = Z_0 + \frac{f}{g}(u\,\Delta y - v\,\Delta x). \tag{14.6}$$

The previous procedure suggests a third alternative utilizing only the height observation, namely, the assumption that the gradient of height of the initial guess field is representative in the neighborhood of the gridpoint, in which case,

$$Z_{G3}^{(1)} = Z_o + \mathbf{r}\cdot\nabla Z_G^{(0)}. \tag{14.7}$$

The best approximation of Z_G may be a linear combination of the various estimates as follows:

$$Z_G = \frac{\sum W_{Gi}Z_{Gi}}{\sum W_{Gi}}, \tag{14.8}$$

where the W_{Gi} are weight factors which may be determined statistically by the method of least squares.

Cressman (1959) modified the methods used prior to that time by using weighted corrections to the initial guess field which are functions of the distance d from the observation to the gridpoint. The weight factors for corrections have the form

$$\begin{aligned} W &= \frac{n^2 - r^2}{n^2 + r^2}, \qquad r \le n, \\ &= 0, \qquad\qquad\quad r > n, \end{aligned} \tag{14.9}$$

where n is a multiple of the grid distance d. The correction procedure is repeated with decreasing values of n.

It should be mentioned that prior to an objective analysis such as described in the foregoing, the observational data are first decoded and then undergo a preanalysis processing which involves various gross-error checks to eliminate or correct erroneous data. Usually included here is a recalculation of upper-air soundings to check for hydrostatic consistency. The various error tolerances depend on the variability of the field as determined from climatology.

14.2 FNWC OBJECTIVE ANALYSIS OF SCALAR DATA

The scheme for the analysis of a two-dimensional scalar field of data F used by the Navy Fleet Numerical Weather Central (FNWC) begins by forming the finite difference Laplacian of the initial guess field F_{ij}:

$$L_{ij} = \nabla^2 F_{ij},$$

where ∇ is the finite difference operator defined earlier in (5.57). Next the L_{ij} field may be smoothed if considered desirable by

$$\bar{L}_{ij} = L_{ij} + k\nabla^2 L_{ij}, \qquad 0 < k < \tfrac{1}{4}. \tag{14.10}$$

Now the observed data are introduced by determining corrections to the initial guess field based on the observed data. Consider an observed value at a point (I, J), where

$$I = i + \Delta i, \qquad J = j + \Delta j, \qquad 1 \geq \Delta i \geq 0, \qquad 1 \geq \Delta j \geq 0.$$

Here (i, j) is a gridpoint and, for simplicity, the grid distance is taken as unity. First interpolate in the initial guess field to obtain the guess value at the point (I, J) using Bessel's central difference formula:

$$\begin{aligned} F_{i',J} = F_{i',j+\Delta j} &= F_{i',j} + \Delta j(F_{i',j+1} - F_{i',j}) \\ &+ \frac{\Delta j(\Delta j - 1)}{4}(F_{i',j+2} - F_{i',j+1} + F_{i',j-1} - F_{i',j}) \\ &+ \frac{\Delta j(\Delta j - 1)(2\Delta j - 1)}{12}(F_{i',j+2} - 3F_{i',j+1} + 3F_{i',j} - F_{i',j-1}), \end{aligned} \tag{14.11}$$

for $i' = i - 1,\ i,\ i + 1,\ i + 2$.

Now repeat the interpolation horizontally with the four values of $F_{i',J}$ to give the value at (I, J), namely, $F_{I,J}$. Next compute the difference between the observed value $O_{I,J}$ at the point (I, J) and the interpolated value:

$$D = O_{I,J} - F_{I,J}. \tag{14.12}$$

At this point an error check is applied to the report. Let T denote a preassigned tolerance based on the type of report and the analysis scan number. Then, if $|D| < T$, the report passes the check. On the other hand, if $|D| \geq T$, the report fails the check and is ignored the rest of that scan. On the first scan, this test serves as a gross-error check by the comparison of the report with the value of the guess field at the point of observation. On subsequent scans the procedure provides a means of cross-checking each report against all other reports insofar as the latter have influenced the analysis through the

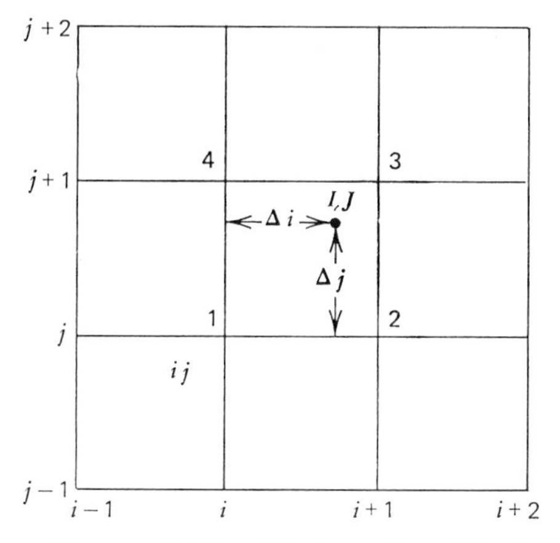

Figure 14.2. Block of gridpoints utilized for interpolating the value of an arbitrary parameter *F* at the point *I, J*.

previous scan. On any given scan a report may remain in the same status as previously or may change from a pass to fail status, or vice versa.

For each of the four gridpoints surrounding the observation, a weight factor is computed which is a function of the distance between the gridpoint and the observation as follows:

$$S_1 = 1 - (\Delta i^2 + \Delta j^2) \text{ if } S_1 \geq 0\text{; otherwise } S_1 = 0.$$

$$S_2 = 1 - [(1 - \Delta i)^2 + \Delta j^2] \text{ if } S_2 \geq 0\text{; otherwise } S_2 = 0.$$

$$S_3 = 1 - [(1 - \Delta i)^2 + (1 - \Delta j)^2] \text{ if } S_3 \geq 0\text{; otherwise } S_3 = 0.$$

$$S_4 = 1 - [\Delta i^2 + (1 - \Delta j)^2] \text{ if } S_4 \geq 0\text{; otherwise } S_4 = 0.$$

This implies that the report does not influence a gridpoint if it is greater than one gridlength away. Then these factors are normalized:

$$S'_k = \frac{S_k}{\sum_{m=1}^{4} S_m}, \qquad k = 1, 2, 3, 4, \tag{14.13}$$

and a reliability coefficient is applied depending on the type of report, giving

$$W_k = RS'_k, \qquad 0 < R < 8. \tag{14.14}$$

Now a weighted difference is computed for each gridpoint:

$$D_k = W_k D, \qquad k = 1, 2, 3, 4,$$

where D is defined earlier in (14.12). These weighted differences D_k are to be applied to each of the gridpoints surrounding the observation. However there may be a number of observations contributing to any particular gridpoint. Consequently the final correction applied will be an appropriate average of the weighted differences contributed from all observations within one mesh length of the gridpoint. For example, if there are N observations that contribute to the modification of the initial guess at a particular gridpoint, say (i, j), the next guess of the value of the scalar function at that gridpoint would be

$$F'_{ij} = F_{ij} + \frac{\sum_{n=1}^{N} W_{ijn} D_n}{\sum_{n=1}^{N} W_{ijn}}. \tag{14.15}$$

The next step consists in holding fixed those values of F that have been adjusted, that is, the F'_{ij}, and then solving the following Poisson equation for a new F field, say, $F_{ij}^{(1)}$, over the remaining gridpoints:

$$\nabla^2 F_{ij}^{(1)} = \bar{L}_{ij},$$

where $\bar{L}_{ij}$ is defined by (14.10). This procedure not only corrects the initial guess values at gridpoints adjacent to the observations but also spreads the influence of these observations into areas without data. The entire process may now be repeated by using the $F_{ij}^{(1)}$ as a starting point.

14.3 THREE-DIMENSIONAL ANALYSIS

It may be recalled from the discussion of multilevel models that, if the static stability parameter became zero or negative, the numerical integration procedure may break down. Hence it is important to ensure that the atmosphere is initially statically stable and remains so. In fact, it is to be expected that under unstable conditions the atmosphere would tend to adjust to a neutral or stable state. If objective analyses are prepared at several levels independently according to one of the two-dimensional techniques discussed earlier, there is no guarantee that the vertical temperature structure will be stable unless suitable precautions are taken. The latter may take various forms, one of which will be discussed briefly.

Holl et al. (1963) achieves this end in an eight-layer model extending from 1,000 mb to 100 mb. The procedure consists of first analyzing a static stability parameter σ defined in terms of the virtual potential temperature θ_v,

$$\sigma = -R p_0^{-\kappa} p^{1+\kappa} \frac{\partial \theta_v}{\partial p}, \qquad \kappa = \frac{R}{c_p}, \tag{14.16}$$

for each of eight layers (defined by interfaces at 775, 600, 450, 350, 275, 225, and 175 mb) based on observed pressure-height data at the ten mandatory levels (1,000, 850, 700, 500, 400, 300, 250, 200, 150, and 100 mb) with the condition that hydrostatic instability is not permitted at any gridpoint. The end product of the technique is to be a relationship between the heights of the mandatory levels on one hand, and on the other, the 1,000-mb height, a virtual temperature parameter, and the static stability parameters for each of the layers indicated in the foregoing. The temperature parameter is chosen to be either the 1,000 to 500-mb or the 1,000 to 300-mb thickness. If (14.16) is integrated through a layer of constant σ and θ_v is replaced by $T_v(p_0/p)^\kappa$, the result is

$$T_v = \sigma(R\kappa)^{-1} - Mgc_p^{-1}p^\kappa, \tag{14.17}$$

where σ and M are constants for the layer. Note that under the assumed constancy of σ, T_v is a linear function of p^κ. Also if $\theta_v = T_v(p_0/p)^\kappa$ is differentiated, dp is replaced from the hydrostatic equation, the equation is integrated through a similar layer, and then T_v is replaced by means of (14.17), the result after some manipulation is

$$z = N + Mp^\kappa - \sigma(g\kappa)^{-1} \ln p. \tag{14.18}$$

Although the static stability is discontinuous from one layer to another, the temperature and height are required to be continuous at the interfaces. Because of the greater density of surface data than upper-air data, the 1,000-mb height has been chosen as a reference height. Then, in essence, a sounding is constructed with the 1,000-mb height, the eight static stabilities, and either the 1,000 to 500-mb thickness or the 1,000 to 300-mb thickness; a sounding which fits the heights of the ten mandatory levels.

In terms of (14.18) and (14.17), there are three constants N, M, σ, which completely specify each of the eight layers, totaling 24 parameters. These are determinable from the ten mandatory heights and the two continuity conditions at each of the seven interfaces. This relationship may be expressed in matrix form as follows:

$$\boldsymbol{\sigma} = D\mathbf{Z}, \tag{14.19}$$

where $\boldsymbol{\sigma}$ is a column vector consisting of the components Z_{10} =1,000-mb height, H = 1,000 to 500-mb thickness (or 1,000 to 300-mb thickness), and the eight static stabilities. The vector $\mathbf{Z}$ is comprised of the ten mandatory heights, and D is a matrix of constants. The temperatures at the mandatory levels are also specified and may be expressed in similar form

$$\mathbf{T}_v = Q\mathbf{Z}. \tag{14.20}$$

It should be noted that these computed temperatures are not the observed values but, according to the authors, yield a representative virtual temperature structure which maintains vertical consistency and is statically stable.

When applying the mass structure model, the data at the mandatory levels in each observed sounding are converted by (14.19) to the 1,000-mb height, the 500-mb height, and the eight static stabilities. Then the 1,000-mb height, 500-mb height, and the static-stability fields are converted to gridpoint values by a two-dimensional objective analysis scheme as described earlier. The first guess for a stability analysis is the previous updated analysis smoothed and adjusted toward climatology in sparse-data areas. The height analysis is obtained by applying the objective analysis routine of Section 14.2 on heights for two passes, then gradients are adjusted to fit observed winds, then two more passes using heights again, and finally two more passes using winds. At this point the 500-mb height field is adjusted where necessary to remove features which are strongly hyperbolic. Now the height values at the 300-mb level are obtained from the inverse of D as follows:

$$\mathbf{Z} = D^{-1}\boldsymbol{\sigma}. \tag{14.21}$$

This 300-mb height field becomes a first guess which is adjusted to fit reported heights and winds similar to the 500-mb analysis procedure described earlier. Then, with H redefined as the 1,000 to 300-mb thickness, height values at all mandatory levels are now obtainable by (14.21). The shift to the 1,000 to 300-mb thickness takes advantage of the much denser coverage of aircraft reports at and near 300 mb and provides for a feedback of this information to the lower levels where such data are scarce. Virtual temperatures are generated for the mandatory levels by the model according to (14.17), and additionally temperatures and heights are available at the interfaces. However, these temperatures may differ considerably from observed values; consequently, a further objective analysis may be carried out by using the constructed temperatures as a first guess.

Experience has shown that further reanalysis of the height fields is necessary at and above 250 mb. Reported heights become increasingly unrepresentative with increasing altitude, much more so than the reported winds. The reanalyses of the higher levels therefore gives additional emphasis on the wind data. However this reanalysis of the higher pressure levels often results in negative static stabilities at various gridpoints. Hence (14.19) is applied to generate the eight static stabilities as defined by the ten mandatory level height fields. These stabilities are now examined for reasonableness and adjusted where necessary to ensure consistency. Then (14.21) is again applied, with $H =$ 1,000 to 300-mb thickness, to produce the final statically stable height fields. The Fleet Numerical Weather Central, Monterey, California, has been using the Holl analysis method with generally satisfactory results.

14.4 INITIALIZATION FOR PRIMITIVE EQUATION FORECASTS

The type of objective analysis described earlier has been generally quite satisfactory for numerical prognosis with the filtered equations (vorticity models); however the momentum (primitive) equations are much more sensitive to the initial state, the boundary conditions, and the differencing scheme.

The conventional analyses of wind and pressure data often contain large imbalances between the pressure and Coriolis forces which can lead to large-amplitude inertial-gravity waves if the latter are permitted in the prediction equations. These imbalances may often be due to observational errors or incorrect analyses. It is apparent that in nature dissipative processes and small-scale phenomena usually lead to a rapid adjustment of the atmosphere to a nearly balanced state with respect to large-scale motions, and large imbalances rarely occur, except those of a cyclostrophic type which are not relevant here. In any event it is unlikely that any imbalances implied by observed wind and pressure data or the objective analyses accurately represent the true state (excepting cyclostrophic types).

The initial conditions used most often for the primitive equations are obtained from the balanced system or are simply geostrophic winds calculated from the pressure-height analyses. This leads to pressure oscillations with periods of several or more hours, which may persist a day or more. In experiments conducted at FNWC, these oscillations were particularly noticeable over mountainous areas. The calculations of initial conditions from the complete balanced system partially allows for cyclostrophic effects but involves lengthy iterative calculations, and even then will not generally represent a truly balanced condition for the system of momentum equations, which include additional physical processes and a different mathematical formulation. As mentioned at the end of Chapter 2, an acceptable numerical integration procedure for the primitive equations must adequately simulate the geostrophic adjustment process in order to bring about the quasi-nondivergent quasi-geostrophic state which characterizes the atmosphere, and then maintain this state while correctly predicting the large-scale flow.

Miyakoda and Moyer (1968) proposed several schemes for initialization, which utilized the prediction equations to step forward two time increments from a balanced state, then backward to the initial time, say τ. During the latter step the divergent portion of the wind is retained while the rotational part is hindcast from time $\tau + 1$ to τ by a vorticity equation for the model. The u and v components of the wind are then recovered from the potential and stream functions and constitute the new value of the velocity with which to repeat the entire procedure. The method did not converge and was abandoned.

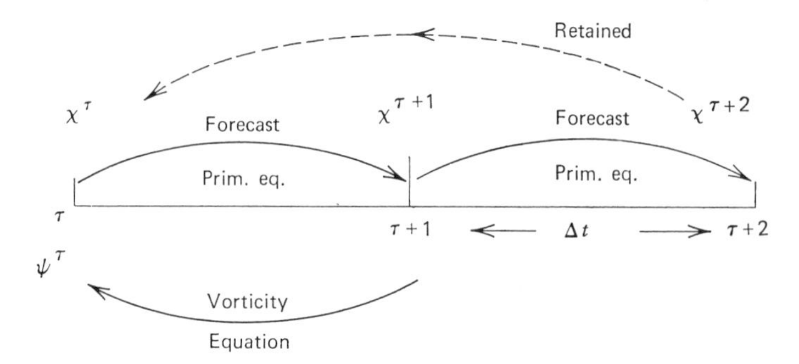

Figure 14.3. Schematic representation of an initialization procedure for a primitive equation prediction model.

A second procedure utilized the Euler "backward" time-differencing technique (see Table 13.1), which was shown to converge for a simple linear system and, moreover, has the desirable property of damping high-frequency modes. In other respects the method was similar to the one described previously, shown schematically in Figure 14.3.

As indicated in the diagram, two steps forward are taken at which time $(\tau + 2)$ the potential function $\chi^{\tau+2}$ is calculated from the predicted velocity components. This value then becomes the next guess value of χ at the initial time τ. From the primitive equations a vorticity equation is formed which is used to make a backward step from $\tau + 1$ to τ to get a new estimate of the rotational part of the wind. A disadvantage of the method is the need to solve Poisson-type equations and a vorticity equation as well.

Nitta and Hovermale (1967) describe several techniques for initialization, which are divided by the authors into two basic categories: *extrapolation methods* and *iteration methods*. The former utilize data from past times whereas the latter utilize data only at the initial time. Each method may be subdivided further according to whether the wind will be made to adjust to a controlled pressure-height field or vice versa. As described earlier (Section 2.8), with quasi-geostrophic motions the wind mainly adjusts to the pressure field when the Rossby deformation length is large, that is, when the characteristic length of the disturbance is large. For smaller scales the mass field tends to adjust more toward the wind field, which might be more appropriate for the tropics where the wind field is considered more reliable than the pressure field.

14.4.1 Extrapolation Methods

The extrapolation method is illustrated in Figure 14.4 where the wind is adjusted to the height field.

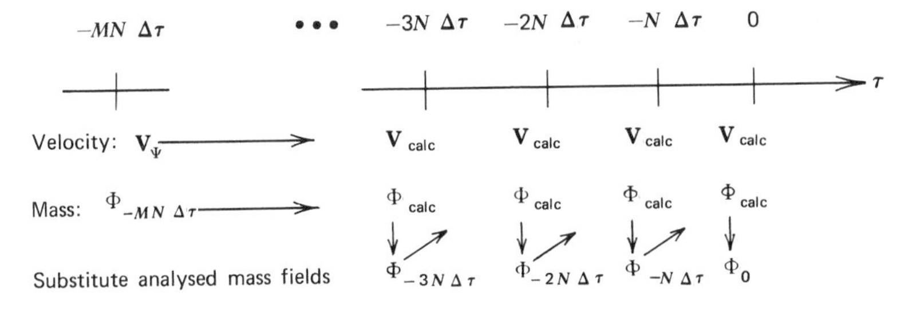

Figure 14.4. An extrapolation technique for determining initial data for a primitive equation prediction model.

Here **V** and Φ represent the wind and mass fields, $\Delta\tau$ is the model time step, N is the number of time steps between recoveries, and M is the number of recoveries. Thus the process begins $MN\,\Delta\tau$ time units prior to the time $\tau = 0$ for which the initial fields are desired to begin the actual numerical forecast. The first estimate of the wind field at $-MN\,\Delta\tau$ may be calculated or computed winds, which together with the analyzed mass field at that time, $\Phi_{-MN\,\Delta\tau}$, permits a forecast for $N\,\Delta\tau$ time steps. Then the mass field is replaced by the analyzed observed field at that time and the process is repeated, and so forth. In this procedure with the wind field may develop serious phase lags due to truncation error. An alternate approach is to solve for the rotational and divergent wind components at each recovery and retain only the forecast divergent part of the wind, which can be combined with the rotational part of the wind V_ψ derived from the Φ_0 field. In this way the phase-lag problem would be eliminated from the largest part of the wind.

Several experiments were carried out by Nitta and Hovermale with data taken from the 17th day of a control run with a two-level model; one experiment was made with a nondivergent wind field by using the basic extrapolation method with $M = 3$, $N = 76$, $\Delta t = 15$ minutes. They also made some experiments with the National Meteorological Center (NMC) primitive equation model. In general the first extrapolation method for both the divergent and rotational wind components proved to be time consuming and inaccurate. The second method, which determined only the divergent portion of the wind by extrapolation, resulted in a forecast almost identical to the NMC operational forecast. Thus it appeared to offer no advantage over conventional initialization; however, the tests were too limited to draw any final conclusions.

14.4.2 Iteration Methods

The iteration methods utilize the observed initial data and then integrate backward and forward in time in some fashion to adjust the wind and

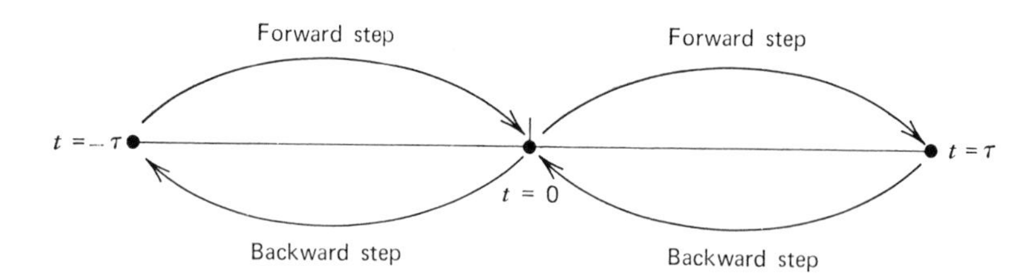

Figure 14.5. Schematic representation of an iterative forward-backward method of initialization for a primitive equation prediction model.

pressure fields until, hopefully, they are representative of the initial large-scale meteorological variables, so that numerical weather forecasts can be made without unrealistic oscillations of the pressure, wind, temperature, etc.

The method of Miyakoda and Moyer cited earlier is an example of an iterative method. A similar technique utilizing forward and backward steps was employed by Nitta and Hovermale (1967) as illustrated in Figure 14.5. Their method utilizes the primitive equations of the model and the Euler two-step time-differencing procedure. After each forward-backward cycle is completed, either the pressure height or the wind field (whichever is considered most accurate) is recovered. The other field is continually adjusted during repeated cycles until a state of large-scale equilibrium is reached, the high-frequency oscillations having been selectively damped. Matsuno (1966) used this method which is now illustrated for the following simple prognostic equations:

$$\begin{aligned} \frac{\partial \mathbf{V}}{\partial t} &= -\nabla\Phi - f\mathbf{k} \times \mathbf{V}, \\ \frac{\partial \Phi}{\partial t} &= -H\nabla \cdot \mathbf{V}. \end{aligned} \tag{14.22}$$

First form a forward time step with the Euler scheme:

$$\begin{aligned} u^* &= u(0) - \alpha\nabla_x\Phi(0) + f\tau v(0), \\ v^* &= v(0) - \alpha\nabla_y\Phi(0) - f\tau u(0), \\ \Phi^* &= \Phi(0) - \alpha H[\nabla_x u(0) + \nabla_y v(0)], \end{aligned} \tag{14.23}$$

$$\begin{aligned} u^{(0)}(\tau) &= u(0) - \alpha\nabla_x\Phi^* + f\tau v^*, \\ v^{(0)}(\tau) &= v(0) - \alpha\nabla_y\Phi^* - f\tau u^*, \\ \Phi^{(0)}(\tau) &= \Phi(0) - \alpha H(\nabla_x u^* + \nabla_y v^*), \end{aligned} \tag{14.24}$$

where $\alpha = \tau/2d$, d is the grid distance, and

$$\nabla_x\Phi = (\Phi_{i+1,j} - \Phi_{i-1,j}).$$

Next a backward time step is taken:

$$\begin{aligned} u^{**} &= u(\tau) + \alpha \nabla_x \Phi(\tau) - f\tau v(\tau), \\ v^{**} &= v(\tau) + \alpha \nabla_y \Phi(\tau) + f\tau u(\tau), \\ \Phi^{**} &= \Phi(\tau) + \alpha H[\nabla_x u(\tau) + \nabla_y v(\tau)], \end{aligned} \tag{14.25}$$

$$\begin{aligned} u^{(1)}(0) &= u(\tau) + \alpha \nabla_x \Phi^{**} - f\tau v^{**}, \\ v^{(1)}(0) &= v(\tau) + \alpha \nabla_y \Phi^{**} + f\tau u^{**}, \\ \Phi^{(1)}(0) &= \Phi(\tau) + \alpha H(\nabla_x u^{**} + \nabla_y v^{**}). \end{aligned} \tag{14.26}$$

At this stage the mass (or wind) field may be set equal to the original value; that is, $\Phi^{(1)}(0)$ is replaced by $\Phi(0)$. It would also be possible to adjust both wind and Φ partially toward their original values. The latter fields normally consist of objective analyses together with a geostrophic or balanced pressure-wind relationship.

This iterative technique was applied in a numerical experiment with the two-layer primitive equation model mentioned earlier and was integrated for 17 days. Then the geopotential fields and rotational wind fields were used as "initial" fields, and the iterative process was applied 150 times. From the "initial guess" of no divergence, a divergent field was recovered that was remarkably similar to the "observed" divergence, and, moreover, the rotational wind components were well preserved during the forward and backward steps. Nevertheless there did remain errors in the wind field, as evidenced by an rms error of about 1.2 m/sec.

As of this time, NMC is using a balanced wind initialization together with 12-hr forecast vertical velocities in the Shuman primitive equation model (personal communication from Dr. Shuman). Pressure oscillations of about 6 mb with a period of approximately 4 hr are produced. The forecast vertical motions make the forecasts somewhat more "noisy" than the use of a zero initial vertical velocity, but the latter gives more systematic errors which is considered more undesirable. A light time smoother is used to form the finished product.

The basic goal of the initialization techniques just described is to determine precisely large-scale wind and pressure fields which are in appropriate balance. Presumably the initialization procedure would follow an objective analysis; the latter, however, has the same goal. Therefore it may be desirable to combine the procedures, if feasible. Nitta and Hovermale and others suggest that the data be carried in terms of the coordinates of the prediction system, that the observations be introduced on the coordinate surfaces directly by the interpolation method discussed earlier, and then final adjustments be made through an iterative method utilizing the Euler two-step time-differencing scheme. The iterative scheme has the flexibility of either allowing the wind to adjust to the pressure field or vice versa, depending on which parameter is considered more reliable.

14.4.3 Initialization for the Navy Primitive Equation Model

The initialization for the NAVY primitive equation model has been primarily based on the linear balance equation. According to LCDR Peter Kesel, USN (personal communication), the total pressure oscillation has been 3 to 5 mb at the onset of integration, then reducing to somewhat less than 3 mb after about 12 hr. This oscillation appears to be a result of (*a*) external gravity waves with periods of 1 to 2 hr, which apparently diminish to negligible values in about 6 hr; (*b*) internal gravity waves with periods of 4 to 8 hr, which diminish but are still evident after 36 hr; and (*c*) inertial oscillations, which diminish but also are present after 36 hr. Some experiments at FNWC with initialization based on the balanced system did not show a significant improvement over the geostrophic start. Dr. F. J. Winninghoff of NPS and FNWC has also experimented with an iterative initialization scheme using the Euler two-step time-differencing, as shown schematically in Figure 14.6. After each time step Δt, the surface pressure and the upper temperatures which define the mass field are restored half-way back to the initial value at $t = 0$. The entire cycle is repeated until the oscillation in the divergence field is negligible. Thus, in adjusting the wind field toward the pressure field, the latter undergoes some change while experiencing "geostrophic adjustment." By the end of the 18 cycles the mass and wind fields generally approximate an equilibrium state suitable for initialization of the numerical forecast. Although the technique appears to give somewhat better results, it is too time-consuming at present for operational application.

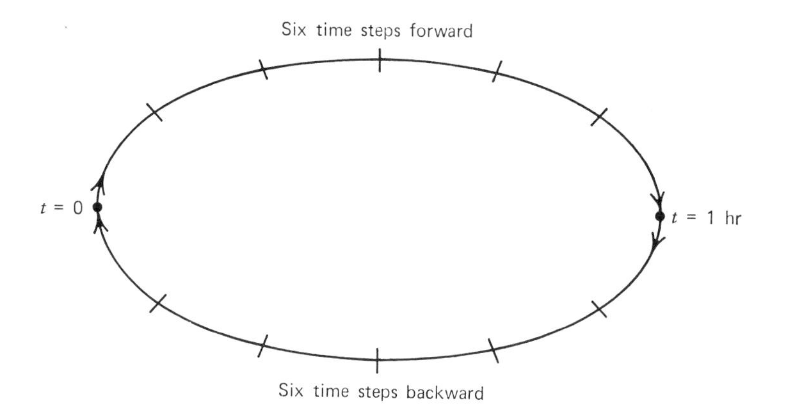

Figure 14.6. Schematic representation of initialization procedure used for the Navy primitive equation prediction model.

14.5 NUMERICAL VARIATIONAL OBJECTIVE ANALYSIS

Sasaki (1958, 1969, 1969) has proposed a sophisticated objective analysis-initialization scheme based on the calculus of variations with the purpose of suppressing the high-frequency noise contained in initial data and yielding dynamically consistent fields in sparse-data areas.

In order to illustrate the basic principle of the variational method, let $F(x, y, y')$ be a twice differentiable function of the three designated variables. Now suppose it is desired to determine the function $y = f(x)$ such that the following integral is a minimum (or maximum):

$$I = \int_a^b F(x, y, y')\, dx, \tag{14.27}$$

where the endpoints a and b are fixed as well as $f(a) = A$ and $f(b) = B$. Following classical theory, suppose there is a small change in y, say from $f(x)$ to $f(x) + \epsilon g(x)$, where $g(x)$ is an arbitrary function for which $g(a) = g(b) = 0$ and ϵ is a parameter. Then there will be corresponding changes δF and δI in F and I, and we may write

$$\delta I = \int_a^b \delta F\, dx, \tag{14.28}$$

where the δ operator denotes the variation in I and F due to the variations δy and $\delta y'$ in y and y'. Therefore

$$\delta F = \frac{\partial F}{\partial y}\delta y + \frac{\partial F}{\partial y'}\delta y'. \tag{14.29}$$

From geometrical considerations or, formally from the definition of $\delta y = \epsilon g(x)$, it may be seen that $\delta(dy/dx) = d(\delta y)/dx$. It follows that

$$F_{y'}\,\delta y' = -\left(\frac{d}{dx}F_{y'}\right)\delta y + \frac{d}{dx}(F_{y'}\,\delta y), \tag{14.30}$$

where the subscripts denote partial derivatives. Integrating this expression over the interval (a, b) will give zero for the last term since δy vanishes at a and b in the case under investigation. As a consequence, substitution of (14.30) and (14.29) into (14.28) gives

$$\delta I = \int_a^b \left(F_y - \frac{d}{dx}F_{y'}\right)\delta y\, dx.$$

If indeed the function $f(x)$ gives a minimum for I, the variation δI must be zero for $y = f(x)$. Moreover since δy is arbitrary, the integral for δI will vanish only if the quantity in parentheses vanishes. Thus a necessary condition for a minimum of I is

$$F_y - \frac{d}{dx} F_{y'} = 0, \tag{14.31}$$

which is generally referred to as the Euler equation or the Euler–Lagrange equation. This result may also be obtained by setting $dI/d\epsilon = 0$ at $\epsilon = 0$. Expanding (14.31) gives

$$F_y - F_{xy'} - F_{yy'}y' - F_{y'y'}y'' = 0, \tag{14.32}$$

which is a second-order differential equation that must be satisfied by $y = f(x)$ for the integral I to be a minimum.

If the function F contains higher derivatives, say y'', the Euler equation is extended to include the term $+d^2F_{y''}/dx^2$. On the other hand, if the integral depends on two functions $y(x)$ and $z(x)$, that is, if $F = F(x, y, z, y', z')$, then there are two Euler equations to be satisfied, (14.31) and a similar one with z and z' in place of y and y'.

Sometimes it is desired to minimize an integral of the form (14.27), subject to a constraint which may be another integral condition, or in the case of two functions, another equation, for example, $G(x, y, z) = 0$. In the latter case a Lagrange multiplier $\lambda(x)$ may be introduced, and the resulting Euler–Lagrange equations are

$$F_y - \frac{d}{dx} F_y - \lambda G_y = 0 \qquad \text{and} \qquad F_z - \frac{d}{dx} F_z - \lambda G_z = 0. \tag{14.33}$$

To illustrate the application of variational theory to meteorological analysis, assume that there are independent functions (Φ_0, T_0) representing the observed geopotential and temperature in the vertical, and it is desired to minimize the differences between the analyzed and observed values subject to the hydrostatic equation as a strict condition. Then an appropriate integral to minimize might be

$$I = \int_0^{P_2} [\alpha(\Phi - \Phi_0)^2 + \beta(T - T_0)^2] \, dP,$$

where α and β are constant weight factors assigned to each variable, and the vertical coordinate is $P = -R \ln (p/p_1)$. The hydrostatic equation can now be written as

$$\frac{d\Phi}{dP} = T,$$

which permits T to be replaced directly by $d\Phi/dP$ in the integrand. The latter is then of the form $F(P, \Phi, \Phi')$, where $\Phi' = d\Phi/dP$.

The Euler equation, $F_\Phi - d(F_{\Phi'})/dP = 0$, which must be satisfied for a minimum value of I, becomes

$$\frac{d^2\Phi}{dP^2} - \frac{\alpha}{\beta}\Phi = -\frac{\alpha}{\beta}\Phi_0 + \frac{dT_0}{dP}. \tag{14.34}$$

Thus the appropriate value of Φ which will minimize I satisfies a second-order differential equation. This equation can be solved numerically, given the values of Φ at $P = 0$ and P_2. After the Φ values have been obtained, the temperature is computed from the hydrostatic equation.

Consider the same problem but assume the observations are known at equally spaced intervals of P denoted by the index k. Then a corresponding sum to be minimized would be

$$I = \sum_k \alpha(\Phi_k - \Phi_{k0})^2 + \beta(T_k - T_{k0})^2.$$

The variation of I is given by

$$\delta I = \sum 2\alpha(\Phi - \Phi_0)\,\delta\Phi + 2\beta(T - T_0)\,\delta T, \tag{14.35}$$

where the index has been omitted to simplify the notation. The hydrostatic equation may be approximated with a central difference as follows:

$$\nabla_P\Phi = T,$$

where

$$\nabla_P\Phi = \frac{\Phi_{k+1} - \Phi_{k-1}}{2\Delta P}. \tag{14.36}$$

Now T may be replaced by $\nabla_P\Phi$ in (14.35), giving

$$\delta I = 2\sum \alpha(\Phi - \Phi_0)\,\delta\Phi + \beta(\nabla_P\Phi - T_0)\nabla_P\,\delta\Phi. \tag{14.37}$$

Next it may be verified that at the interior points

$$\begin{aligned} \sum \beta\nabla_P\Phi\nabla_P\,\delta\Phi &= -\sum \beta\,\delta\Phi\nabla_P(\nabla_P\Phi), \\ \sum T_0\nabla_P\,\delta\Phi &= -\sum \delta\Phi\nabla_P T_0, \end{aligned} \tag{14.38}$$

which corresponds to the integration by parts of (14.30).

When these results are substituted into (14.37) and the variational principle is applied by requiring the coefficient of $\delta\Phi$ to vanish, the result is

$$\nabla_P^2\Phi - \frac{\alpha}{\beta}\Phi = -\frac{\alpha}{\beta}\Phi_0 + \nabla_P T_0,$$

which is a finite difference equivalent of (14.34). This equation is a one-dimensional elliptic Helmholtz type equation for Φ and can be solved by relaxation, given the upper and lower boundary values. The central differences are not usable at the boundaries and must be replaced by forward and backward differences which will satisfy the conditions required by (14.38).

Consider next a somewhat more realistic analysis problem where it is desired to analyze fields of Φ, T, u, and v, subject to the hydrostatic and geostrophic laws. Then an appropriate quantity to be minimized might be

$$\int [\alpha(\Phi - \Phi_0)^2 + \beta(T - T_0)^2 + \gamma(fv - fv_0)^2 + \gamma(fu - fu_0)^2]\, dx\, dy\, dP.$$

The hydrostatic and geostrophic conditions can be strictly imposed by replacing T by Φ_P, as before, fv by Φ_x, and fu by $-\Phi_y$. Then the integrand is of the form $F(x, y, P, \Phi, \Phi_x, \Phi_y, \Phi_P)$, and the Euler-Lagrange equation is

$$F_\Phi - \frac{\partial}{\partial x}\frac{\partial F}{\partial p} - \frac{\partial}{\partial y}\frac{\partial F}{\partial q} - \frac{\partial}{\partial P}\frac{\partial F}{\partial r} = 0,$$

where $p = \Phi_x$, $q = \Phi_y$, $r = \Phi_P$. For the preceding integral the Euler equation to be solved for Φ is

$$\gamma(\Phi_{xx} + \Phi_{yy}) + \beta\Phi_{PP} - \alpha\Phi = -\alpha\Phi_0 + \beta\frac{\partial T_0}{\partial P} + \gamma\mathbf{k}\cdot\nabla\times f\mathbf{V}_0.$$

An analogous finite difference equation may be obtained by minimizing a sum similar to the integral.

If it is further desired to impose the thermal wind relationship on the analysis, terms of the form $\xi[\Phi_{Px} - f(\partial v_0/\partial P)]^2$ and $\xi[\Phi_{Py} + f(\partial u_0/\partial P)]^2$ should be added to the integrand. Then the Euler equation must be augmented by several third- and fourth-order partial derivatives, namely,

$$-\xi\left(\Phi_{PPxx} + \Phi_{PPyy} - \frac{\partial}{\partial x}\frac{f\,\partial^2 v_0}{\partial P^2} + \frac{\partial}{\partial y}\frac{f\,\partial^2 u_0}{\partial P^2}\right),$$

which would complicate the problem considerably.

In any event, after the geopotential field has been obtained by solution of the Euler equation, the temperature and wind fields are calculated from the hydrostatic and geostrophic relationships, which have been strictly imposed in this example. If it is desired that these conditions only hold approximately, say for the geostrophic relations, then the following additional terms would be added to the integrand:

$$\mu(fv - \Phi_x)^2 + \mu(fu + \Phi_y)^2.$$

In this case the Φ, u, and v must be treated as separate functions and there would be three simultaneous Euler–Lagrange equations to be solved for these variables.

The preceding example may be considered as an objective analysis technique oriented toward initialization for the filtered equations. Sasaki (1969) has illustrated a recent version of his analysis technique with a barotropic model directed toward initialization for the primitive equations:

$$\frac{\partial \mathbf{V}}{\partial t} + \mathbf{V} \cdot \nabla \mathbf{V} + f\mathbf{k} \times \mathbf{V} + \nabla \Phi = 0,$$
$$\frac{\partial \Phi}{\partial t} + \nabla \cdot \Phi \mathbf{V} = 0. \tag{14.39}$$

The high-frequency components may be suppressed by requiring *near* steadiness as a condition on the analysis, namely,

$$\frac{\partial \mathbf{V}}{\partial t} \doteq 0, \qquad \frac{\partial \Phi}{\partial t} \doteq 0. \tag{14.40}$$

The objective analyses of the u, v, and Φ from gridpoint "observations" $\tilde{u}$, $\tilde{v}$, and $\tilde{\Phi}$ are obtained by minimizing the following sum:

$$\sum_{ij} \{\tilde{\alpha}(\Phi - \tilde{\Phi})^2 + \tilde{\gamma}(u - \tilde{u})^2 + \tilde{\gamma}(v - \tilde{v})^2 + \alpha_t(\nabla_t \Phi)^2 + \gamma_t(\nabla_t u)^2 + \gamma_t(\nabla_t v)^2$$
$$+ \alpha[(\nabla_x \Phi)^2 + (\nabla_y \Phi)^2] + \gamma_u[(\nabla_x u)^2 + (\nabla_y u)^2] + \gamma_v[(\nabla_x v)^2 + (\nabla_y v)^2]\}. \tag{14.41}$$

The ∇'s are finite difference operators which are defined in (14.44). The obvious purpose of the terms involving $\tilde{\alpha}$ and $\tilde{\gamma}$ is to minimize the differences between the analyzed fields and the observations, analogous to the method of least squares, while the α_t, γ_t terms provide for the condition of quasi-steadiness (14.40). The $\alpha, \gamma_u, \gamma_v$ terms are introduced to make the corresponding Euler–Lagrange equations elliptic and solvable by relaxation methods, but they also serve as low-pass filters to suppress short wavelengths. The dynamic requirement of quasi-steadiness (14.40), which is incorporated directly in the sum to be minimized, is expressible in finite difference form as follows:

$$\nabla_t \Phi = -\nabla_x(\langle\Phi\rangle u) - \nabla_y(\langle\Phi\rangle v),$$
$$\nabla_t u = f\langle v\rangle - \nabla_x \Phi - (u\overline{\nabla}_x u + \langle v\rangle \overline{\nabla}_y u), \tag{14.42}$$
$$\nabla_t v = -f\langle u\rangle - \nabla_y \Phi - (\langle u\rangle \overline{\nabla}_x v + v\overline{\nabla}_y v).$$

The notation used in (14.42) is defined in (14.44) with the aid of Figure 14.7. Utilizing the expressions for $\nabla_t \Phi$, etc., in (14.44), the variation is taken by holding the α's and γ's constant as usual. Equating the coefficients of the variations $\delta\Phi$, δu, and δv to zero as required for a minimum leads to the

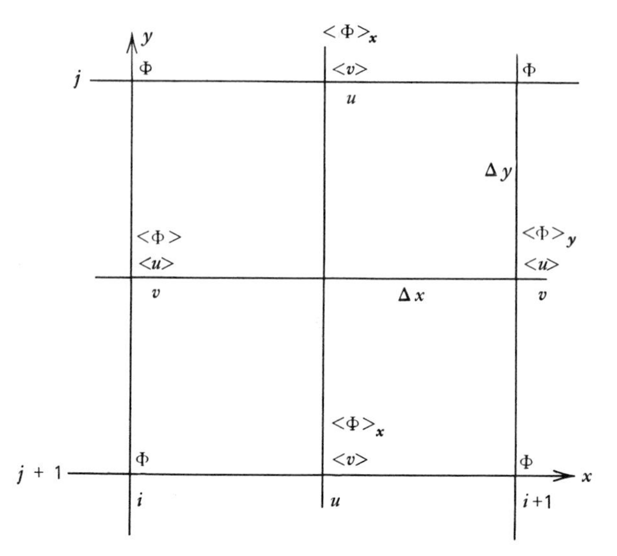

Figure 14.7. Specification of meteorological parameters at gridpoints for Sasaki's (1969) variational method of objective analysis.

following simultaneous set of Euler–Lagrange equations:

$$
\begin{aligned}
&\tilde{\alpha}(\Phi - \tilde{\Phi}) + \gamma_t(\nabla_x\nabla_t u + \nabla_y\nabla_t v) + \alpha_t(\langle u\nabla_x\nabla_t\Phi\rangle_x + \langle v\nabla_y\nabla_t\Phi\rangle_y) - \alpha\nabla^2\Phi = 0,\\
&\tilde{\gamma}(u - \tilde{u}) + \alpha_t\langle\Phi\rangle_x\nabla_x\nabla_t\Phi - \gamma_t f\langle\nabla_t v\rangle\\
&+ \gamma_t[\overline{\nabla}_x(u\nabla_t u) + \overline{\nabla}_y(\langle v\rangle\nabla_t u) - \nabla_t u\,\overline{\nabla}_x u - \langle\nabla_t v\overline{\nabla}_x v\rangle] - \gamma_u\nabla^2 u = 0, \qquad (14.43)\\
&\tilde{\gamma}(v - \tilde{v}) + \alpha_t\langle\Phi\rangle_y\nabla_y\nabla_t\Phi + \gamma_t f\langle\nabla_t u\rangle\\
&+ \gamma_t[\overline{\nabla}_x(\langle u\rangle\nabla_t v) + \overline{\nabla}_y(v\nabla_t v) - \langle\nabla_t u\,\overline{\nabla}_y u\rangle - \nabla_t v\overline{\nabla}_y v] - \gamma_v\nabla^2 v = 0.
\end{aligned}
$$

The various finite difference operators are defined in terms of the staggered grid in Figure 14.7 as follows:

$$
\begin{aligned}
\langle\Phi\rangle_x &= \tfrac{1}{2}(\Phi_{ij} + \Phi_{i+1,j}), \qquad \langle\Phi\rangle_y = \tfrac{1}{2}(\Phi_{ij} + \Phi_{i,j+1}),\\
\langle\Phi\rangle &= \tfrac{1}{4}(\Phi_{ij} + \Phi_{i+1,j} + \Phi_{i,j+1} + \Phi_{i+1,j+1}),\\
\nabla_t\Phi &= \frac{1}{2\Delta t}(\Phi_{n+1} - \Phi_{n-1}),\\
\nabla_x\Phi &= \frac{1}{2\Delta x}(\Phi_{i+1,j} - \Phi_{ij}), \qquad (14.44)\\
\overline{\nabla}_x\Phi &= \frac{1}{4\Delta x}(\Phi_{i+1,j} - \Phi_{i-1,j}),\\
\nabla^2\Phi_{ij} &= \frac{1}{\Delta x^2}(\Phi_{i+1,j} + \Phi_{i-1,j} - 2\Phi_{ij}) + \frac{1}{\Delta y^2}(\Phi_{i,j-1} + \Phi_{i,j+1} - 2\Phi_{ij}).
\end{aligned}
$$

Similar definitions apply to the y direction which increases in the direction of decreasing j index. The solutions of (14.43) for u, v, and Φ are obtained by a convergent relaxation method based on a linearized set of these equations. A geostrophic initial field is shown to be a suitable first guess for the iterative technique. Sasaki applied the method to the analysis of 500-mb height and wind data of hurricane Dora, 1200 GCT, September 8, 1964. Part of the data were deliberately not used, and later were compared to the analysis as a measure of the accuracy of the analysis technique. Some of the weight factors used were:

$$\tilde{\alpha} = \alpha_t = \frac{1}{(200 \text{ m}^2/\text{sec}^2)^2},$$

$$\tilde{\gamma} = \gamma_t = \frac{1}{(10 \text{ m/sec})^2}.$$

The denominators are the squares of the assumed amplitudes of geopotential and wind speed anomalies. The objective analyses obtained generally compared favorably with a subjective analysis except near the hurricane center due to the paucity of data.

14.5.1 Further Remarks on Initialization and Prediction

In the preceding discussion it was tacitly assumed that all observed data were synchronous; hence the principal objective of the analysis and initialization procedure was to specify the values of the dependent variables over a three-dimensional grid at the observation time. Then with these grid values as initial data, the prediction equations are integrated forward in time to yield a forecast of the meteorological variables at a future time. It has been indicated, of course, that the initial specification must be done with great care so that the data fields adequately represent the quasi-geostrophic, quasi non-divergent large-scale, low frequency meteorological disturbances. If this is not done properly, large, high frequency inertial-gravity oscillations will be generated which can give unrealistic pressure changes and completely obscure the meteorological forecast. In nature, as a consequence of the geostrophic adjustment mechanism, most of the energy of the atmosphere is found in the low-frequency motions and little in the inertial-gravity waves.

In actual practice considerable meteorological data is not sychronous with the regular observation times, and the analysis-initialization procedures described here are not specifically designed to incorporate such off-time data. This problem is becoming more acute as increasing amounts of meteorological data are now available from satellites and other sources which are essentially continuously measuring devices. In fact, any global prediction system will have to depend heavily on satellite measurements. Consequently, a better analysis-prediction system must be devised which can more or less continuously incorporate new data.

Naturally, a better and more complete specification of the initial data and a finer resolution may be expected to improve the accuracy of the forecasts. Additionally, a better simulation of the physical processes is needed, particularly the parameterization of the sub-grid scale phenomena of turbulence, convection, radiation, etc., which transfer heat, momentum and moisture and strongly influence the large-scale atmospheric circulation. In fact, not only does the earth's surface exercise a great influence on the motions in the atmosphere through the exchange of heat, momentum, and moisture, but the atmosphere, in turn, has a strong controlling influence on the ocean circulation.

In the not very distant future, coupled atmosphere-ocean global models will be used for prediction purposes. The feasibility of such a system is, of course, dependent on an increase in the computing and data retrieval speed of the computers by several orders of magnitude, however such computers have already been designed and will be operational by the latter half of this decade.

14.6 SMOOTHING AND FILTERING†

In various places in the text it has been mentioned that because of numerical errors, computational instabilities, etc., there may be a spurious growth of short waves, which if left unchecked could blow up and obscure a meteorological forecast. Even if a catastrophic instability does not arise it is usually desirable to eliminate very high wave number components for aesthetic reasons. In spectral models short waves can be automatically removed by systematically omitting wave numbers larger than some arbitrarily chosen value. In the finite-difference integration schemes, explicit or implicit diffusion can be included in the difference equations to suppress the high wave numbers. Nevertheless, systematic explicit smoothing is very often included in operational prediction schemes, at least for the final products.

In order to gain some insight about smoothing techniques, consider the simple one-dimensional three-point operator

$$\bar{f}_j = (1 - S)f_j + \frac{S}{2}(f_{j+1} + f_{j-1}), \tag{14.45}$$

where $x = j\,\Delta x$ and S is a constant which may be negative. If the operator is applied to the harmonic form

$$f = Ae^{ikx}, \tag{14.46}$$

† The treatment here follows an article by R. Shapiro, "Smoothing, Filtering, and Boundary Effects", Reviews in Geophysics and Space Physics, **8**, pp. 359–387, 1970.

where the wave number $k = 2\pi/L$ and A is a constant which may be complex, the result is

$$\bar{f} = Rf. \tag{14.47}$$

Here R, referred to as the *response* function, is given by

$$R = 1 - S(1 - \cos k\,\Delta x) = 1 - 2S\sin^2 \pi\,\Delta x/L. \tag{14.48}$$

It is evident that this particular smoothing operator does not affect the wave number nor the phase (provided $R \geq 0$) of the original wave but only its amplitude. If $|R| > 1$, the wave is *amplified* by the operation, while $|R| < 1$ results in *damping*. Note that a negative R would be undesirable here since the operation would then produce a 180° phase shift from the original wave.

It may also be shown that the average value of the smoothed function approaches that of the original function as the domain increases. From (14.45) it follows that

$$\frac{1}{2J+1}\sum_{j=-J}^{J}\bar{f}_j = \frac{1}{2J+1}\sum_{j=-J}^{J} f_j + \frac{S}{2(2J+1)}\sum_{j=-J}^{J}(f_{j+1} - 2f_j + f_{j-1})$$

$$= \frac{1}{2J+1}\sum f_j + \frac{S}{4J+2}(f_{J+1} - f_J - f_{-J} + f_{-J-1}).$$

It may now be seen that the last term will approach zero as J approaches infinity, which establishes the desired result.

Consider next some special cases. With $S = \frac{1}{2}$ (14.45) becomes

$$\bar{f}_j = f_j + \tfrac{1}{4}(f_{j+1} - 2f_j + f_{j-1}) = \tfrac{1}{4}(f_{j+1} + 2f_j + f_{j-1}),$$

or

$$\bar{f}_j = f_j + \tfrac{1}{4}\nabla_x^2 f_j, \tag{14.49}$$

where

$$\nabla_x^2 f_j = f_{j+1} - 2f_j + f_{j-1}. \tag{14.50}$$

From (14.48) the response function with $S = \frac{1}{2}$ is

$$R(\tfrac{1}{2}) = 1 - \tfrac{1}{2}(1 - \cos k\,\Delta x) = 1 - \sin^2 \pi\,\Delta x/L = \cos^2 \pi\Delta x/L. \tag{14.51}$$

Now for $L = 2\Delta x$, $R = 0$, hence *two-gridlength waves will be removed* by the smoother (14.49). As pointed out in Chapter 12, wavelengths less than $2\Delta x$ cannot be resolved on a grid with a spacing of Δx, hence any such harmonics will appear in longer wavelengths through aliasing. Moreover according to (14.51), wavelengths larger than $2\Delta x$ will be damped by the smoother. For example, with $L = 10\Delta x$,

$$R(\tfrac{1}{2}) = \cos^2 \pi/10 = 0.905,$$

which is less than 10% reduction in amplitude. However if this smoothing operator is applied ten times the amplitude is reduced to 0.37 of the original value, and after one hundred applications a wavelength of $10\Delta x$ will have essentially vanished. For such repetitions of the simple operator (14.49) the response function is obviously

$$R^m(\tfrac{1}{2}) = \cos^{2m} \pi\, \Delta x/L. \tag{14.52}$$

Consider next the case $S = 1$, for which $\bar{f}_j = (f_{j+1} + f_{j-1})/2$. Then the response function turns out to be

$$R(1) = \cos 2\pi\, \Delta x/L, \tag{14.53}$$

which results in a 180° phase change for wavelengths between $4\Delta x/3$ and $4\Delta x$. With two applications of this operator the phase change is prevented since $R^2(1) = \cos^2 2\pi\, \Delta x/L$. This result is similar to (14.52) except that it applies to twice the grid distance.

It is apparent that repeated applications of the simple smoother (14.49) would be undesirable because of excessive damping of even medium and long waves. In fact, it would be desirable to leave waves longer than several gridlengths relatively unaffected. It is possible to accomplish this objective by a judicious combination of several smoothers which results in an effective filter for a specified wave band.

According to (14.48) the response function corresponding to a series of smoothers of the form (14.45) with coefficients $S_1, S_2, \ldots, S_n$ would be

$$R_n(S, k) = R_1 R_2 \ldots R_n = \prod_{i=1}^{n} (1 - 2S_i \sin^2 k\, \Delta x/2). \tag{14.54}$$

Consider the case with just two smoothing operators. Since it is generally desirable to remove two-gridlength noise, one of the two operators will be (14.53), hence $S_1 = \frac{1}{2}$. The second constant S_2 is undetermined as yet. Thus the response function may be written

$$R_2(S_2, k) = (1 - \sin^2 k\, \Delta x/2)(1 - 2S_2 \sin^2 k\, \Delta x/2) \tag{14.55}$$

It follows that $R_2 = 0$ for $k = \pi/\Delta x$, i.e., $L = 2\Delta x$. Moreover R approaches 1 as k approaches 0, i.e., as $L \to \infty$; so that infinitely long waves remain unaffected. In order to specify other properties of the response function (14.55), the first and second derivatives of R_2 with respect to k may be calculated with the results

$$\frac{dR_2}{dk} = -\Delta x \sin \frac{k\, \Delta x}{2} \cos \frac{k\, \Delta x}{2} (1 + 2S_2 \cos k\, \Delta x) \tag{14.56}$$

$$\frac{d^2R_2}{dk^2} = (\Delta x)^2[S_2 \sin^2 k\, \Delta x - \tfrac{1}{2} \cos k\, \Delta x(1 + 2S_2 \cos k\, \Delta x)]. \tag{14.57}$$

It is evident from (14.56) that the extrema, for which $dR_2/dk = 0$, occur at the following values of k:

$$k\,\Delta x/2 = 0,\ \pi/2,\ \pi,\ 3\pi/2,\ 2\pi,\ \ldots, \tag{14.58}$$

and also at the particular wave number k^* for which

$$\cos k^*\,\Delta x = -1/2S_2. \tag{14.59}$$

The values of k in (14.58) correspond to the wavelengths $L = \infty$, $2\Delta x$, Δx, $2\Delta x/3$, $\Delta x/2$, etc., and the corresponding values of R_2 are either 1 or 0. When the relation (14.59) is substituted into (14.57) it may be seen that R_2 is a maximum or a minimum according to whether S_2 is negative or positive, respectively. The corresponding value of R_2 at k^* is found to be

$$R_2 = -\frac{1}{8S_2}(2S_2 - 1)^2. \tag{14.60}$$

If there is to be no amplification of any wave, then R_2 cannot exceed 1. Substituting the maximum value $R_2 = 1$ into (14.60) gives the value $S_2 = -\frac{1}{2}$. On the other hand, if phase changes are to be avoided, the minimum value of the response function R_2 at k^* must be zero. Now for $R_2 = 0$, (14.60) gives $S_2 = \frac{1}{2}$. However, the use of this value in (14.55) merely corresponds to $m = 2$ in (14.52) and produces further damping of all wavelengths larger than $2\Delta x$. Since the objective is to restore the amplitudes of wavelengths larger than $2\Delta x$, the value $S_2 = -\frac{1}{2}$ is chosen with the resulting response

$$\begin{aligned} R_2(k) &= (1 - \sin^2 k\,\Delta x/2)(1 + \sin^2 k\Delta x/2) \\ &= (1 - \sin^4 k\,\Delta x/2). \end{aligned} \tag{14.61}$$

For values of $S_2 < -\frac{1}{2}$, there will be amplification of some wavelengths. If $2S_2 = -(1 + \epsilon)$, where $\epsilon > 0$, then

$$R_2(L) > 1 \qquad \text{when} \qquad \sin^2 \pi\,\Delta x/L < \epsilon/(1 + \epsilon).$$

This results in amplification of waves longer than a certain critical value and also certain short waves which are not resolvable on the grid system, but by aliasing can feed energy into longer waves. In general this is undesirable but may be acceptable under some circumstances.

If it is desired to restore the amplitude of a particular wavelength L' to the original value, the value of $k' = 2\pi/L'$ can be substituted into (14.55) with $R_2(S_2, k') = 1$. The appropriate value of S_2 is then found to be

$$S_2 = -\tfrac{1}{2}\sec^2 \pi\,\Delta x/L'. \tag{14.62}$$

The procedure described here for restoring the amplitude of wavelengths larger than $2\Delta x$ with the resulting response function (14.55) can be extended to more operators. A logical extension might be to find a constant S_3 corresponding to the response function

$$R_3 = (1 - \sin^4 k\,\Delta x/2)(1 - 2S_3 \sin^2 k\,\Delta x/2) \tag{14.63}$$

Proceeding as before, it is found that $S_3 = \pm i/4$. As a consequence, these values must be used as a complex conjugate pair in order for the smoothed result to remain real. Thus (14.63) should have had the form

$$R_4 = (1 - \sin^4 k\,\Delta x/2)(1 - 2S_3 \sin^2 k\,\Delta x/2)(1 + 2S_3 \sin^2 k\,\Delta x/2)$$

and the procedure now leads to $S_3 = \pm i/2$. The resulting response function is

$$\begin{aligned} R_4 &= (1 - \sin^4 k\,\Delta x/2)(1 - i \sin^2 k\,\Delta x/2)(1 + i \sin^2 k\,\Delta x/2) \\ &= (1 - \sin^8 k\,\Delta x/2) \end{aligned} \tag{14.64}$$

This procedure can be generalized to yield an operator which will filter two-gridlength "noise"; and, with neither amplification nor phase change of any wave, result in arbitrarily small damping of longer wavelengths. An operator with such a response function is referred to as a *low-pass filter*. For an n'th order operator consisting of constants $S_1 = \frac{1}{2}, S_2, S_3, \ldots, S_n$, there are two real smoothing elements and $(n - 2)$ elements comprised of complex conjugate pairs with n restricted to the values

$$n = 2^j, \qquad j = 0, 1, 2, \ldots \tag{14.65}$$

For $j \geq 1$, there are $(j - 1)$ conjugate pairs. For example, where $n = 8$, $j = 3$, and there are two conjugate pairs $\pm\frac{1}{2}i^{1/2}, \pm\frac{1}{2}ii^{1/2}$, in addition to the previous set corresponding to (14.64), namely, $S_1 = \frac{1}{2}, S_2 = -\frac{1}{2}, S_3 = \pm i/2$. The resulting response function is easily found to be

$$R_8 = (1 - \sin^{16} k\,\Delta x/2). \tag{14.66}$$

For a wavelength of $4\Delta x$, the response function is $R_4 = 0.93750$ and $R_8 = 0.99609$; while for $L = 3\Delta x$, $R_8 = 0.89989$.

In progressively obtaining the smoothing coefficients corresponding to the general response function $(1 - \sin^{2n} k\,\Delta x/2)$, the coefficients corresponding to the factor $(1 - \sin^n k\,\Delta x/2)$ would have already been determined; hence only those associated with the factor $(1 + \sin^n k\,\Delta x/2)$ would have to be found. These may be constructed for any value of n from the roots of the equation

$$(2S)^{n/2} = -1 \tag{14.67}$$

For example, with $n = 1$, $S = \frac{1}{2}$, with $n = 2$, $S = -\frac{1}{2}$, and with $n = 4$, $S = \pm i/2$, and so forth.

14.6.1 Two-Dimensional Smoothers

The smoothing procedures described previously for one dimension can be easily extended to two dimensions. For example, the simple smoother (14.45) can be applied sequentially to the two directions or simultaneously, namely,

$$\bar{f}_{ij}^{i,j} = \overline{\bar{f}_{ij}^{i}}^{j} = \overline{\bar{f}_{ij}^{j}}^{i} \tag{14.68}$$

$$\bar{f}_{ij}^{ij} = \tfrac{1}{2}(\bar{f}_{ij}^{i} + \bar{f}_{ij}^{j}). \tag{14.69}$$

The sequential application (14.68) clearly represents a 9-point operator, while (14.69) is a 5-point operator. Note that the order of applying the operators is immaterial. Expanding (14.68) and (14.69) gives

$$\begin{aligned}\bar{f}_{ij}^{i,j} &= f_{ij} + \frac{S}{2}(1 - S)\nabla^2 f_{ij} \\ &\quad + \frac{S^2}{4}(f_{i-1\,j-1} + f_{i-1\,j+1} + f_{i+1\,j-1} + f_{i+1\,j+1})\end{aligned} \tag{14.70}$$

$$\begin{aligned}\bar{f}_{ij}^{ij} &= f_{ij} + \frac{S}{4}(f_{i-1\,j} + f_{i+1\,j} + f_{i\,j-1} + f_{i\,j+1} - 4f_{ij}) \\ &\equiv f_{ij} + \frac{S}{4}\nabla^2 f_{ij}.\end{aligned} \tag{14.71}$$

If the simple harmonic function $Ae^{i(kx+ly)}$ is substituted into (14.70) and (14.71), it is easily found that the corresponding response functions are, respectively,

$$R(k, 1) = (1 - 2S\sin^2 k\,\Delta x/2)(1 - 2S\sin^2 l\,\Delta y/2) \tag{14.72}$$

$$R(k1) = 1 - S(\sin^2 k\,\Delta x/2 + \sin^2 l\,\Delta y/2) \tag{14.73}$$

The response function (14.72) corresponding to the 9-point operator (14.70) has the same form as (14.48) and hence has the advantage that the technique used earlier to generalize the one-dimensional operators is applicable.

14.6.2 Bandpass Filters

The emphasis in the discussion thus far has been on low-pass filters which remove the undesirable short wavelength components. However, this filter may be utilized to isolate and retain the short wavelengths. This is easily accomplished by substracting the filtered field from the original function, which is equivalent to a *high-bandpass* filter with the response function $(1 - R_n)$.

Linear combinations of operators may also be used to dampen or isolate a particular band of wavelengths, the result being a *bandpass filter*. Thus the response function of a bandpass filter would be $R_a(f) - R_b(f)$, provided $a < b$. Particular bands can be isolated by appropriate choices of a and b.

14.6.3 Boundary Effects

No mention has been made thus far about the effects of boundaries on the smoothed functions. However, it is evident from the simple smoothing operator (14.45) that a boundary value will influence the smoothed value at the first interior point after one application. With repeated applications of the smoothing operator the influence of the boundary value will propagate inward one gridpoint per application. The final result after a large number of applications depends on the form of the smoothing operator and the type of boundary conditions.

When zero boundary conditions are imposed, a non-zero mean value of the function may be decreased and spurious growth may occur for certain long-wave components at the coefficient $S_2 < -\frac{1}{2}$. This problem can sometimes be avoided by insuring that the function has a zero mean value over the domain. Of course, another way of avoiding this problem is to use an operator that does not amplify any wave.

In general, when dealing with finite domains, periodic boundary conditions are ideal, according to Shapiro. However, if the imposition of such periodicity is not feasible, a non-amplifying smoothing operator is probably best.

The following table after Shapiro shows the response of the low-pass filter (14.54) as a function of wavelength. The values are shown for a single application and for 1024 applications of an n-component smoother, $n = 1$ through $n = 8$.

Table 14.1 The response of an ideal low-pass filter as a function of $L/\Delta x$ and the number n of smoothing elements after a single application and after 1024 applications. (after Shapiro, 1970).

1 Application

$L/\Delta x$	$n \to 1$	2	4	8	16	32	64
2	0	0	0	0	0	0	0
3	0.25	0.44	0.68	0.90	1.00	1.00	1.00
4	0.50	0.75	0.94	1.00	1.00	1.00	1.00
6	0.75	0.94	1.00	1.00	1.00	1.00	1.00
10	0.90	0.99	1.00	1.00	1.00	1.00	1.00
15	0.96	1.00	1.00	1.00	1.00	1.00	1.00
20	0.98	1.00	1.00	1.00	1.00	1.00	1.00
50	1.00	1.00	1.00	1.00	1.00	1.00	1.00

Table 14.1. (*contd.*)

1024 Applications

$L/\Delta x$	$n \rightarrow 1$	2	4	8	16	32	64
2	0	0	0	0	0	0	0
3	0.00	0.00	0.00	0.00	0.00	0.90	1.00
4	0.00	0.00	0.00	0.02	0.98	1.00	1.00
6	0.00	0.00	0.02	0.98	1.00	1.00	1.00
10	0.00	0.00	0.92	1.00	1.00	1.00	1.00
15	0.00	0.15	1.00	1.00	1.00	1.00	1.00
20	0.00	0.54	1.00	1.00	1.00	1.00	1.00
50	0.02	0.98	1.00	1.00	1.00	1.00	1.00

appendix

Examples of Numerical Analyses and Forecasts

In this appendix some examples of computerized objective meteorological analyses and forecasts are presented. The charts displayed here were obtained from the Navy Fleet Numerical Weather Central, (FNWC) Monterey, California through the courtesy of LCDR P. G. Kesel, USN and LTJG P. H. Hildebrand, USNR.

CASE I

Figure A.1 (page 282) shows the FNWC surface analysis with isobars drawn at 4-mb intervals for the Northern Hemisphere at 00Z (Greenwich Civil Time), 30 January 1970, based on 4,426 reports and displayed on a polar stereographic map projection. A 63 by 63 gridpoint mesh has been used with a grid distance of 381 km, true at 60 degrees north latitude. Figure A.2 (page 283) shows the corresponding chart 36 hours later at 12Z, 31 January 1970.

Figures A.3 through A.7 are 36-hour sea-level pressure forecasts made from the 00Z, 30 January 1970 chart with five different versions of the Navy primitive equation model. The basic equations are described in Section 13.7; the principal modifications for moisture and radiation are discussed in Chapter 9, and the method of convective parameterization is discussed in Section 10.4. Figure A.3 (page 284) which will be termed the "control" forecast has idealized (heavily smoothed) mountains, linear balanced initial conditions, the "sponge" (or "restoration") boundary conditions mentioned in Section 5.4, diabatic heating including sensible heating from the surface, latent heat release from the both large-scale vertical ascent and parameterized convection, and solar and terrestrial radiation. Figure A.4 (page 285) shows a forecast for the same period with all conditions the same for the control model except that all forms of diabatic heating were omitted. Figure A.5 (page 286) is similar to the control forecast except that geostrophic initial winds are used. The forecast in Figure A.6 (page 287) utilizes two-way cyclic boundary conditions instead of the restoration boundaries, but otherwise it is identical to the control model. Figure A.7 (page 288) differs from the control run only with respect to orography which has considerably less smoothing than the idealized mountains used in the control run. The terrain heights for these more realistic mountains were derived by first blending the mean values derived by L. Berkofsky (AFCRL) from one degree latitude squares and modal heights obtained from one degree squares by L. C. Clarke of FNWC. This blend was then averaged over each square of the FNWC Northern Hemisphere 63 by 63 grid and finally filtered to remove two-gridlength noise.

No attempt will be made here to discuss all of the detailed differences between the various forecasts. However, merely to illustrate how significantly the diabatic heating, initialization, orography, boundary conditions, etc., can influence a forecast, the subsequent behavior of the cyclone with a central pressure of 1004.8 mb south of Japan at 00Z, 30 January 1970 (Figure A.1) will be examined. During the next thirty-six hours this low pressure center moved northeastward and deepened to about 965 mb, as may be seen from the verifying analysis, Figure A.2.

The 36-hour forecast with the control model (Figure A.3) predicted a central pressure of 962 mb for this cyclone, which agrees very closely with the observed value. On the other hand, the adiabatic model (Figure A.4) deepened the cyclone to 991 mb, which is only about one third of the observed decrease in central pressure. Similarly, the forecast with geostrophic initial conditions (Figure A.5) which predicted a central pressure of 981 mb for the Japan low, did not produce adequate intensification. The boundaries were evidently too far removed from this cyclone to yield significant differences between the control forecast and the forecast with the cyclic conditions shown in Figure A.6, which gave a central value of 961 mb. The different orography used in the A.7 forecast did have some effect on the Japan low and gave a slightly higher central pressure of 968 mb. Note also that these higher mountains gave the best forecast for the high pressure system over the western United States.

It is evident from these results that substantial differences in the forecasts can occur with the variations of a basic model described here. In this instance, the higher mountains with the diabatic model gave a better forecast over the Pacific northwest than the idealized mountains; the balanced initial winds

Table A.1. A comparison of the forecast central pressure of the Japan cyclone with various versions of the Navy primitive equation prediction model.

Figure	Description	Central Pressure (mb) of the Japan Cyclone
A.1	Initial analysis	1005
A.2	Verifying analysis	965
A.3	Control forecast	962
A.4	Adiabatic forecast	991
A.5	Geostrophic initial winds	981
A.6	Cyclic boundary condition	961
A.7	Higher mountains	968

and diabatic heating gave a better forecast of the Japan low than geostrophic initial winds or adiabatic flow. Table A.1 summarizes the results of the central pressure forecasts of the Japan cyclone with the various versions of the Navy primitive equation prediction model.

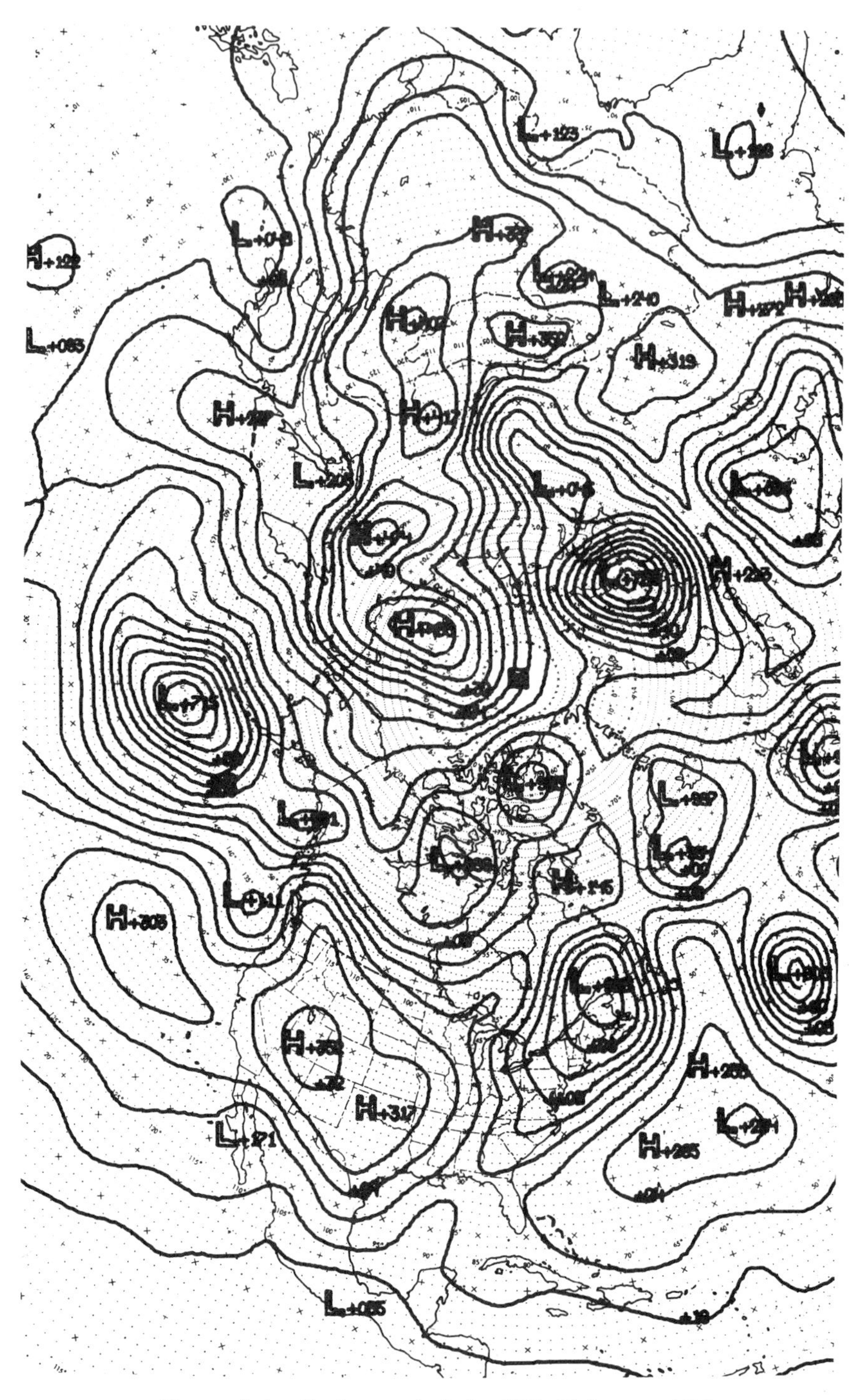

Figure A.1. Surface analysis for 00Z, 30 January 1970.

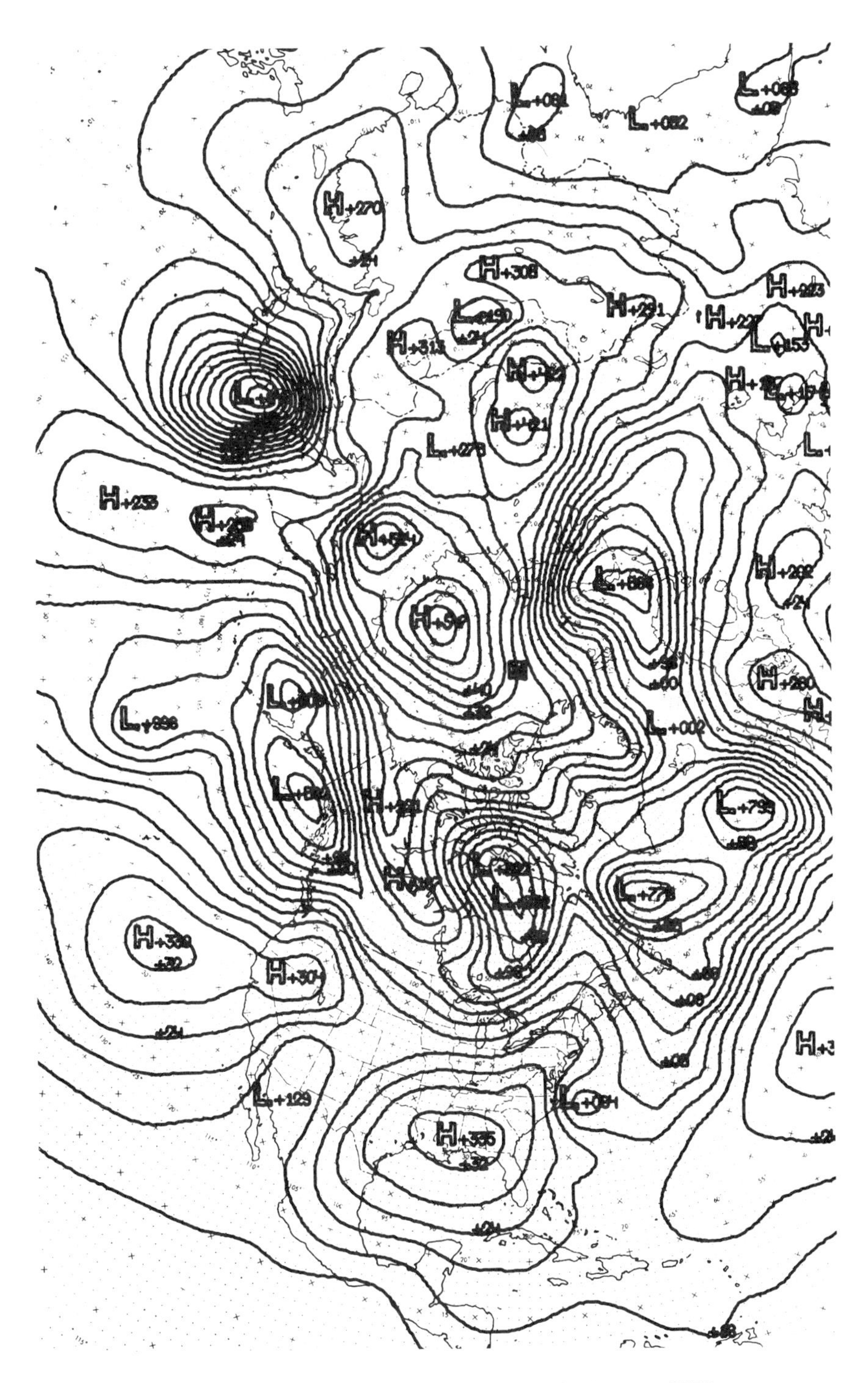

Figure A.2. Surface analysis for 12Z, 31 January 1970.

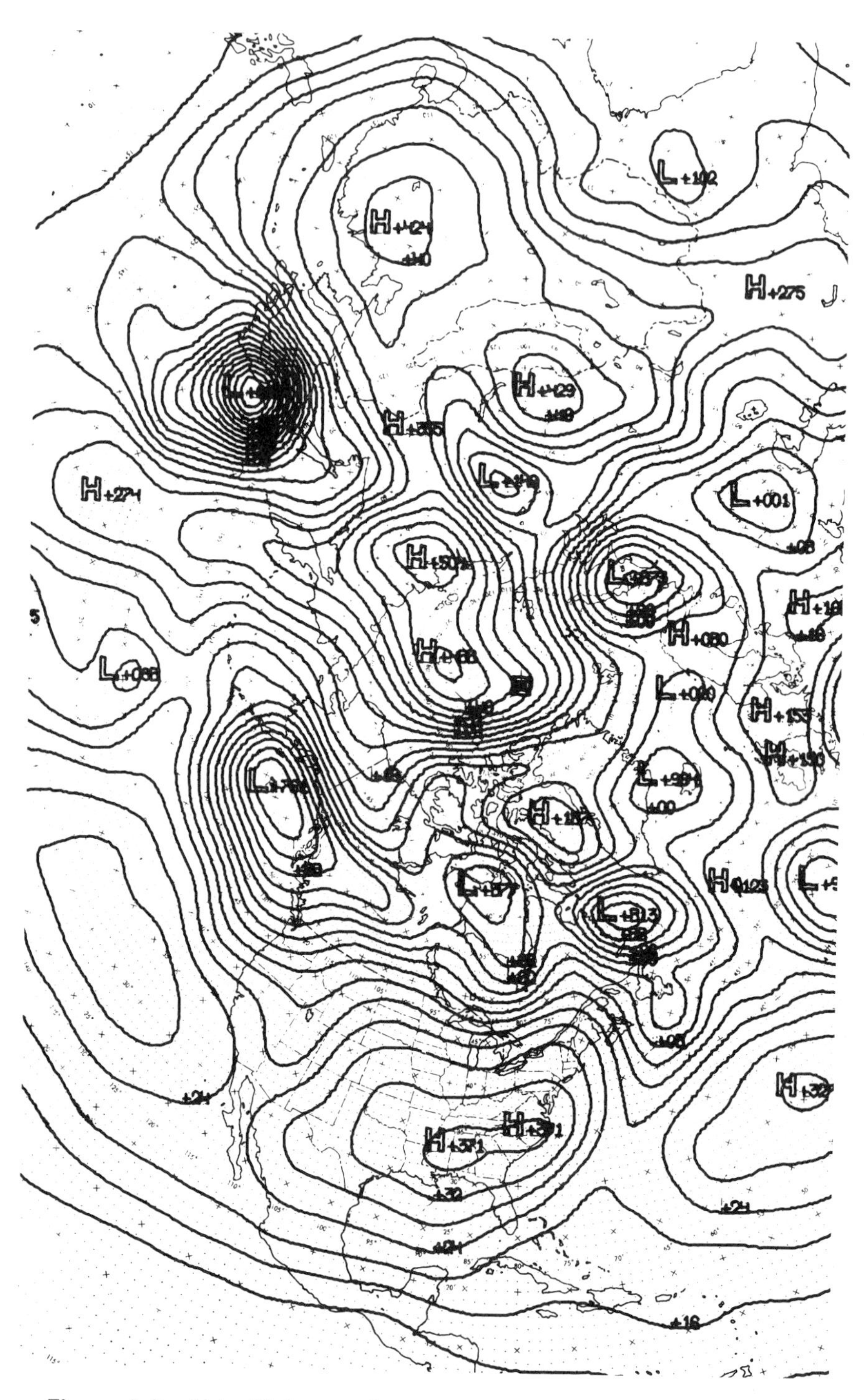

Figure A.3. 36-hr PE forecast (control), verifying at 12Z, 31 January 1970.

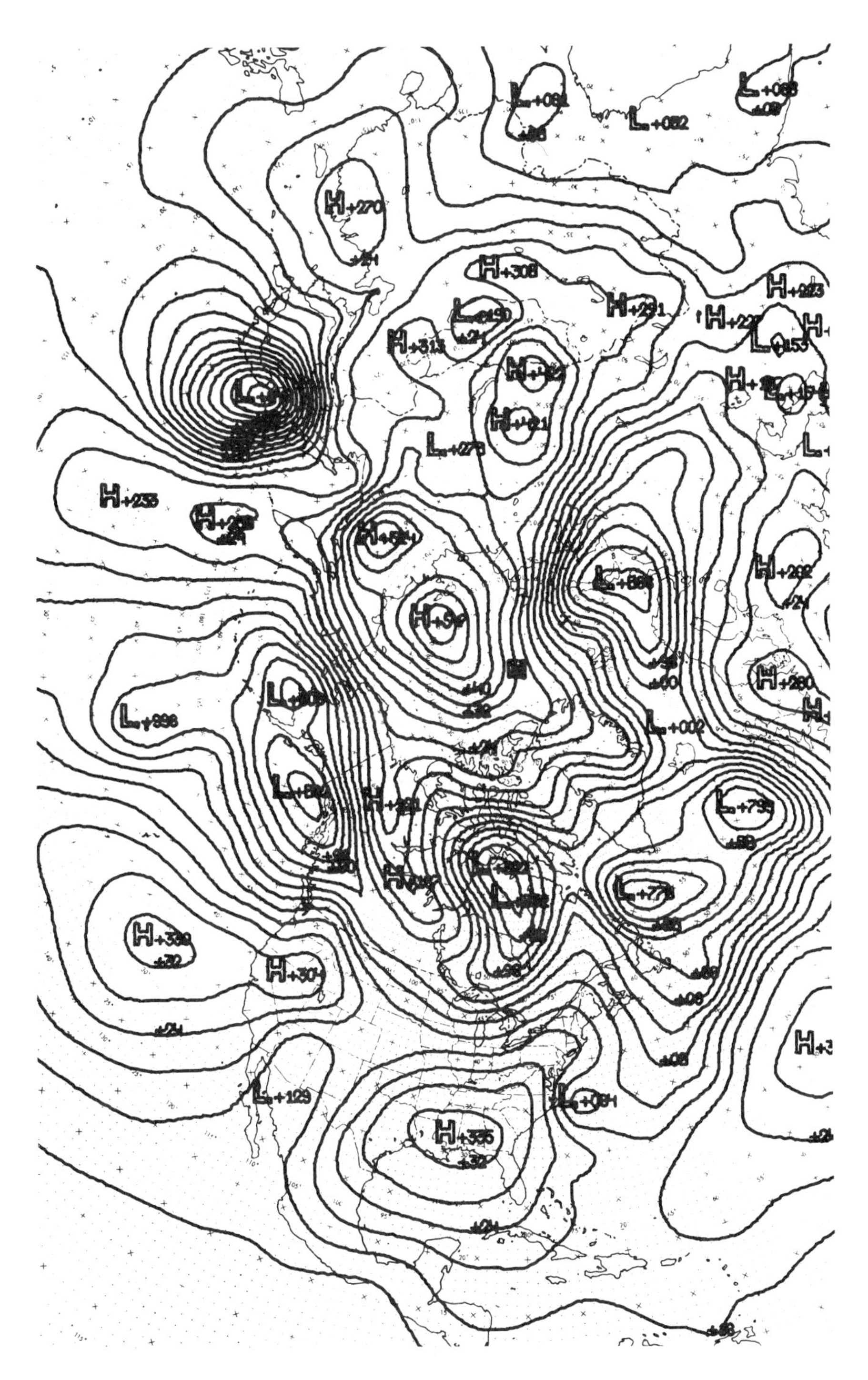

Figure A.2. Surface analysis for 12Z, 31 January 1970.

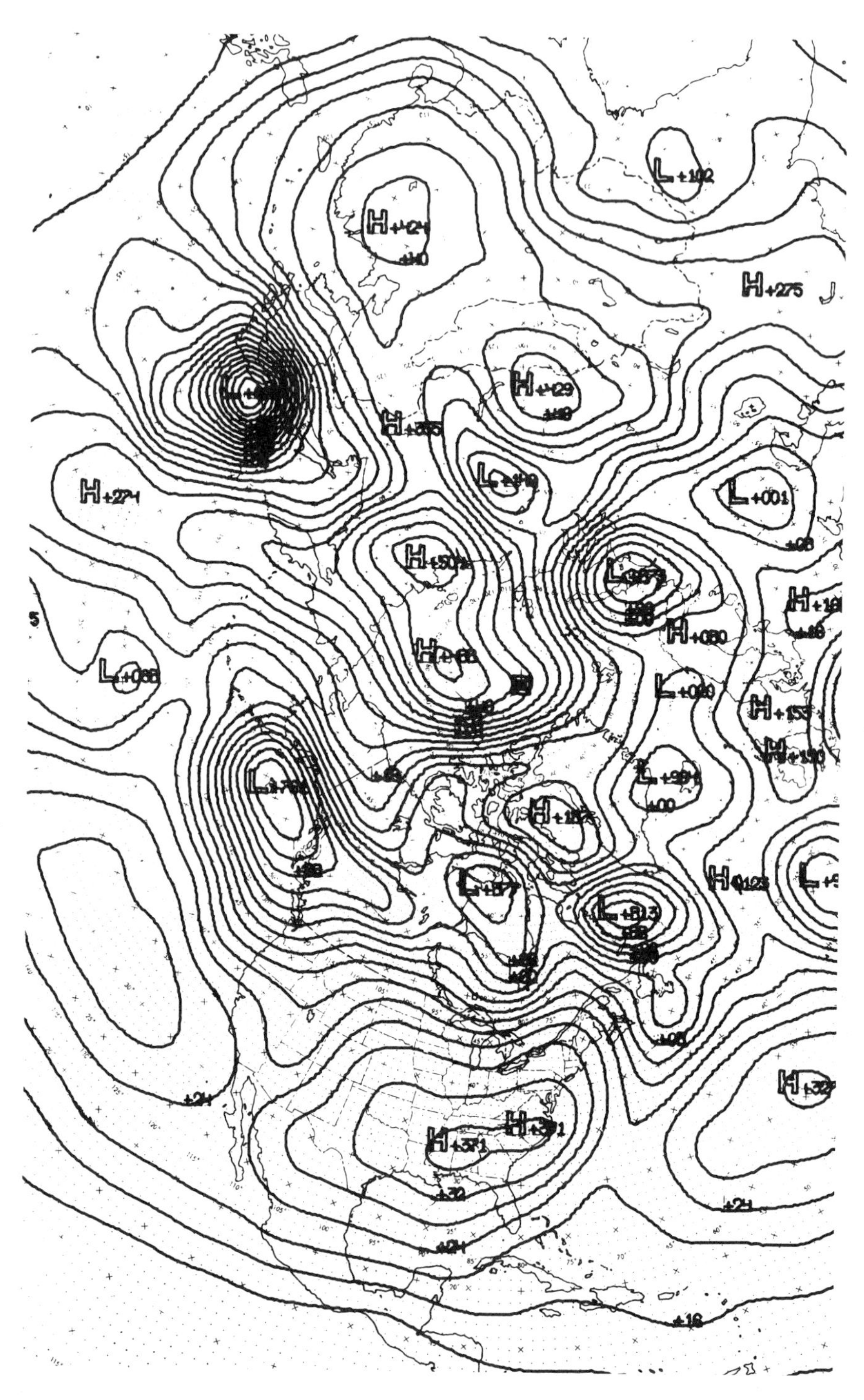

Figure A.3. 36-hr PE forecast (control), verifying at 12Z, 31 January 1970.

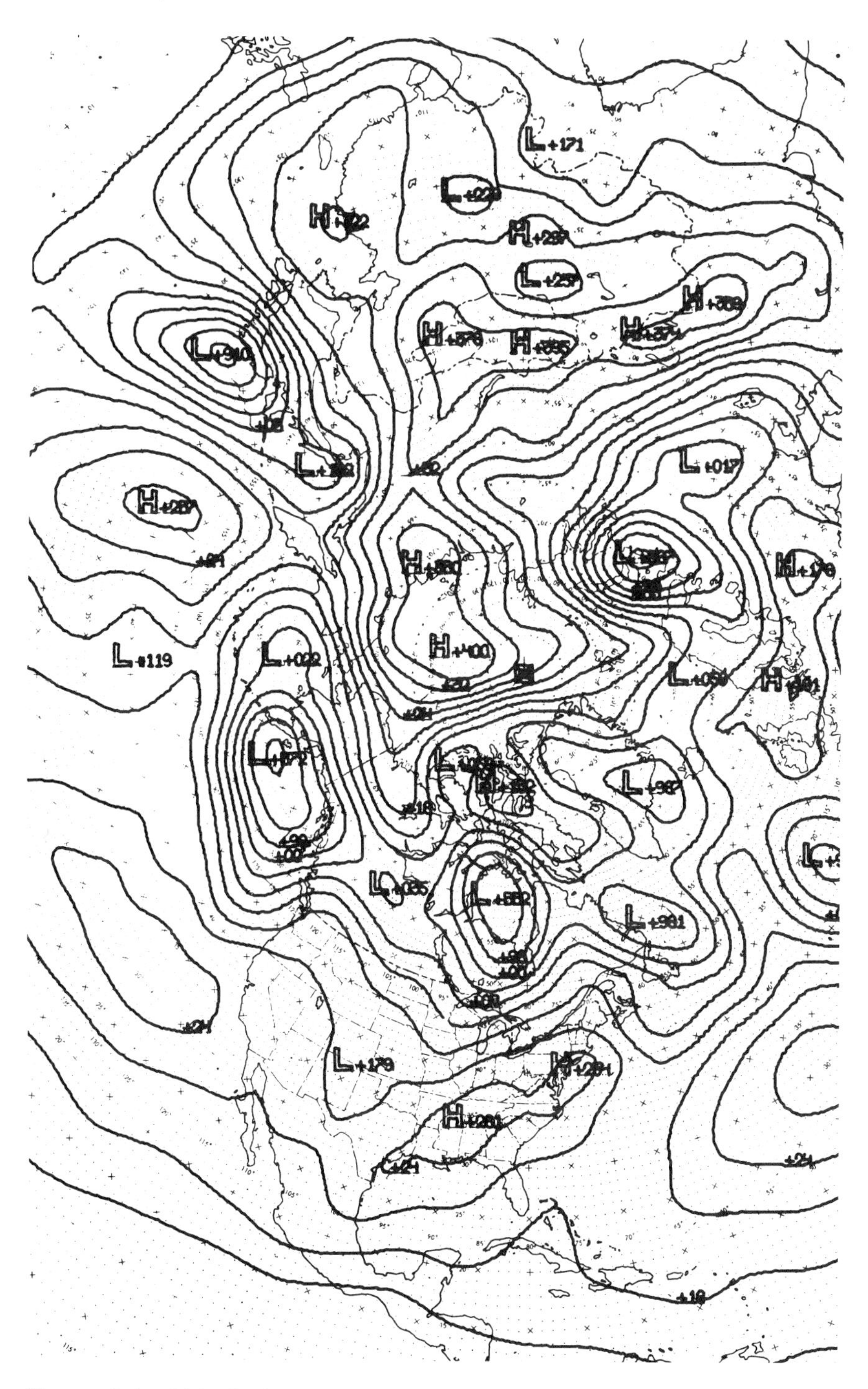

Figure A.4. 36-hr PE forecast (adiabatic), verifying at 12Z, 31 January 1970.

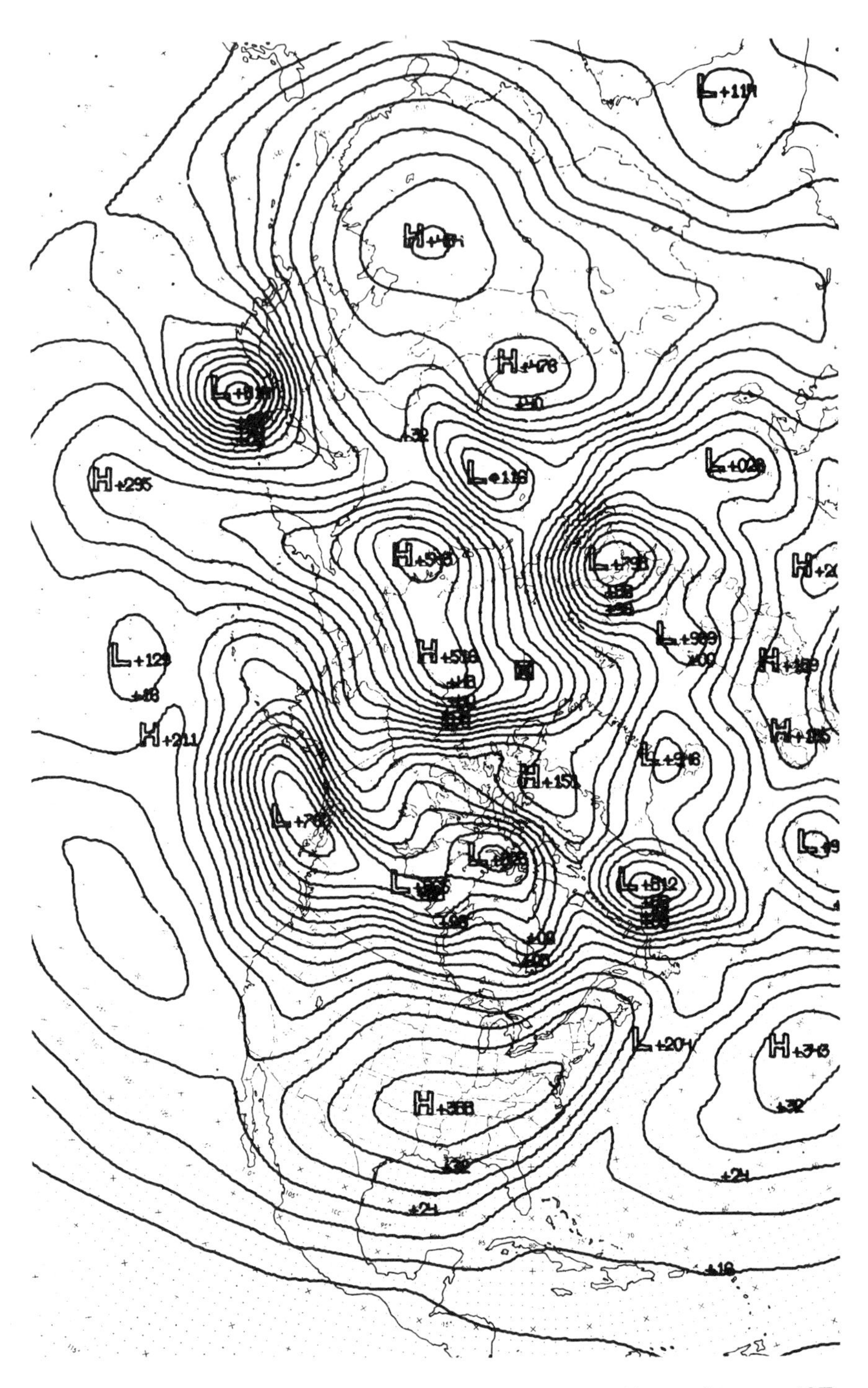

Figure A.5. 36-hr PE forecast (geostrophic initial winds), verifying at 12Z, 31 January 1970.

Figure A.6. 36-hr PE forecast (restoration boundaries), verifying at 12Z, 31 January 1970.

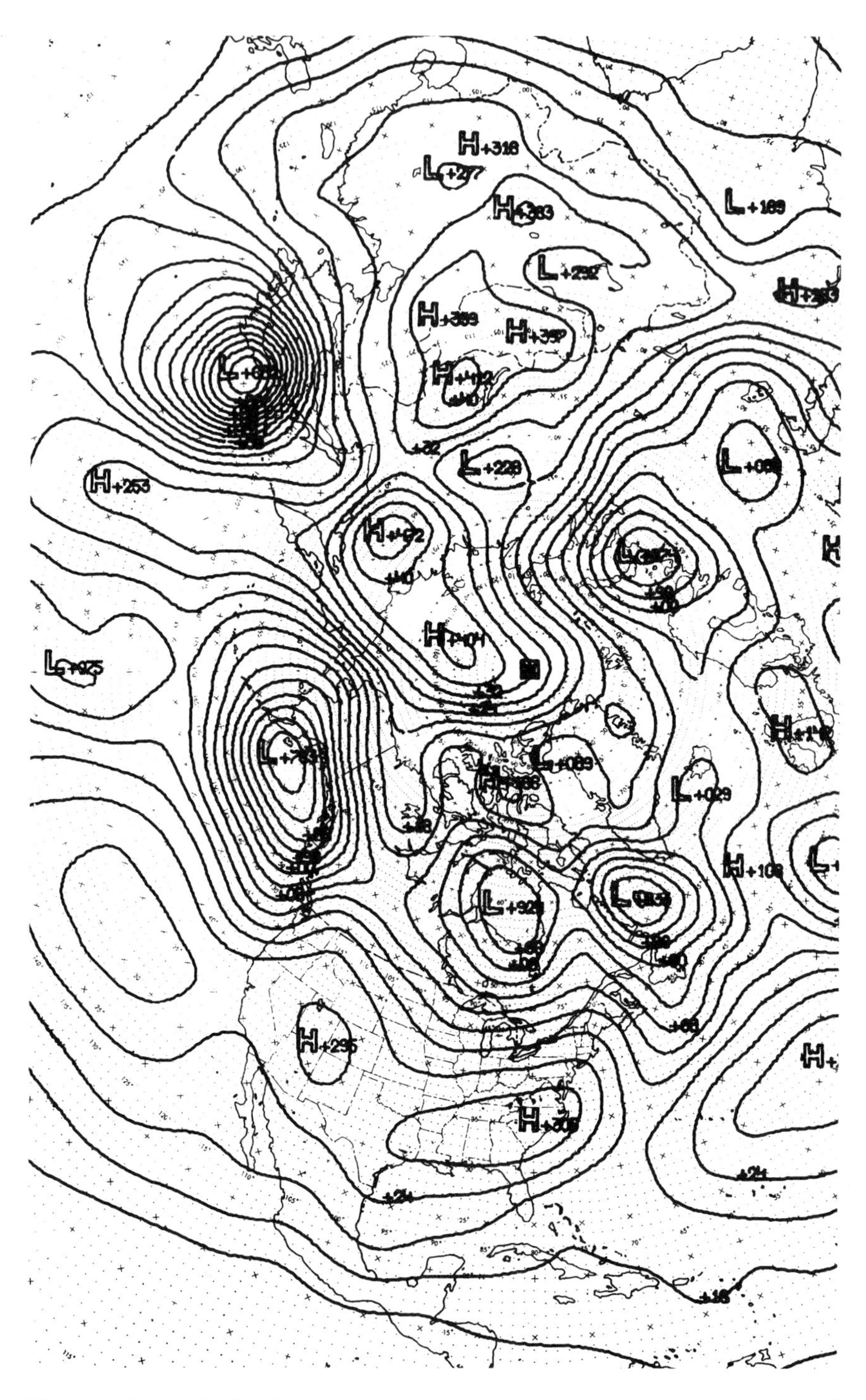

Figure A.7. 36-hr PE forecast (Berkofsky-Clarke mountains), verifying at 12Z, 31 January 1970.

CASE II

Figures A.8 through A.19 show a more complete example of a winter forecast which was picked at random from the files of FNWC. Although the analyses and forecasts were prepared for the entire Northern Hemisphere, as in Case I, only an area encompassing most of North America and portions of adjacent oceans is displayed. The point 80N, 100W is located at the center of the upper border of each figure.

Figures A.8 and A.9 (page 292) are the surface analyses for 00Z, 26 January 1971 and 12Z, 27 January 1971; isobars are drawn at 4-mb intervals. The major features of the earlier map, Figure A.8, are a rather weak low pressure center (cyclone) just southwest of the Great Lakes, a high pressure center west of California with a ridge extending inland over the southwestern states, a cold high pressure area over western Canada and Northern Alaska, and a deep cyclone in the Gulf of Alaska. During the next 36 hours the cyclone initially located near the Great Lakes moved into the New England area and intensified markedly, with the central pressure dropping to about 966 mb. Behind this cyclone, cold air has spread southward over the Great Plains resulting in a high pressure cell centered over the Midwest with a prominent ridge extending into the Gulf of Mexico. On the other hand, the high pressure system off the West Coast and the cyclone in the Gulf of Alaska have remained nearly stationary, with the cyclone filling about a dozen mb.

Figure A.10 (page 293) shows a 36-hour surface forecast from the initial conditions at 00Z, 26 January 1971, which is verified at 12Z, 27 January 1971. The prediction scheme used is essentially a one-parameter (the so-called barotropic) 500-mb filtered model combined with an advective 500–1000 mb thickness forecast (SLP). The prediction equations are similar to Equations (6.14) and (7.7) in the text, but include some empirical refinements which improve their skill. The predicted movement of the Great Lakes low was very good, however the system deepened only to about 986 mb, which was not nearly enough. The Gulf of Alaska storm was kept at about the same intensity and was erroneously moved eastward over Western Canada with a trough extending into the northwestern United States, which did not verify well at all. The high pressure area over the western and central United States was forecast fairly well, but was too prominent over the Minnesota area and not nearly intense enough over the northwestern United States and the Gulf of Mexico.

Figure A.11 (page 293) is the corresponding surface forecast made with the Navy five-layer primitive equation model. It may be seen that in general with respect to the three pressure systems previously discussed, this prediction is better, although even here, the Gulf of Alaska cyclone has intruded somewhat too far inland, the anticyclone centered over the United

States-Canadian border is too far north, and the ridge over the Gulf of Mexico is too strong. In fact, the current PE model generally tends to overdevelop anticyclones somewhat and revisions to the model are being made to correct this deficiency.

Figures A.12 through A.15 are the corresponding set of 500-mb charts; contours are drawn at 60 meter intervals. On the initial chart A.12 (page 294) there is a broad trough just west of the Great Lakes, a ridge dominates the west coast of the United States and Canada, and a strong trough extends south-southwestward from the Gulf of Alaska. Thirty-six hours later (Figure A.13 (page 294) the trough over the United States has moved to the East Coast and has intensified markedly, while the West Coast ridge and the Gulf of Alaska trough have remained nearly stationary. The barotropic forecast (Figure A.14, page 295) moves the central United States trough eastward too slowly and does not intensify it nearly enough. On the other hand the ridge over the western United States is erroneously moved well inland. The primitive equation forecast (Figure A.15, page 295) is again considerably better in most respects than the barotropic forecast.

Figures A.16, A.17 (page 296), and A.18 (page 297) show the initial 300-mb analysis, the verifying analysis, and the 36-hr primitive equation forecast. The eastward movement and deepening of the trough affecting the eastern half of the United States are predicted quite accurately, as well as the more or less stationary ridge-trough system dominating the Pacific Coast and Gulf of Alaska. A filtered barotropic forecast is not available since this model is utilized only at 500 mb.

Just a few of the many products of a multilevel baroclinic primitive equation forecast have been displayed here. Since the momentum equations are used, horizontal winds are explicitly predicted along with temperature, moisture, and surface pressure. In addition, vertical velocities, precipitation, and the heights of upper pressure surfaces are calculated. Figure A.19 (page 297) gives the predicted precipitation (in inches) for the 36-hour period beginning 00Z, 26 January 1971 and ending 12Z, 27 January 1971. Included here is precipitation resulting from large-scale vertical motions and from parameterized convection. Quite heavy precipitation is predicted for the region affected by the intense storm in northeastern United States and lesser amounts of precipitation are predicted in the Pacific Northwest.

Table A.2 gives both the root-mean-square errors (RMSE) of the SLP vs the primitive equation sea-level pressure forecasts and the barotropic vs the primitive equation 500-mb forecasts for Case II.

Note that the primitive equation sea-level forecasts are better than the advective-barotropic forecasts for every period throughout the 72 hours. Although the 24 and 36-hr verifications were not available at 500 mb because of a computer error, the mean errors at 500 mb are apparently the same up to

Table A.2. Root-mean-square errors (RMSE) for the SLP-BARO and the primitive equation (PE) forecasts of sea-level pressure (mb) and the 500-mb height (meters) for intervals up to 72 hours starting from 00Z, 26 January 1971.

Forecast	Sea Level Pressure RMSE		500-mb Height RMSE	
Period (hrs)	SLP (mb)	PE	BARO (meters)	PE
12	3.3	2.9	28	29
24	5.2	3.9	—	—
36	7.1	5.1	—	—
48	8.9	6.1	56	56
60	10.6	6.7	103	68
72	11.9	7.4	116	75

48 hours for the two models; however the PE forecasts are markedly better for the 60 and 72-hour forecasts. While this is only a single example of forecasts by the two methods, they have been compared systematically at FNWC for six months or more now and the example shown here is indeed representative. The primitive equation forecasts have been definitely better in terms of RMSE. Of course, the reader need hardly be reminded that the RMSE by itself is not an adequate measure of forecasting skill and should be supplemented by other means such as a comparison of the forecast and actual movements and changes of intensity of the individual pressure systems.

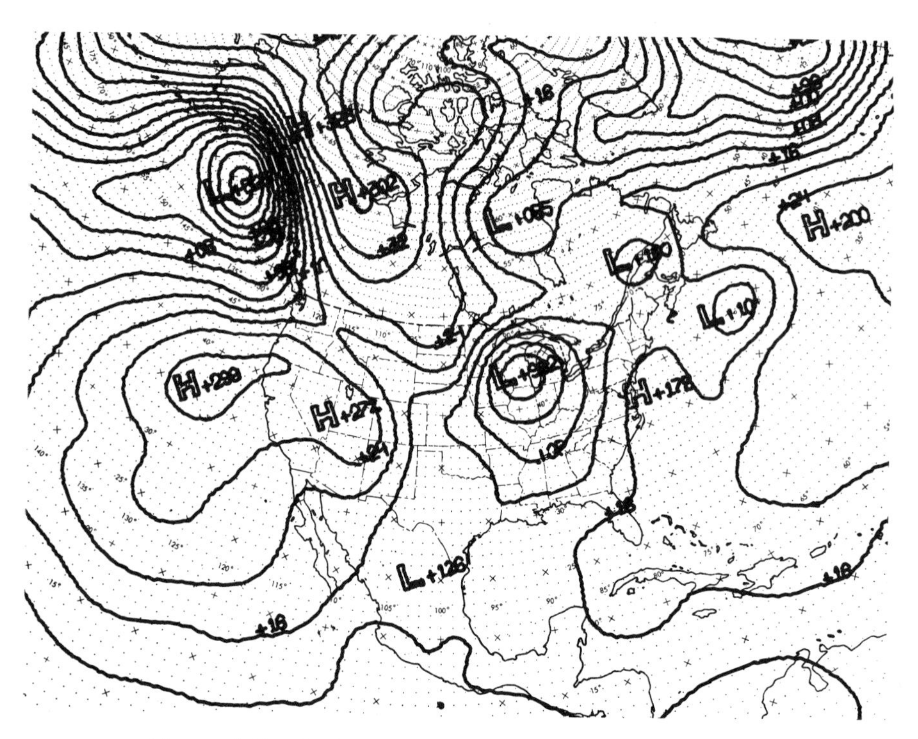

Figure A.8. Surface analysis for 00Z, 26 January 1971.

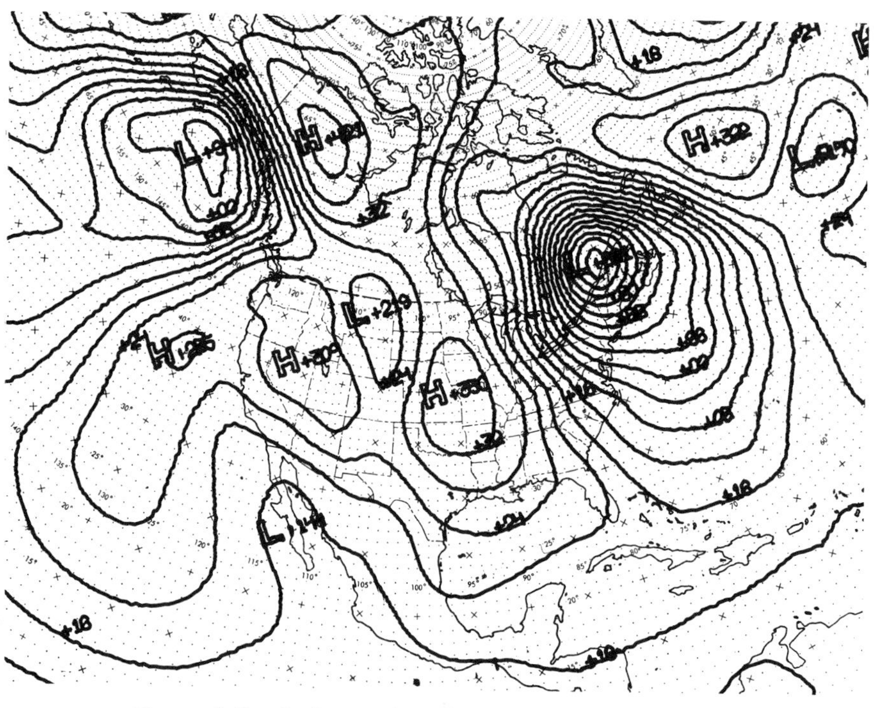

Figure A.9. Surface analysis for 12Z, 27 January 1971.

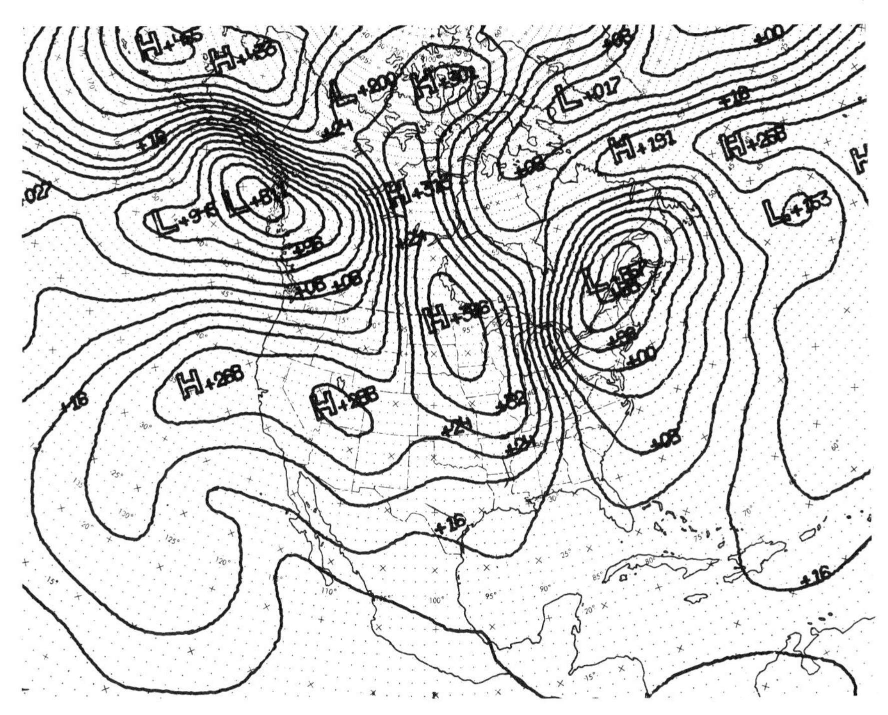

Figure A.10. 36-hr SLP surface forecast, verifying at 12Z, 27 January 1971.

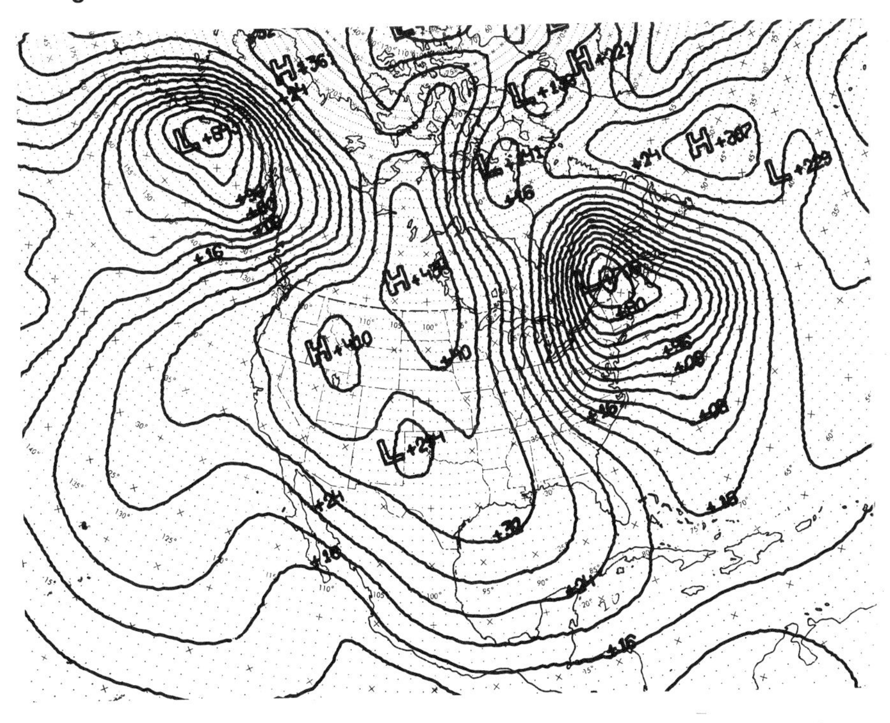

Figure A.11. 36-hr PE surface forecast, verifying at 12Z, 27 January 1971.

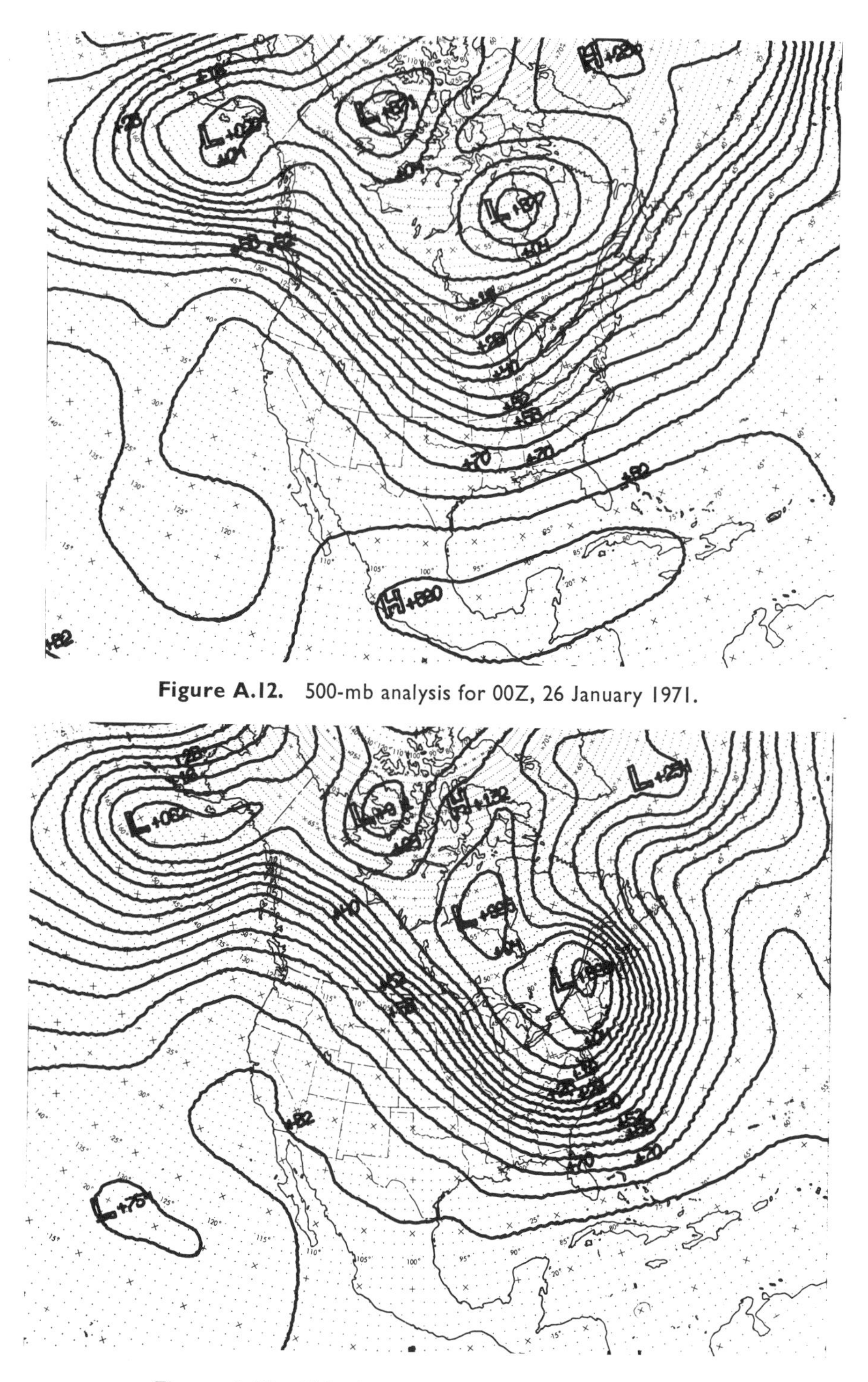

Figure A.12. 500-mb analysis for 00Z, 26 January 1971.

Figure A.13. 500-mb analysis for 12Z, 27 January 1971.

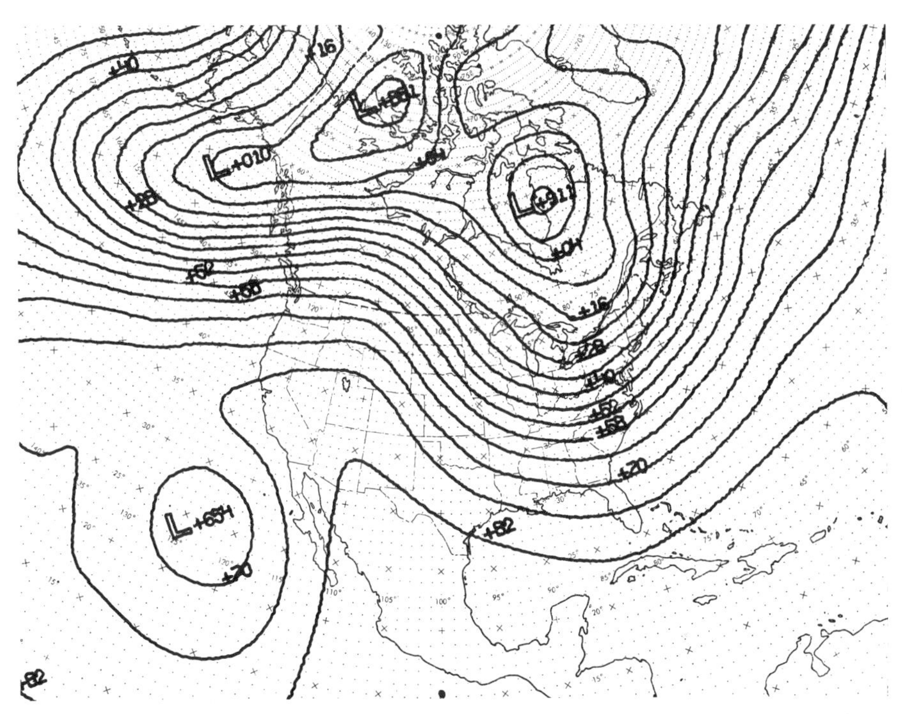

Figure A.14. 36-hr barotropic 500-mb forecast, verifying at 12Z, 27 January 1971.

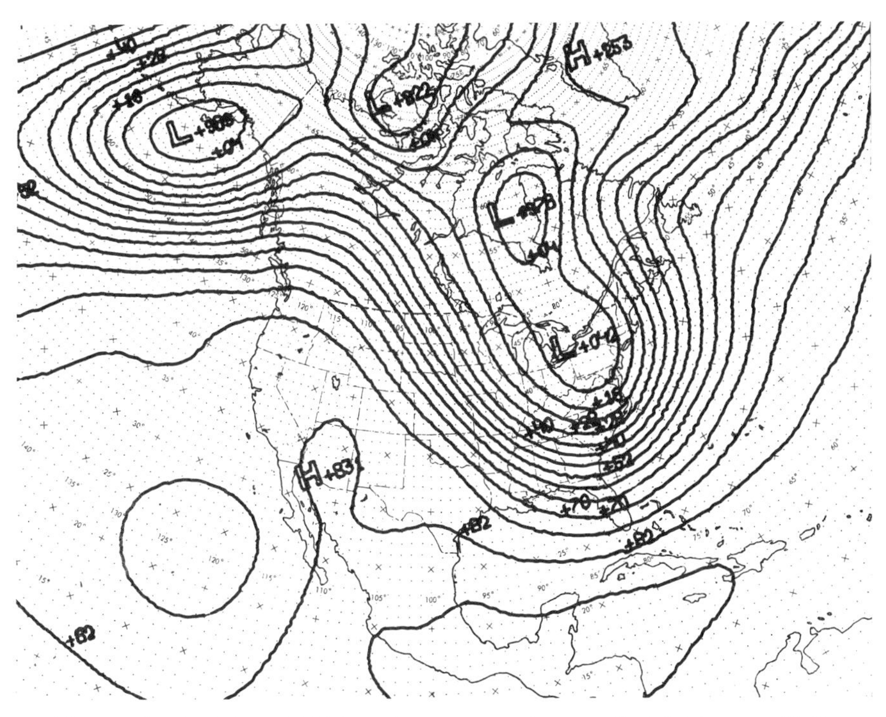

Figure A.15. 36-hr PE 500-mb forecast, verifying at 12Z, 27 January 1971.

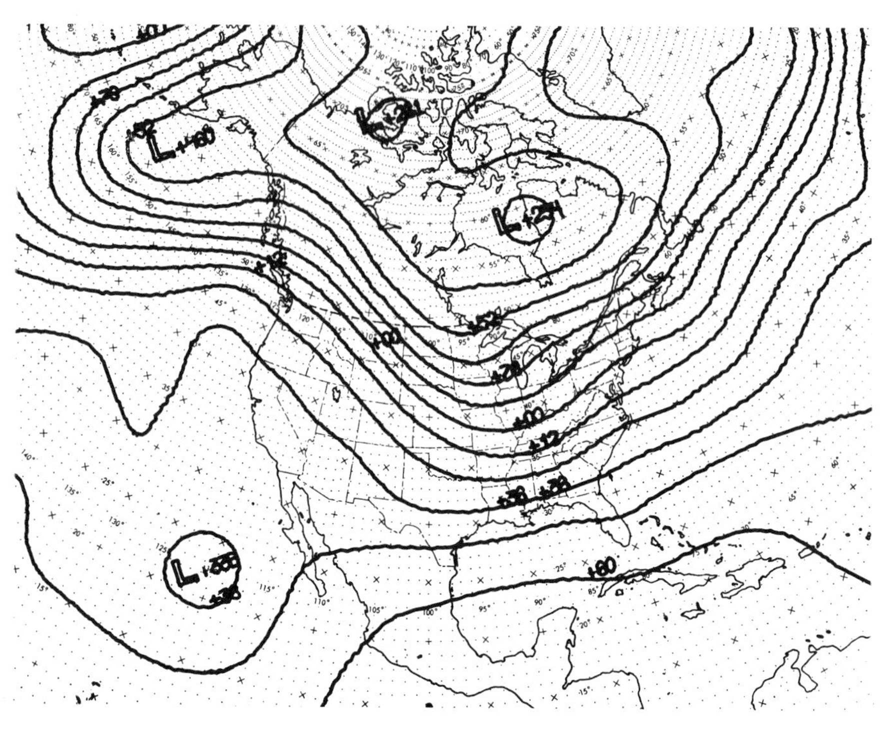

Figure A.16. 300-mb analysis for 00Z, 26 January 1971.

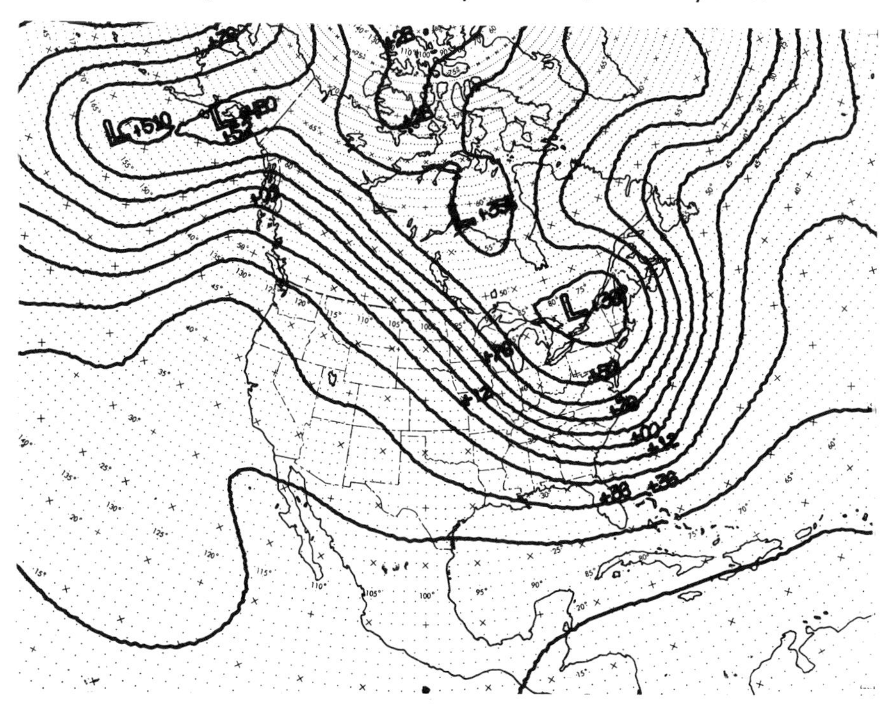

Figure A.17. 300-mb analysis for 12Z, 27 January 1971.

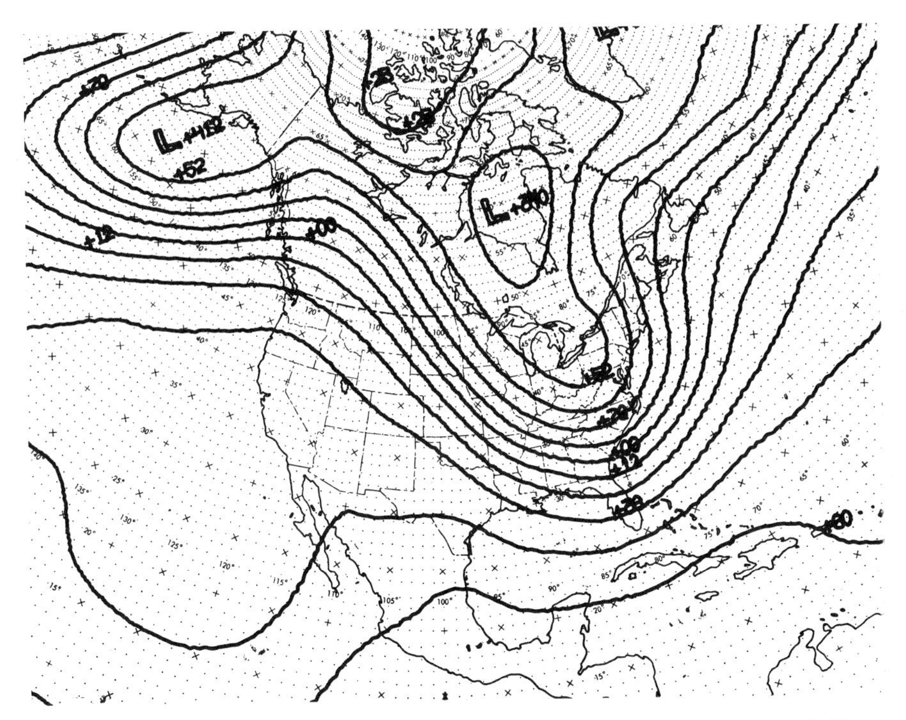

Figure A.18. 36-hr PE 300-mb forecast, verifying at 12Z, 27 January 1971.

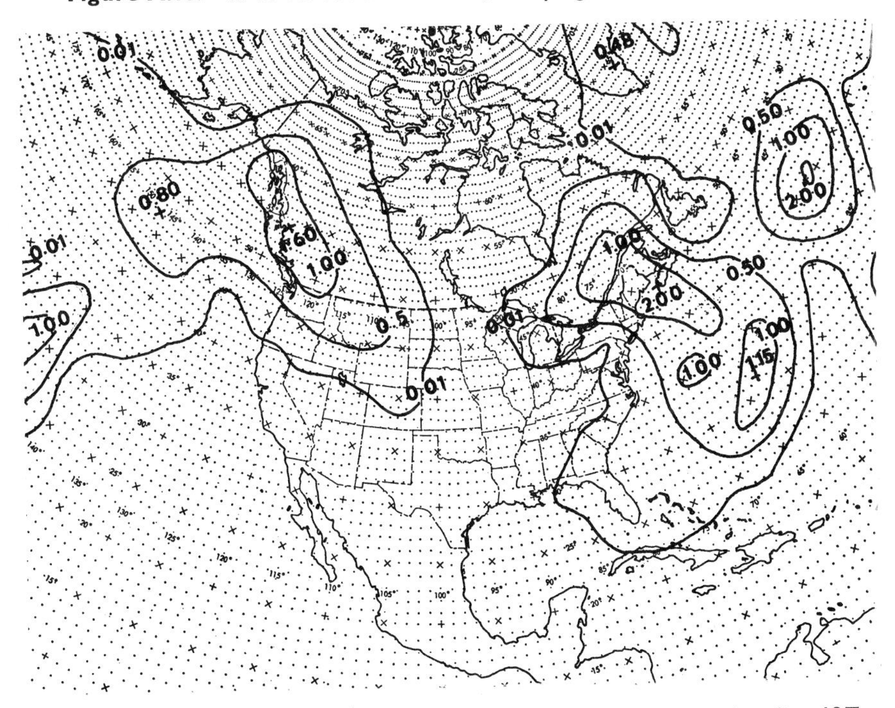

Figure A.19. PE precipitation forecast (inches) for the 36-hr period ending 12Z, 27 January 1971.

References

Advances in Numerical Weather Prediction, 1965–1966 *Seminar Series*, sponsored by Travelers Research Center, Inc.

Arakawa, A., "Nongeostrophic Effects in Baroclinic Prognostic Equations." *Proc. Intern. Symp. Numerical Weather Prediction, Tokyo* (1960).

———, "The Life Cycle of Large-Scale Disturbances." Paper delivered at the Fourth Numerical Prediction Conference, Los Angeles, Calif., Jan. 31–Feb. 3, 1962.

———, "Computational Design for Long-Term Numerical Integrations of the Equations of Atmospheric Motion." *J. Computational Phys.*, **1,** 119–143 (1966).

Arakawa, A., A. Katayama, and Y. Mintz, "Numerical Simulation of the General Circulation of the Atmosphere." *Proc. WMO (IUGG Symp. National Weather Prediction), Tokyo*, 1968.

Arnason, G., "A Convergent Method for Solving the Balance Equation." *Joint Numerical Weather Prediction Unit, Tech. Note*, Suitland, Md. (1957).

Aubert, E., "On the Release of Latent Heat as a Factor in Large-Scale Atmospheric Motions." *J. Meteor.* **14,** 527–542 (1957).

Baer, F., "Integration of the Spectral Vorticity Equation." *J. Atmos. Sci.*, **21,** 260–276 (1964).

———, "Studies in Low-Order Spectral Systems." *Atmos. Sci. Paper no.* 129, Dept. of Atmos. Sci., Colorado State University (1968).

Baer, F., and G. W. Platzman, "A Procedure for Numerical Integration of the Spectral Vorticity Equation." *J. Meteor.*, **18,** 393–401 (1961).

Barteau, C. L. and W. D. Groscup, "A Five-Level Baroclinic Prediction Model." M.S. thesis, U.S. Naval Post Graduate School, Monterey, Calif., (1963), 15 pp.

Baumhefner, D. P., "Global Real Data Forecasts with the NCAR Two-Layer General Circulation Mode." *Mon. Wea. Rev.*, **98,** 92–99 (1970).

Bergthorsson, P. and B. Doos, "Numerical Weather Map Analysis." *Tellus*, **7,** 329–340 (1955).

Berkofsky, L., "A Three Parameter Baroclinic Numerical Prediction Model." *J. Meteor.* **13,** 102–111 (1956).

Bjerknes, V., "Das Problem von der Wettervorhersage, betrachtet vom Standpunkt der Mechanik und der Physik." *Meteor. Z.*, **21,** 1–7 (1904).

Bjerknes, V., "Wettervorhersage." *Meteor. Z.*, Braunschweig, **36,** 68 pp. (1919).

Blinova, E. N. and I. A. Kibel, "Hydrodynamical Methods of the Short- and Long-Range Weather Forecasting in the U.S.S.R." *Tellus*, **9,** 447–463 (1957).

Bolin, B., "The Adjustment of a Non-Balanced Velocity Field Towards Geostrophic Equilibrium in a Stratified Fluid." *Tellus*, **5,** 373–385 (1953).

Bonwell, G. R. R. and F. H. Bushby, "A Case Study of Frontal Behavior Using a 10-Level Primitive Equation Model." *Quart. J. Roy. Meteor. Soc.*, **96,** 287–296 (1970).

Brown, A. John Jr., "A Numerical Investigation of Hydrodynamic Instability and Energy Conversions in a Quasi-Geostrophic Atmosphere." *J. Atmos. Sci.*, **26,** no. 3, 352–375 (1969).

Brunt, D., *Physical and Dynamical Meteorology*. Cambridge University Press, 1939, 428 pp.

Burger, A., "Scale Considerations of Planetary Motions of the Atmosphere." *Tellus*, **10,** 195–205 (1958).

Bushby, F. H. and M. S. Timpson, "A 10-Level Model and Frontal Rainfall." *Quart. J. Roy. Meteor. Soc.*, **93,** 1–17 (1967).

Bushby, F. H. and M. Hinds, "Computation of Tendencies and Vertical Motion With a Two-Parameter Model of the Atmosphere." *Quart. J. Roy. Meteor. Soc.*, **80,** 16–25 (1954).

Bushby, F. H. and V. M. Huckle, "The Use of a Stream Function in a Two-Parameter Model of the Atmosphere." *Quart. J. Roy. Meteor. Soc.*, **82,** 409–418 (1956).

Cahn, A., "An Investigation of the Free Oscillations of a Simple Current System." *J. Meteor.*, **2,** 113–119 (1945).

Carstensen, L. P. and G. E. Lawniczak, Jr., "Routines for a Vertically Consistent Analysis and Pattern Decomposition." Joint meeting of the American Geophysical Union and the American Meteorological Society, Washington, D.C., April 1964.

Charney, J. G., "The Dynamics of Long Waves in a Baroclinic Westerly Current." *J. Meteor.*, **4,** 135–162 (1947).

———, "On the Scale of Atmospheric Motions." *Geofys. Publikasjoner*, **17,** (1948), 17 pp.

———, "On the Scale of Atmospheric Motions for Baroclinic Flow in a Closed Region." *Geofys. Publikasjoner* **17,** no. 2, (1949).

———, "On a Physical Basis for Numerical Prediction of Large-Scale Motions in the Atmosphere." *J. Meteor.*, **6,** 371–385 (1949).

———, "Numerical Prediction of Cyclogenesis." *Proc. Natl. Acad. Sci., U.S.*, **40,** no. 2, 99–110 (1954).

———, "The Use of the Primitive Equations of Motion in Numerical Prediction." *Tellus*, **7,** 22–26 (1955).

———, A Note on Large-Scale Motions in the Tropics. *J. Atmos. Sci.*, **20,** 607–609 (1963).

———, "The Feasibility of a Global Observation and Analysis Experiment." *Natl. Acad. Sci.*, *Publ.* 1290, National Research Council, Oct. 1965.

Charney, J. G., "The Feasibility of a Global Observation System." Report of panel on International Meteorological Programs, Committee on Atmospheric Sciences, *Natl. Acad. Sci.*, sec. I, 1965, 48 pp. (Also see *Bull. Amer. Meteor. Soc.*, **47,** no. 3, 1966).

———, "Integration of the Primitive and Balance Equations." *Proc. Intern. Symp. Numerical Weather Prediction, Tokyo; Meteor. Soc. Japan*, 131–152 (1962).

———, "A Further Note on Large-Scale Motions in the Tropics." *J. Atmos. Sci.*, **26,** 182–185 (1969).

Charney, J. G. and A. Eliassen, "A Numerical Method for Predicting the Perturbations on the Middle Latitude Westerlies." *Tellus*, **1,** 38–54 (1949).

———, "On the Growth of the Hurricane Depression." *J. Atmos. Sci.*, **21,** no. 1, 68–75 (1964).

Charney, J. G., R. Fjortoft, and J. von Neumann, "Numerical Integration of the Barotropic Vorticity Equations." *Tellus*, **2,** no. 4, 237–254 (1950).

Charney, J. G., B. Gilchrist, and F. Shuman, "The Prediction of General Quasi-geostrophic Motions." *J. Meteor.*, **13,** 489–499 (1956).

Charney, J. G. and N. A. Phillips, "Numerical Integration of the Quasigeostrophic Equations for Barotropic and Simple Baroclinic Flows." *J. Meteor.*, **10,** 71–99 (1953).

Cressman, G. P., "Barotropic Divergence and Very Long Atmospheric Waves." *Mon. Wea. Rev.*, **86,** 293–297 (1958).

———, "A Three-Level Model Suitable for Daily Numerical Forecasting." *Natl. Meteor. Center, U.S. Weather Bureau, Tech. Memo.*, no. 22, (1963).

———, "Improved Terrain Effects in Barotropic Forecasts." *Mon. Wea. Rev.*, **88,** 327–342 (1960).

Cressman, G. P. and H. A. Bedient, "An Experiment in Automatic Data Processing." *Mon. Wea. Rev.*, **85,** 333–340 (1957).

Crowley, W. P., "Numerical Advection Experiments." *Mon. Wea. Rev.*, **96,** 1–11, (1968).

Danard, M. B., "A Quasi-Geostrophic Model Incorporating the Effects of Release of Latent Heat." *J. Appl. Meteor.*, **5,** no. 1, 85–93 (1966).

———, "A Simple Method of Including Longwave Radiation in a Tropospheric Numerical Prediction Model." *Mon. Wea. Rev.*, **97,** 77–85, (1969).

Doos, B. and M. Eaton, "Upper Air Analysis Over Ocean Areas." *Tellus*, **9,** 184–194 (1957).

Eliassen, A., "The Quasi-State Equations of Motion With Pressure as Independent Variable." *Geofys. Publikasjoner*, **17,** no. 3 (1949), 44 pp.

———, "Simplified Dynamic Models of the Atmosphere, Designed for the Purpose of Numerical Prediction." *Tellus*, **4,** 145–156 (1952).

———, "A Procedure for Numerical Integration of the Primitive Equations of the Two-Parameter Model of the Atmosphere." *Sci. Rept.* no. 4, Dept. of Meteorology, UCLA (1956).

———, "On the Use of a Material Layer Model of the Atmosphere in Numerical Prediction." *Proc. Intern. Symp. Numerical Weather Prediction, Tokyo; Meteor. Soc. Japan*, 207–211 (1962).

Ellsaesser, H. W., "Evaluation of Spectral Versus Grid Methods of Hemispheric Numerical Weather Prediction." *UCRL 7865-T, Rev.*, **2,** (1964).

Fischer, G., "A Survey of Finite-Difference Approximations to the Primitive Equations." *Mon. Wea. Rev.*, **93,** no. 1, 1–10 (1965).

Fjortoft, R., "On the Changes in the Spectral Distribution of Kinetic Energy for Two-Dimensional, Nondivergent Flow." *Tellus*, **5,** 225–230 (1953).

Fleet Numerical Weather Facility, "General Environmental Computer Product Catalog." *FNWF Tech. Note* no. 11 (1966).

Gambo, K., "The Scale of Atmospheric Motions and the Effect of Topography on Numerical Weather Prediction in the Lower Atmosphere." *Papers in Meteorology and Geophys*, **8,** no. 1, 1–24 (1957).

Gates, W. L., "On the Truncation Error, Stability and Convergence of Difference Solutions of the Barotropic Vorticity Equation." *J. Meteor.*, **16,** 556–568 (1959).

———, "The Stability Properties and Energy Transformations of the Two-Layer Model of Variable Static Stability." *Tellus*, **13,** 460–471 (1961).

———, "A Numerical Study of Transient Rossby Waves in a Wind-Driven Homogeneous Ocean." *Rand Corp. Rept.*, March 1967.

Gates, W. L. and C. A. Riegel, "A Study of Numerical Errors in the Integration of Barotropic Flow on a Spherical Grid." *J. Geoph. Res.*, **67**, no. 2, 773–784 (Feb. 1962).

———, "Comparative Numerical Integration of Simple Atmospheric Models on a Spherical Grid." *Tellus*, **15,** no. 4, 406–423 (Nov. 1963).

Gilchrist, B. and G. P. Cressman, "An Experiment in Objective Analysis." *Tellus*, **6,** 309–318 (1954).

Gadd, A. J. and Keers, J. F., "Surface Exchanges of Sensible and Latent Heat in a 10-Level Model Atmosphere." *Quart. J. Roy. Meteor. Soc.*, **96,** 297–308 (1970).

Green, J. S. A., "A Problem in Baroclinic Stability." *Quart. J. Roy. Meteor. Soc.*, **86,** 237–251 (1960).

Hague, S. M. A., "The Initiation of Cyclonic Circulation in a Vertically Unstable Stagnant Air Mass." *Quart. J. Roy. Met. Soc.*, **78** (1952).

Haltiner, G. J., "The Effects of Sensible Heat Exchange on the Dynamics of Baroclinic Waves." *Tellus*, **19,** no. 2 (1967).

———, "Computer Applications in Environmental Prediction." *Proc. 8th Navy Sci. Symp.*, **1,** 54–75 (1964).

Haltiner, G. J., G. Arnason, and J. Frawley, "Higher Order Geostrophic Wind Approximations." *Mon. Wea. Rev.*, **90** (1962).

Haltiner, G. J. and D. E. Caverly, "The Effect of Friction on the Growth and Structure of Baroclinic Waves." *Quart. J. Roy. Meteor. Soc.*, **91** (1965).

Haltiner, G. J., L. C. Clarke, and G. F. Lawniczak, "Computation of Large-Scale Vertical Velocity." *J. Appl. Meteor.*, **1,** 1–7 (1962).

Haltiner, G. J. and F. L. Martin, *Dynamical and Physical Meteorology*. McGraw-Hill Book Co., 1957, 470 pp.

Haurwitz, B., "The Motion of Atmospheric Disturbances on the Spherical Earth." *J. Marine Res.* (Sears Foundation), **3,** 254–267 (1940).

Hinkelmann, K., *Ein numerisches Experiment mit den primitiven Gleichungen.* C. G. Rossby Memorial Volume, Stockholm, 1959.

Holl, M. J., J. P. Bibbo, and J. R. Clark, "Linear Transforms for State-parameter Structure." *Meteor. Intern. Tech. Rept.* (1964).

Holton, J. R., "On the Theory of Easterly Waves." *Proc. AMS Symp. Tropical Meteorology, Honolulu* (1970).

Holton, J. R., J. M. Wallace, and J. A. Young, "A Note on Boundary Layer Dynamics and the ITCZ." Unpublished manuscript.

Houghton, D. and W. Washington, *On Global Initialization of the Primitive Equations.* Part 1, NCAR Manuscript, 1968.

Howard, J. N., D. L. Burch, and D. Williams, "Near-Infrared Transmission Through Synthetic Atmospheres." Geophys. Res. Papers, no. 40, *AFCRC-TR-55-213*, (1955), 244 pp.

Hubert, W. E., "Hurricane Trajectory Forecasts From a Nondivergent Nongeostrophic Barotropic Model." *Mon. Wea. Rev.*, **85,** 83–87 (1957).

———, "Operational Forecasts of Sea and Swell." *Proc. First U.S. Navy Symp. Military Oceanogr.*, 113–124 (1964).

Joseph, J. H., "Calculation of Radiative Heating in Numerical General Circulation Modes." *Numerical Simulation of Weather and Climate. Tech. Rept.* no. 1, Dept. of Meteorology, UCLA (1966).

Kasahara, A., "On Certain Finite-Difference Models for Fluid Dynamics." *Mon. Wea. Rev.*, **93,** no. 1, 27–31 (Jan. 1965).

Kasahara, A. and T. Asai, "Effects of an Ensemble of Convective Elements on the Large Scale Motions of the Atmosphere." *J. Meteor. Soc. Japan*, ser. II, **45,** no. 4 (1967).

Kasahara, A., and W. M. Washington, "NCAR Global General Circulation Model of the Atmosphere," *Mon. Wea. Rev.*, **95,** 389–402 (1967).

Knighting, E., D. E. Jones, and M. K. Hinds, "Numerical Experiments in the Integration of the Meteorological Equations of Motion." *Quart. J. Roy. Meteor. Soc.*, **84,** 91–107 (1958).

Krishnamurti, T. N., "An Experiment in Numerical Prediction in the Equatorial Latitudes." *Quart. J. Roy. Meteor. Soc.*, **95,** 594–620 (1969).

Kuo, H. L., "Dynamical Aspects of the General Circulation and the Stability of Zonal Flow." *Tellus*, **3,** 268–284 (1951).

———, "The Stability Properties and Structure of Disturbances in a Baroclinic Atmosphere." *J. Meteor.*, **10,** 235–243 (1953).

———, "On Formation and Intensification of Tropical Cyclones Through Latent Heat Release by Cumulus Convection." *J. Atmos. Sci.*, **22,** 40–63 (1965).

———, "Dynamical Instability of Two-Dimensional Nondivergent Flow in a Barotropic Atmosphere." *J. Meteor.*, **6,** 105–122 (1949).

Kuo, H. L. and J. Nordo, "Integration of Four-Level Prognostic Equations Over the Hemisphere." *Tellus*, **11,** no. 4, 412–424 (Nov. 1959).

Kurihara, Y., "On the Use of Implicit and Iterative Methods for the Time Integration of the Wave Equation." *Mon. Wea. Rev.*, **93,** no. 1, 33–46 (Jan. 1965).

———, "Numerical Integration of the Primitive Equations on a Spherical Grid." *Mon. Wea. Rev.*, **93,** no. 7 (July 1965).

Kwizak, M. and A. J. Robert, "A Semi Implicit Scheme for Grid Point Atmospheric Models of the Primitive Equations." *Mon. Wea. Rev.*, **99,** *No.* 1, 32–36, 1971.

Laevastu, T., "Synoptic Scale Heat Exchange and Its Relations to Weather." *FNWF Tech. Note* no. 7 (1965).

Landis, Robert C. and Dale F. Leipper, "Effects of Hurricane Betsy upon Atlantic Ocean Temperature Based upon Radio-Transmitted Data." *J. Appl. Meteor.*, **7,** no. 4, 554–562 (1968).

Langlois, W. E. and H. C. W. Kwok, "Description of the Mintz-Arakawa Numerical General Circulation Model." *UCLA Department of Meteorology Tech. Rept.* no. 3 (1969).

Lax, P. and R. Richtmyer, "Survey of the Stability of Linear Finite Difference Equations." *Communs. Pure and Appl. Math.*, **9,** 267–293 (1956).

Leipper, D. F., "Observed Ocean Conditions and Hurricane Hilda 1964." *J. Atmos. Sci.*, **24,** 182–186 (1967).

Leith, C. E., "Numerical Simulation of the Earth's Atmosphere." *Rept. UCRL 7986-T* (1964), 40 pp.

———, "Numerical Simulation of the Earth's Atmosphere." in *Methods in Computational Physics*, Vol. 4, Applications in Hydrodynamics, Academic Press, 1965.

Lilly, D. K., "On the Computational Stability of Numerical Solutions of Time-Dependent Nonlinear Geophysical Fluid Dynamics Problems." *Mon. Wea. Rev.*, **93,** no. 1, 11–26 (1965).

———, "On the Theory of Disturbances in a Conditionally Unstable Atmosphere." *Mon. Wea. Rev.*, **88,** no 1 (1960).

Lorenz, E. N., "Available Potential Energy and the Maintenance of the General Circulation." *Tellus*, **7,** 157–167 (1955).

———, "Maximum Simplication of the Dynamic Equations." *Tellus*, **12,** no. 3, 243–254 (Aug. 1960).

———, "Energy and Numerical Weather Prediction." *Tellus*, **12,** no. 4, 364–373 (1960).

———, "The Predictability of Hydrodynamic Flow." *Trans. N.Y. Acad. Sci.*, ser. II, **25,** 409–432 (1963).

———, "Deterministic Nonperiodic Flow." *J. Atmos. Sci.*, **20,** no. 2, 130–141 (1963).

Manabe, S. and F. Moller, "On the Radiative Equilibrium and Heat Balance of the Atmosphere." *Mon. Wea. Rev.*, **89,** 503–532 (1961).

Manabe, S., J. Smagorinsky, and R. F. Strickler, "Simulated Climatology of a General Circulation Model With a Hydrologic Cycle." *Mon. Wea. Rev.*, **93,** no. 12, 769–798 (Dec. 1965).

Manabe, S. and R. F. Strickler, "On the Thermal Equilibrium of the Atmosphere With a Convective Adjustment." *J. Atmos. Sci.*, **21,** no. 4, 361–385 (1964).

Matsuno, T., "Numerical Integration of the Primitive Equations by a Simulated Backward Difference Method." *J. Meteor. Soc. Japan*, **44,** 76–84 (1966).

———, "A Finite Difference Scheme for Time Integrations of Oscillatory Equations with Second Order Accuracy and Sharp Cutoff for High Frequency." *J. Meteor. Soc. Japan*, **44,** 86–88 (1966).

———, "False Reflection of Waves at a Boundary due to the Use of Finite Difference." *J. Meteor. Soc. Japan*, **44,** 145–157 (1966).

Mintz, Y., "Very Long-Term Global Integration of the Primitive Equations of Atmospheric Motion." *WMO Tech. Note* no. 66 (*WMO-IUGG Symp. on Res. and Dev.*), 1965: Aspects of Long-Range Forecasting, Boulder, Colo., (1964.)

Miyakoda, K., "On a Method of Solving the Balance Equation." *J. Meteor. Soc. Japan*, **34,** no. 6, 364–367 (Dec. 1956). Reprinted in *Vortex*, **2** (June 1957).

———, "Contribution to the Numerical Weather Prediction Computation With Finite Difference." *Japanese J. Geophys.*, **3,** 75–190 (1962).

Miyakoda, K. and R. W. Moyer, "A Method of Initialization for Dynamical Weather Forecasting." *Tellus*, **20,** 115–128 (1968).

Miyakoda, K., J. Smagorinsky, R. F. Strickler, and G. D. Hembree, "Experimental Extended Predictions With a Nine-Level Hemispheric Model." *Mon. Wea. Rev.*, **97,** no. 1, 1–76 (1969).

Nitta, T., and Hovermale, J. B., "On Analysis and Initialization for the Primitive Forecast Equations." *Weather Bureau Tech. Memo*, NMC–42 (1967).

Nitta, T., and Yanai, M., "A Note on the Barotropic Instability of the Tropical Easterly Current." *J. Meteor. Soc. Japan*, **47,** no. 2 (1968).

Obukhov, A., "K. Voprosu O Geostrofichesko Vetra." *Izv. Akad. Nauk SSSR, Ser Geograf. Geofiz*, **13,** 281–306 (1949).

Ogura, "Y., Frictionally Controlled, Thermally Driven Circulations in a Circular Vortex With Application to Tropical Cyclones." *J. Atmos. Sci.*, **21,** no. 6, 610–621 (Nov. 1964).

Ogura, Y. and J. G. Charney, "A Numerical Model of Thermal Convection in the Atmosphere." *Proc. Intern. Symp. Numerical Weather Prediction, Tokyo,* 1960; *Meteor. Soc. Japan*, 431–452 (1962).

Okland, H., "An Experiment in Cyclogenesis Prediction by a Two-Level Model." *Mon. Wea. Rev.*, **93,** 663–672 (1965).

Ooyama, K., "A Dynamic Model for the Study of Tropical Cyclone Developments." Dept. of Meteor. and Oceanog., N.Y. Univ. (1963), 26 pp.

———, "A Dynamical Model for the Study of Tropical Cyclone Development." *Geofisica Intern.* (*Mexico*), **4,** 187–198 (1964).

———, "Numerical Simulation of the Life Cycle of Tropical Cyclones." *J. Atmos. Sci.*, **26,** 3–40 (1969).

Panofsky, H., "Objective Weather Map Analysis." *J. Meteor.*, **6,** 386–392 (1949).

Pedlosky, Joseph, "Baroclinic Instability in Two Layer Systems." *Tellus*, **15,** 20–25 (1963).

———, "The Stability of Currents in the Atmosphere and the Ocean." *J. Atmos. Sci.*, **21,** 201–219, 342–353 (1964).

Petterssen, S., D. L. Bradbury, and K. Pedersen, "The Norwegian Cyclone Models in Relation to Heat and Cold Sources." *Geophys. Publ. Geophysica Norvegica*, **24,** no. 9, 243–280 (1962).

Phillips, N. A., *An Example of Nonlinear Computational Instability*. Rossby Memorial Volume, the Rockefeller Inst. Press, 1959.

———, "A Simple Three-Dimensional Model for the Study of Large-Scale Extra-Tropical Flow Patterns." *J. Meteor.*, **8,** 381–394 (1951).

Phillips N. A., "The General Circulation of the Atmosphere: A Numerical Experiment." *Quart. J. Roy. Meteor. Soc.*, **82** (1956), 123 pp.

———, "A Map Projection System Suitable for Large-Scale Numerical Weather Prediction." *J. Meteor. Soc. Japan*, 75th Anniv. Vol., 262–267 (1957).

———, "A Coordinate System Having Some Special Advantages for Numerical Forecasting." *J. Meteor.*, **14,** 184–185 (1957).

———, "An Example of Non-Linear Computational Instability." In *The Atmosphere and the Sea in Motion*, pp. 501–504, Rockefeller Inst. Press in association with Oxford Univ. Press, 1959, 509 pp.

———, "Numerical Integration of the Primitive Equations on the Hemisphere." *Mon. Wea. Rev.*, **87,** 109–120 (1959).

———, "On the Problem of Initial Data for the Primitive Equations." *Tellus*, **12,** 121–126 (1960).

———, "Numerical Weather Prediction." *Advances in Computers* (F. Alt, editor), vol. 1, Academic Press, 1960.

———, "Geostrophic Motion." *Rev. Geophys.*, **1,** no. 2, 123–176 (May 1963).

———, "Numerical Integration of the Hydrostatic System of Equations With a Modified Version of the Eliassen Finite-Difference Grid." *Proc. Intern. Symp. Numerical Weather Prediction*, *Tokyo*, 1960; *Meteor. Soc. Japan*, 109–120 (1962).

Platzman, G. W., "The Computational Stability of Boundary Conditions in Numerical Integration of the Vorticity Equation." *Arch. Meteor. Geophys. u. Bioklimatol.*, ser. A7, 29–40 (1954).

———, "The Lattice Structure of the Finite-Difference Primitive and Vorticity Equations." *Mon. Wea. Rev.*, **86,** 285–292 (1958).

———, "The Spectral Form of the Vorticity Equation." *J. Meteor.*, **17,** 635–644 (1960).

Platzman, G. W. and F. Baer, "Numerical Integration of the Spectral Vorticity Equation." Univ. of Chicago, Dept. of Meteorology (1959).

Reed, R. J., "Experiments in 1000-mb Prognosis." *Nat. Meteor. Center*, *Weather Bureau*, *ESSA*, *Tech. Memo.*, no. 26, U.S. Department of Commerce (1963).

Renard, R. J. and L. C. Clarke, "Experiments in Numerical Objective Frontal Analysis." Paper presented at the 45th Annual American Meteorology Society Meeting in New York City (Jan. 27, 1965).

Richardson, L. F., *Weather Prediction by Numerical Process.* Cambridge Univ. Press, London, 1922, 236 pp. Reprinted by Dover.

Richtmyer, R. D., "A Survey of Difference Methods for Non-Steady Fluid Dynamics." *NCAR Tech. Notes*, nos. 63-2, Boulder, Colo., 1963, 25 pp.

Richtmyer, R. D. and K. W. Morton, "*Difference Methods for Initial-Value Problems*," Wiley-Interscience, 399 pages (1967).

Riehl, H. and J. S. Malkus, "Some Aspects of Hurricane DAISY." *Tellus*, **13** (1961).

Robert, A. J., "The Integration of a Low Order Spectral Form of the Primitive Meteorological Equations." *J. Meteor. Soc. Japan*, **44,** no. 5, 237–245 (1966).

———, "Integration of a Spectral Barometric Model for Global 500 mb." *Mon. Wea. Rev.*, **96,** no. 2 (1968).

Rosenthal, Stanley L., "Numerical Experiments with a Multilevel Primitive Equation Model Designed to Simulate the Development of Tropical Cyclones—Experiment 1." *Tech. Memo ERLTM-NHRL*, **82,** U.S. Dept. of Commerce, ESSA (1969).

Rossby, C. G., "On the Mutual Adjustment of Pressure and Velocity Distribution in Certain Simple Current Systems, II." *J. Marine Res.* (Sears Foundation), 239–263 (1938).

———, "Planetary Flow Patterns in the Atmosphere"' *Quart. J. Roy. Meteor. Soc.*, **66,** 68–87 (1940).

———, "On the Propagation of Frequencies and Energies in Certain Types of Oceanic and Atmospheric Waves." *J. Meteor.*, **2,** 187–204 (1945).

———, "On the Dispersion of Planetary Waves in a Barotropic Atmosphere." *Tellus*, **1,** 54–58 (1949).

Rossby, C. G., et al., "Relations Between Variations in the Intensity of the Zonal Circulation of the Atmosphere and the Displacements of the Semi-Permanent Centers of Action." *J. Marine Res.* (Sears Foundation), **2,** 38–55 (1939).

Sadourny, R., A. Arakawa, and Y. Mintz, "Integration of the Nondivergent Barotropic Vorticity Equation With an Icosahedral-Hexagonal Grid for the Sphere." *Mon. Wea. Rev.*, **96,** 351-356 (1968).

Saltzman, B., "Spectral Statistics of the Wind at 500 Mb." *J. Atmos. Sci.*, **19,** no. 2, 195–206 (Mar. 1962).

Saltzman, B. and S. Teweles, "Further Statistics on the Exchange of Kinetic Energy Between Harmonic Components of the Atmospheric Flow." *Tellus*, **16,** 432–435 (1964).

Sasaki, Y., "An Objective Analysis Based on the Variational Method." *J. Meteor. Soc. Japan*, **36,** 77–78 (1958).

———, "Numerical Variational Objective Analysis Formulated under the Constraints Implied by Long Wave System and Quasi-Steady State." Unpublished manuscript, Univ. of Okla, 1969.

———, "Proposed Inclusion of Time Variation Terms, Observational and Theoretical, in Numerical Variational Objective Analysis." *J. Meteor. Soc. Japan*, **47,** 115–124 (1969).

Shuman, F. G., "A Method for Solving the Balance Equation." *Joint Numerical Weather Prediction Unit*, *Tech. Memo.* no. 6 (1955), 12 pp.

———, "A Method of Designing Finite-Different Smoothing Operators to Meet Specification." *Joint Numerical Weather Prediction Unit Tech. Memo.* no. 7 (1955), 14 pp.

———, "Predictive Consequences of Certain Physical Inconsistencies in the Geostrophic Barotropic Model." *Mon. Wea. Rev.*, **85,** no. 7, 229–234 (July 1957).

———, "Numerical Methods in Weather Prediction: II. Smoothing and Filtering." *Mon. Wea. Rev.*, **85,** 357–361 (1957).

———, "Numerical Experiments With the Primitive Equations." *Proc. Intern. Symp. Numerical Weather Prediction*, *Tokyo* (Nov. 7–13, 1960); *Meteor. Soc. Japan* (1962), 656 pp.

Shuman, F. G. and J. B. Hovermale, "An Operational Six-Layer Primitive Equation Model." *J. Appl. Meteor.*, **7,** no. 4 (1968).

Shuman, F. G. and L. W. Vanderman, "Difference System and Boundary Conditions for the Primitive Equation Barotropic Forecast." *Mon. Wea. Rev.*, **94,** 329–335 (1966).

Silberman, I., "Planetary Waves in the Atmosphere." *J. Meteor.*, **11,** 27–34 (1954).

Simons, T. S., "A Three-dimensional Spectral Prediction Equation." *Atmos. Sci. Paper* no. 127, Dept. of Atmos. Sci., Colorado State University (1968).

Smagorinsky, J., "The Dynamical Influence of Large-Scale Heat Sources and Sinks on the Quasi-Stationary Mean Motions of the Atmosphere." *Quart. J. Roy. Meteor. Soc.*, **79,** 342–366 (1953).

———, "On the Inclusion of Moist Adiabatic Processes in Numerical Prediction Models." *Berichte Deutsch. Wetterdienstes*, **5,** no. 38, 82–90 (1956).

———, "On the Numerical Integration of the Primitive Equations of Motion for Baroclinic Flow in a Closed Region." *Mon. Wea. Rev.*, **86,** 457–466 (1958).

———, "On the Application of Numerical Methods to the Solution of Systems of Partial Difference Equations Arising in Meteorology." In *Frontiers of Numerical Mathematics* (R. E. Langer, editor), University of Wisconsin Press, 1960.

———, "On the Dynamical Prediction of Large-Scale Condensation by Numerical Methods." Physics of Precipitation, *Geophysical Mono.*, no. 5, American Geophysical Union, 71–78 (1960).

———, "A Primitive Equation Model Including Condensation Processes." *Proc. Intern. Symp. Numerical Weather Prediction, Tokyo*, 1960; *Meteor. Soc. Japan* (March 1962).

———, "General Circulation Experiments With the Primitive Equations, I, The Basic Experiment." *Mon. Wea. Rev.*, **91,** 99–165 (1963).

———, "Some Aspects of the General Circulation." *Quart. J. Roy. Meteor. Soc.*, **90,** 1–14 (1964).

Smagorinsky, J. and G. O. Collins, "On the Numerical Prediction of Precipitation." *Mon. Wea. Rev.*, **83,** no. 3, 53–68 (March 1955).

Smagorinsky, J., S. Manabe, and J. L. Holloway, "Numerical Results From a Nine-Level General Circulation Model of the Atmosphere." *Mon. Wea. Rev.*, **93,** no. 12, 727–768 (Dec. 1965).

Smagorinsky, J. and Staff Members, "Prediction Experiments With a General Circulation Model." *Proc. Intern. Symp. Dynamics of Large Scale Processes in the Atmosphere (IAMAP/WMO), Moscow*, 1965.

Syono, S. and M. Yamasaki, "Stability of Symmetrical Motions Driven by Latent Heat Release by Cumulus Convection Under the Existence of Surface Friction." *J. Meteor. Soc. Japan*, sec. II, **44,** no. 6 (1966).

Teweles, S., "Spectral Aspects of the Stratosphere Circulation During the IGY." *MIT Rept.* no. 8 (Jan. 7, 1963).

Thompson, P. D., "On the Theory of Large-Scale Disturbance in a Two-Dimensional Baroclinic Equivalent of the Atmosphere." *Quart. J. Roy. Meteor. Soc.*, **79,** 51–69 (1953).

———, "A Theory of Large-Scale Disturbances in Nongeostrophic Flow." *J. Meteor.*, **13,** 251–261 (1956).

Thompson, P. D., "Uncertainty of Initial State as a Factor in the Predictability of Large-Scale Atmospheric Flow Patterns." *Tellus*, **9**, 275–295 (1957).

———, *Numerical Weather Analysis and Prediction*. The Macmillan Co., 1961, 170 pp.

Thompson, P. D. and W. L. Gates, "A Test of Numerical Prediction Methods Based on the Barotropic and Two-Parameter Baroclinic Method." *J. Meteor.*, **13**, 127–141 (1956).

Wiin-Nielsen, A., "On Certain Integral Constraints for the Time Integration of Baroclinic Models." *Tellus*, **11**, 45–60 (1959).

Wiin-Nielsen, A., J. A. Brown, and M. Drake, "On Atmospheric Energy Conversion Between the Zonal Flow and the Eddies." *Tellus*, **15**, 261–279 (1963).

———, "Further Studies of Energy Exchange Between the Zonal Flow and the Eddies." *Tellus*, **16**, 168–180 (1964).

Williams, R. T., "A Theoretical Investigation of the Structure of Easterly Waves.," Unpublished manuscript, Naval Postgraduate School, 1969.

Winninghoff, F. and A. Arakawa, "Numerical Simulation of the Geostrophic Adjustment Process." *Mon. Wea. Rev.*, **99** (1971).

Wolff, P. M., "The Error in Numerical Forecasts Due to Retrogression of Ultra-Long Waves." *Mon. Wea. Rev.*, **86**, 245–250 (1958).

Wolff, P. M., L. P. Carstensen, and T. Laevastu, "Analyses and Forecasting of Sea Surface Temperature." *FNWF Tech. Note* no. 8 (1965), 30 pp.

Yamasaki, M., "Numerical Simulation of Tropical Cyclone Development with the Use of Primitive Equations." *J. Meteor. Soc. Japan*, **46**, no. 3 (1968).

———, "A Tropical Cyclone Model with Parameterized Partition of Released Latent Heat." *J. Meteor. Soc. Japan*, **46**, no. 3, 202–214 (1968).

———, "Large-Scale Disturbances in a Conditionally Unstable Atmosphere in Low Latitudes." *Papers in Meteor. Geophys*, **20** (1970).

Yang, C. H., "Nonlinear Aspects of Large Scale Motions in the Atmosphere." *Univ. of Michigan Tech. Rept.*, 1967.

Index